Springer Finance

Springer Finance

Springer Finance is a programme of books aimed at students, academics and practitioners working on increasingly technical approaches to the analysis of financial markets. It aims to cover a variety of topics, not only mathematical finance but foreign exchanges, term structure, risk management, portfolio theory, equity derivatives, and financial economics.

Kerry Back

A Course in Derivative Securities

Introduction to Theory and Computation

Springer

Kerry Back
Department of Finance
Mays Business School
Texas A&M University
306 Wehner Building
College Station, TX 77843-4218
USA
e-mail: *kback@mays.tamu.edu*

Mathematics Subject Classification (2000): 91B28, 91B70, 9104, 65C05, 65M06, 60G44, 6004
JEL Classification: G13, C63

ISBN-13 978-3-642-06474-6   e-ISBN-13 978-3-540-27900-6

Springer is a part of Springer Science+Business Media

springeronline.com

© Springer-Verlag Berlin Heidelberg 2010
Printed in The Netherlands

Cover design: *design & production*, Heidelberg

Printed on acid-free paper   41/3142sz - 5 4 3 2 1 0

To my parents, Roy and Verla.

Preface

This book is an outgrowth of notes compiled by the author while teaching courses for undergraduate and masters/MBA finance students at Washington University in St. Louis and the Institut für Höhere Studien in Vienna. At one time, a course in Options and Futures was considered an advanced finance elective, but now such a course is nearly mandatory for any finance major and is an elective chosen by many non-finance majors as well. Moreover, students are exposed to derivative securities in courses on Investments, International Finance, Risk Management, Investment Banking, Fixed Income, etc. This expansion of education in derivative securities mirrors the increased importance of derivative securities in corporate finance and investment management.

MBA and undergraduate courses typically (and appropriately) focus on the use of derivatives for hedging and speculating. This is sufficient for many students. However, the seller of derivatives, in addition to needing to understand buy-side demands, is confronted with the need to price and hedge. Moreover, the buyer of derivatives, depending on the degree of competition between sellers, may very likely benefit from some knowledge of pricing as well. It is "pricing and hedging" that is the primary focus of this book. Through learning the fundamentals of pricing and hedging, students also acquire a deeper understanding of the contracts themselves. Hopefully, this book will also be of use to practitioners and for students in Masters of Financial Engineering programs and, to some extent, Ph.D. students in finance.

The book is concerned with pricing and hedging derivatives in frictionless markets. By "frictionless," I mean that the book ignores transaction costs (commissions, bid-ask spreads and the price impacts of trades), margin (collateral) requirements and any restrictions on short selling. The theory of pricing and hedging in frictionless markets stems of course from the work of Black and Scholes [6] and Merton [51] and is a very well developed theory. It is based on the assumption that there are no arbitrage opportunities in the market. The theory is the foundation for pricing and hedging in markets with frictions (i.e., in real markets!) but practice can differ from theory in important ways if the frictions are significant. For example, an arbitrage opportunity in a frictionless market often will not be an arbitrage opportunity for a trader who moves the market when he trades, faces collateral requirements, etc. This book has nothing to say about how one should deviate from the benchmark frictionless theory when frictions are important. Another important omission from the book is jump processes—the book deals exclusively with binomial and Brownian motion models.

The book is intended primarily to be used for advanced courses in derivative securities. It is self-contained, and the first chapter presents the basic financial concepts. However, much material (functioning of security exchanges, payoff diagrams, spread strategies, etc.) that is standard in an introductory book has not been included here. On the other hand, though it is not an introductory book, it is not truly an advanced book on derivatives either. On any of the topics covered in the book, there are more advanced treatments available in book form already. However, the books that I have seen (and there are indeed many) are either too narrow in focus for the courses I taught or not easily accessible to the students I taught or (most commonly) both. If this book is successful, it will be as a bridge between an introductory course in Options and Futures and the more advanced literature. Towards that end, I have included cites to more advanced books in appropriate places throughout.

The book includes an introduction to computational methods, and the term "introduction" is meant quite seriously here. The book was developed for students with no prior experience in programming or numerical analysis, and it only covers the most basic ideas. Nevertheless, I believe that this is an extremely important feature of the book. It is my experience that the theory becomes much more accessible to students when they learn to code a formula or to simulate a process. The book builds up to binomial, Monte Carlo, and finite-difference methods by first developing simple programs for simple computations. These serve two roles: they introduce the student to programming, and they result in tools that enable students to solve real problems, allowing the inclusion of exercises of a practical rather than purely theoretical nature. I have used the book for semester-length courses emphasizing calculation (most of the exercises are of that form) and for short courses covering only the theory.

Nearly all of the formulas and procedures described in the book are both derived from first principles and implemented in Excel VBA. The VBA programs are in the text and in an Excel workbook that can be downloaded free of charge at www.kerryback.net. I use a few special features of Excel, in particular the cumulative normal distribution function and the random number generator. Otherwise, the programs can easily be translated into any other language. In particular, it is easy to translate them into MATLAB, which also includes a random number generator and the cumulative normal distribution function (or, rather, the closely related "error function") as part of its basic implementation. I chose VBA because students (finance students, at least) can be expected to already have it on their computers and because Excel is a good environment for many exercises, such as analyzing hedges, that do not require programming. An appendix provides the necessary introduction to VBA programming.

Viewed as a math book, this is a book in applied math, not math proper. My goal is to get students as quickly as possible to the point where they can compute things. Many mathematical issues (filtrations, completion of filtrations, formal definitions of expectations and conditional expectations, etc.) are entirely ignored. It would not be unfair to call this a "cookbook" approach. I try to explain intuitively why the recipes work but do not give proofs or even formal statements of the facts that underlie them.

I have naturally taken pains to present the theory in what I think is the simplest possible manner. The book uses almost exclusively the probabilistic/martingale approach, both because it is my preference and because it seems easier than partial differential equations for students in business and the social sciences to grasp. A sampling of some of the more or less distinctive characteristics of the book, in terms of exposition, is:

- Important theoretical results are highlighted in boxes for easy reference; the derivations that are less important and more technical are presented in smaller type and relegated to the ends of sections.
- Changes of numeraire are introduced in the first chapter in a one-period binomial model, the probability measure corresponding to the underlying as numeraire being given as much emphasis as the risk-neutral measure.
- The fundamental result for pricing (asset prices are martingales under changes of numeraire) is presented in the first chapter, because it does not need the machinery of stochastic calculus.
- The basic ideas in pricing digital and share digitals, and hence in deriving the Black-Scholes formula, are also presented in the first chapter. Digitals and share digitals are priced in Chap. 3 before calls and puts.
- Brownian motion is introduced by simulating it in discrete time. The quadratic variation property is emphasized, including exercises that contrast Brownian motion with continuously differentiable functions of time, in order to motivate Itô's formula.
- The distribution of the underlying under different numeraires is derived directly from the fundamental pricing result and Itô's formula, bypassing Girsanov's theorem (which is of course also a consequence of Itô's formula).
- Substantial emphasis is placed on forwards, synthetic forwards, options on forwards and hedging with forwards because these have many applications in fixed income and elsewhere—a simple but characteristic example is valuing a European option on a stock paying a known cash dividend as a European option on the synthetic forward with the same maturity.
- Following Margrabe [50] (who attributes the idea to S. Ross) the formula for exchange options is derived by a change of numeraire from the Black-Scholes formula. Very simple arguments derive Black's formula for forward and futures options from Margrabe's formula and Merton's formula for stock options in the absence of a constant risk-free rate from

Black's formula. This demonstrates the equivalence of these important option pricing formulas as follows:

$$
\begin{aligned}
\text{Black-Scholes} &\Longrightarrow \text{Margrabe} \\
&\Longrightarrow \text{Black} \\
&\Longrightarrow \text{Merton} \\
&\Longrightarrow \text{Black-Scholes}
\end{aligned}
$$

- Quanto forwards and options are priced by first finding the portfolio that replicates the value of a foreign security translated at a fixed exchange rate and then viewing quanto forwards and options as standard forwards and options on the replicating portfolio.
- The market model is presented as an introduction to the pricing of fixed-income derivatives. Forward rates are shown to be martingales under the forward measure by virtue of their being forward prices of portfolios that pay spot rates.
- In order to illustrate how term structure models are used to price fixed-income derivatives, the Vasicek/Hull-White model is worked out in great detail. Other important term structure models are discussed much more briefly.

Of course, none of these items is original, but in conjunction with the computational tools, I believe they make the "rocket science" of derivative securities accessible to a broader group of students.

The book is divided into three parts, labeled "Introduction to Option Pricing," "Advanced Option Pricing," and "Fixed Income." Naturally, many of the chapters build upon one another, but it is possible to read Chaps. 1–3, Sects. 7.1–7.2 (the Margrabe and Black formulas) and then Part III on fixed income. For a more complete coverage, but still omitting two of the more difficult chapters, one could read all of Parts I and II except Chaps. 8 and 10, pausing in Chap. 8 to read the definitions of baskets, spreads, barriers, lookbacks and Asians and in Chap. 10 to read the discussion of the fundamental partial differential equation.

I would like to thank Mark Broadie, the series editor, for helpful comments, and especially I want to thank my wife, Diana, without whose encouragement and support I could not have written this. She mowed the lawn—and managed everything else—while I typed, and that is a great gift.

College Station, Texas   *Kerry Back*
April, 2005

Contents

Appendices

Part I

Introduction to Option Pricing

1

Asset Pricing Basics

This chapter introduces the "change of numeraire" (or "martingale") method for valuing derivative securities. The method is introduced in a binomial model and then extended to more general (continuum of states) models. Computations in the more general model require the continuous-time mathematics that will be presented in Chap. 2. We will begin with a brief description of the basic derivatives (calls and puts) and some other financial concepts. More detailed descriptions can be found in any of the many introductory books on derivative securities (e.g., [37] or [49]).

It should be noted that the pricing and hedging results in this book are not tied to any particular currency. However, for specificity (and as a consequence of the author's habit) the discussion will generally be in terms of dollars. Multiple currencies are addressed in Chap. 6.

1.1 Fundamental Concepts

Longs, Shorts, and Margin

In financial markets, the owner of an asset is said to be "long" the asset. If person A owes something to person B, the debt is an asset to person B but a liability to person A. One also says that person A is "short" the asset. For example, if someone borrows money and invests the money in stocks, then the individual is short cash and long stocks.

One must invest some of one's own money when borrowing money to buy stocks. For example, an individual could invest $600, borrow $400, and buy $1000 of stock. The $600 is called the "margin" posted by the investor, and buying stocks in this way is called buying "on margin." The investor, or the portfolio, is also said to be "levered," because buying $1000 of stock with only a $600 investment amplifies the risk and return per dollar of investment. On a percentage basis, we would say the account has 60% margin, the 60% being the ratio of the equity (assets minus liabilities = $1000 of stock minus $400 debt) to the assets ($1000 of stock). If the value of the stock drops sufficiently far, then it may become doubtful whether the investor can repay the $400. In this case, the investor must either sell the stock or invest more of his own funds (i.e., he receives a "margin call"). In other words, in actual markets there are "margin requirements," that specify a minimum percent margin an investor must have initially (when borrowing money) and a minimum percent margin the investor must maintain.

Rather than borrowing money to buy stocks, an investor can do the opposite—he can borrow stocks to buy money. In this case, "buying money" means selling the borrowed stocks for cash. Such an investor will be short stocks and long cash. This is called "short selling" (or, more briefly, "shorting") stocks. For example, suppose individual A borrows 100 shares of stock from individual B and then sells them to individual C. Both B and C are long the 100 shares and A is short, so the net long position is $2 \times 100 - 100$, which is the original 100 shares that B was long. A short seller of stocks must pay to the lender of the stocks any dividends that are paid on the stock. In our example, both B and C own the 100 shares so both expect to receive dividends. The company will pay dividends only to C, and A must pay the dividends to B.

Of course, investors always wish to buy low and sell high. The usual method is to buy stocks and hope they rise. An investor who short sells also wishes to buy low and sell high, but he reverses the order—he sells first and then hopes the stocks fall. The risk is that the stocks will instead rise, which will increase the value of his liability (short stock position) without increasing the value of his assets (long cash position), thus putting him "under water." To shield the lender of the stocks from this risk, a short seller must also invest some of his own funds, and this amount is again called the investor's margin. For example, an investor might invest $600, and borrow and sell $1000 of stock. In this case, the investor will be long $1600 cash and short $1000 worth of stock. His equity is $600 and his percent margin is calculated as $600/$1000 = 60%. Again, there are typically both initial and maintenance margin requirements. An additional feature of short selling for small individual investors is that they typically will not earn interest on the proceeds of the short sale (the $1000 cash obtained from selling stocks in the above example).

In this book, we will assume there is a single risk-free rate at which one can both borrow and lend. Moreover, we will assume that investors earn this rate on margin deposits, including the proceeds of short sales (and including any margin that may be required when buying and selling forward and futures contracts). Thus, investors gain from buying on margin if the asset return is sure to exceed the risk-free rate, and they gain from short selling if the return on an asset is sure to be below the risk-free rate. These assumptions are not reasonable for small individual investors, but they are fairly reasonable for institutional investors. We will assume that no asset has a return that is certain to be above the risk-free rate nor certain to be below the risk-free rate, because institutional investors could "arbitrage" such guaranteed high-return or guaranteed low-return assets.

Calls and Puts

Call and put options are the basic derivative securities and the building blocks of many others. A derivative security is a security the value of which depends upon another security. A call option is the right to buy an asset at a pre-specified price. The pre-specified price is called the exercise price, the strike price, or simply the strike. We will often call the asset a "stock," but there are options on many other types of assets also, and everything we say will be applicable to those as well.[1] The asset to which the call option pertains is called the "underlying asset," or, more briefly, the "underlying." If the market value of the asset exceeds the exercise price, then we say the call option is "in the money." Buying a call option is a way to bet on the upside of the underlying asset.

A put option is the right to sell an asset at a pre-specified (exercise, strike) price. Buying a put is a way to bet on an asset price becoming low (similar to shorting). A put option is in the money if the exercise price exceeds the value of the asset. Both puts and calls are potentially valuable and hence the buyer of a put or call must pay the seller.

A long put option provides insurance to someone who is long the underlying asset, because it guarantees that the asset can always be sold at the strike price of the put (of course, it can be sold at the market price, if that is higher than the strike of the put). Symmetrically, a long call option provides insurance to someone who is short the underlying asset. The terminology in option markets reflects the parallels between options and insurance contracts. In particular, the seller of an option is said to "write" the option and the compensation (price) he receives from the buyer is called the option "premium," just as an insurance company writes insurance contracts in exchange for premium income. Calculating the price at which one should be willing to trade an option is the main topic of this book.

It is important to recognize the different situations of someone who is short a call option and someone who is long a put. Both positions are bets on the downside of the asset. Both the investor who is short a call and the investor who is long a put may eventually sell the underlying asset and receive the exercise price in exchange. However, the investor who is long a put has an option to sell the asset at the exercise price and the investor who is short a call has an *obligation* to sell the asset at the exercise price, should the counterparty choose to exercise the call. Thus, the investor who is long a put will be selling at the exercise price when it is profitable to do so, whereas the investor who is short a call will be selling at the exercise price when it is

---

[1] One caveat is that by "asset" we mean something that can be stored; thus, for example, electricity is, practically speaking, not an asset.

unprofitable. The buyer of a put must pay the premium to the seller; he then profits if the asset price is low, with his maximum possible profit being quite large (the maximum value is attained when the market value of the underlying asset reaches zero). In contrast, the seller of a call receives premium income, and the premium is his maximum possible profit, whereas his potential losses are unbounded. Thus, these are very different positions.

Individuals who sell calls usually sell out-of-the-money covered calls. "Covered" means that they own the underlying asset and can therefore deliver the underlying if the call is exercised without incurring any further expense—they experience only an "opportunity cost" in delivering it for less than the market price.[2] A call being out of the money implies that the price of the underlying must rise before the call would be exercised against the seller; thus, the seller of an out-of-the money covered call still has some potential for profit from the underlying. In addition, of course, the seller receives the premium income from the call. Institutions often follow this strategy also, using the premium income to "enhance" their return from the underlying. One can hedge a short call without owning a full share of the underlying asset, if one is able to rebalance the hedge over time. Calculating such hedges is another of the principal topics of this book.

In a certain sense, option markets are zero-sum games. The profit earned by one counterparty to an option transaction is a loss suffered by the other. However, options can allow for an increase in the welfare of all investors by improving the allocation of risk. A producer who must purchase a certain input may buy a call option, giving him the right to buy the input at a fixed price. This caps his expense. The seller of the call now bears the risk that the input price will be high—in this case, the option will be exercised and he will be forced to sell at a price below the market price. It may be that the seller is in a better position to bear the risk (for example, he may have less of the risk in his portfolio) and the option transaction may thereby improve the allocation of risks across investors. The similarity to insurance should be apparent.

Quite complex bets or hedges can be created by combining options. For example, a long call and put with the same strike price is called a "straddle." Such a portfolio is (almost) always in the money. It is in fact a bet on volatility—a big move in the underlying asset value away from the exercise price will lead to either the call or put having a high value. Another important example of an option portfolio is a "collar." A collar consists of a long put and a short call, or a short call and a long put, with the options having the same maturity. As mentioned before, a long put provides insurance to someone who is long the underlying asset. Selling a call provides premium income that can

---

[2] In contrast, one who sells a call without owning the underlying is said to sell a "naked call." The seller of a naked call, or the seller of a put, must post margin, just like a short seller of stocks, in order to ensure that he can meet his obligation. However, this does not apply to sellers of covered calls.

be used to offset the cost of the put (the most popular type of collar is a zero-cost collar: a collar in which the premium of the call is equal to the premium of the put). The cost of selling a call for an owner of the underlying is that it sells off the upside of the underlying asset—if the value of the asset exceeds the strike price of the call, then the call will be exercised and the underlying asset must be delivered for the strike price (rather than the higher market price). Thus, one can purchase the downside insurance provided by a long put by selling part of the upside potential of the asset, rather than paying the cost of the insurance out of pocket. There are many other examples of option portfolios that could be given.

Some puts and calls are traded on exchanges. In this case, the exchange clearinghouse "steps between" the buyer and seller and becomes the counterparty to both the buyer and seller. This eliminates the risk that the seller might default on his obligation when the buyer chooses to exercise his option. If the owner of an option chooses to exercise, the clearinghouse randomly chooses someone who is short the option to fulfill the obligation. Most exchange traded options are never exercised, because any gain on a long contract can be captured by selling the contract at the market price, thus cancelling the position. Obviously, however, the right to exercise is essential, because it determines the market price. Puts and calls are also transacted "over the counter," which means that they are private contracts of the counterparties. Moreover, puts and calls are embedded in many other financial instruments. A prosaic but important example is that most homeowners have the right to pay off their mortgages early. This means they have call options on their mortgages, with exercise price equal to the remaining mortgage principal. Similarly, callable bonds can be redeemed early by the company issuing them, convertible bonds have embedded call options on the company's stock (which are exercised by "converting" the bonds) and there are many, many other examples. Puts and calls also exist outside financial markets. For example, a company may begin manufacturing a new product at a small scale; if the product is successful, the scale can be expanded. In this case, the company buys a call option on large-scale production with the premium being the cost of launching small-scale production. Adapting the methods developed for financial options to value such "real options" is an important and growing field.

Exercise Policies for Calls and Puts

It may be rational to exercise a call if the asset value exceeds the exercise price. Thus, denoting the price of the asset by S and the exercise price by K, the owner of a call option can profit by $S - K$ dollars by exercising the option when $S > K$. If $S < K$, exercise would be irrational. Thus, the payoff to the owner of the call option is[3] $\max(0, S - K)$. It has been said that timing is

---

[3] We use the standard notation: $\max(a, b)$ denotes the larger of a and b and $\min(a, b)$ denotes the smaller.

everything, and the timing here should be made clearer. The simplest type of option is called a "European" option. A European option has a finite lifetime and can only be exercised at its maturity date. For a European call option, the exercise strategy just described is the optimal one, with S representing the asset price at the maturity date of the option. Equally, if not more, important are "American" options, which can be exercised at any time before maturity.

For an American call option, the exercise strategy just described is the optimal one at the maturity date, but it may also be optimal to exercise prior to maturity. Let K denote the exercise price, T the date the option matures, and $S(t)$ the price of the underlying asset at date $t \leq T$. The "intrinsic value" of the call option at date t is defined to be $\max(0, S(t) - K)$. One would of course never exercise unless the intrinsic value is positive—i.e., unless the option is in the money. Moreover, if the asset does not pay a dividend (or other type of cash flow) prior to the option maturity then one should not exercise in any circumstances prior to maturity. This is captured in the saying: "calls are better alive than dead." Exercise being suboptimal is equivalent to the value of the option exceeding the intrinsic value.

The principle that calls on non-dividend-paying assets are better alive than dead follows from two facts: (i) it is generally a good thing (in financial markets as well as in life) to keep one's options open, and (ii) early exercise implies early payment of the exercise price and hence foregone interest. The usual protest that is heard when this statement is made is that one should surely exercise if he expects the stock price to plummet, because by exercising (and then selling the stock acquired) one can lock in the current stock price rather than waiting for it to fall, in which case the option will surely be worth less. This intuition is a reasonable one, but it ignores the fact that the investor could short sell the stock if he expects it to plummet—he doesn't need to exercise the option to lock in the current stock price. In fact, shorting the stock and retaining the option is always better than exercising, assuming the underlying asset does not pay a dividend.

Specifically, suppose an investor considers exercising at date t. As an alternative to exercising early, consider shorting the stock at date t and retaining the option. This is always better than exercising at date t, because the short position can be "covered" (the stock can be purchased and returned to the lender to cancel the short position) at cost K at date T by exercising the option, and paying K at date T is better than paying it at date t, given that interest rates must be nonnegative. To be more precise, note that exercise at date t produces $S(t) - K$ dollars at date t. On the other hand, retaining the option, shorting the stock at date t, and covering the short either by exercising the option or buying the stock in the market (whichever is cheaper) produces $S(t)$ dollars at date t and

$$\max(0, S(T) - K) - S(T) = \max(-S(T), -K) = -\min(S(T), K) \geq -K$$

dollars at date T. If $S(T) > K$, one has $-K$ dollars at date T, in which case retaining the option has been superior due to the time value of money.

Furthermore, if $S(T) < K$, the strategy of retaining the option and shorting the stock produces $-S(T) > -K$ dollars at date T, so retaining the option is superior due both to flexibility (waiting until T to decide whether to exercise turns out to be better than committing at date t) and because of the time value of money.[4]

Early exercise of a call option can be optimal when the underlying asset pays a dividend. The above analysis does not apply in this case, because paying the dividend to the lender of the stock is an additional cost for the strategy of retaining the option and shorting the stock. If the dividend is so small that it cannot offset the time value of money on the exercise price, then early exercise will not be optimal. In other cases, deriving the optimal exercise strategy is a complicated problem that we will first begin to study in Chap. 5.

A European put option will be exercised at its maturity T if the price $S(T)$ of the underlying asset is below the exercise price K. In general, the value at maturity can be expressed as $\max(0, K - S(T))$. Early exercise of an American put can be optimal, regardless of whether the underlying pays a dividend. While it is valuable to keep one's options open (for puts as well as calls) the time value of money works in the opposite direction for puts. Early exercise of a put option implies early receipt of the exercise price, and it is better to receive cash earlier rather than later. In general, whether early exercise is optimal depends on how deeply the option is in the money—if the underlying asset price is sufficiently low, then it will be fairly certain that exercise will be optimal, whether earlier or late; in this case, one should exercise earlier to earn interest on the exercise price. How low it should be to justify early exercise depends on the interest rate (a higher rate makes the time-value-of-money issue more important, leading to earlier exercise) and the volatility of the underlying asset price (a lower volatility reduces the value of keeping one's options open, leading also to earlier exercise). We will begin to study the valuation of American puts in Chap. 5 also.

----

[4] Recall that we are assuming investors earn interest on the proceeds of short sales; otherwise, the $S(t)$ dollars earned from exercising the option and selling the stock will be worth more than the $S(t)$ dollars earned from shorting the stock. In this case, early exercise could be optimal. However, assuming institutional investors can earn interest on the proceeds of shorts, such investors should prefer owning the option and shorting the stock to exercising. This means they should bid up the price of the option to the point where it exceeds the value $S(t) - K$ of exercise. If this is the case, then an investor who cannot earn interest on the proceeds of shorts should simply sell the option in the market rather than exercise it. Thus, a sufficient condition for calls to be "better alive than dead" is that there be some investors who can earn interest on the proceeds of shorts. This type of reasoning is possible for each situation in this book where the assumption of earning interest on margin deposits is important, and we will not deal with it in this much detail again.

Compounding Interest

During most of the first two parts of the book (the only exception being Chap. 7) we will assume there is a risk-free asset earning a constant rate of return. For simplicity, we will specify the rate of return as a continuously compounded rate. For example, if the annual rate with annual compounding is r_a, then the corresponding continuously compounded rate is r defined as $r = \log(1 + r_a)$, where "log" denotes the natural logarithm function. This means that the gross return over a year (one plus the rate of return) is $e^r = 1 + r_a$. More generally, an investment of x dollars for a time period of length T (we measure time in years, so, e.g., a six-month investment would mean $T = 0.5$) will result in the ownership of xe^{rT} dollars at the end of the time period.

Expressing the interest rate as a continuously compounded rate enables us to avoid having to specify in each instance whether the rate is for annual compounding, semi-annual compounding, monthly compounding, etc. For example, the meaning of an annualized rate r_s for semi-annual compounding is that an investment of x dollars will grow over a year to $x(1 + r_s/2)^2$. The equivalent continuously compounded rate is defined as $r = \log(1 + r_s/2)^2$, and in terms of this rate we can say that the investment will grow in six months to $xe^{0.5r}$ and that it will grow in one year to xe^r. We can interpret this rate as being continuously compounded because compounding n times per year at an annualized rate of r results in \$1 growing in a year to $(1 + r/n)^n$ and

$$\lim_{n \to \infty} \left(1 + \frac{r}{n}\right)^n = e^r \ .$$

To develop pricing and hedging formulas for derivative securities, it is a great convenience to assume that investors can trade continuously in time. This requires us to assume also that returns are computed continuously. In the case of a risk-free investment of $x(t)$ dollars at any date t at a continuously compounded rate of r, we will say that the interest earned in "an instant dt" is $x(t)r\, dt$ dollars. This is only meaningful when we accumulate the interest over a non-infinitesimal period of time. So consider investing $x(0)$ dollars at time 0 and reinvesting interest in the risk-free asset over a time period of length T. Let $x(t)$ denote the account balance at date t, for $0 \le t \le T$. The change in the account balance in each instant is the interest earned, so we have $dx(t) = x(t)r\, dt$. The real meaning of this equation is that $x(t)$ satisfies the differential equation

$$\frac{dx(t)}{dt} = x(t)r \ ,$$

and it is well known (and easy to verify) that the solution is

$$x(t) = x(0)e^{rt} \ ,$$

leading to an account balance at the end of the time period of $x(T) = x(0)e^{rT}$. Thus, the statement that "the interest earned in an instant dt is $x(t)r\, dt$" is equivalent to the statement that interest is continuously compounded at the rate r.

In the last part of the book, we will drop the assumption that the risk-free asset earns a constant rate of return. In this case, we will still generally assume that there is a risk-free asset for very short-term investments (i.e., for investments with infinitesimal durations!). We will let $r(t)$ denote the risk-free rate for an instantaneous investment at date t. This means that an investment of $x(t)$ dollars at date t in the risk-free asset earns interest in an instant dt equal to $x(t)r(t)\,dt$. Consider again an investment of $x(0)$ dollars at date 0 in this instantaneously risk-free asset with interest reinvested and let $x(t)$ denote the account balance at date t. Then $x(t)$ must satisfy the differential equation

$$\frac{dx(t)}{dt} = x(t)r(t) \ .$$

The solution of this differential equation is

$$x(t) = x(0)\exp\left(\int_0^t r(s)\,ds\right) \ .$$

The expression $\int_0^t r(s)\,ds$ can be interpreted as a continuous sum over time of the rates of interest $r(s)$ earned at times s between 0 and t. If these rates are all the same, say equal to r, then $\int_0^t r(s)\,ds = rt$ and our compounding factor $\exp\left(\int_0^t r(s)\,ds\right)$ is e^{rt} as before.

1.2 State Prices in a One-Period Binomial Model

To introduce the concepts that will be discussed in the remainder of the chapter, we will consider in this and the following section the following very simple framework. There is a stock with price S today (which we will call date 0). At the end of some period of time of length T, the stock price will take one of two values: either S_u or S_d, where $S_u > S_d$. If the stock price equals S_u we say we are in the "up" state of the world, and if it equals S_d we say we are in the "down" state. The stock does not pay a dividend. There is also a risk-free asset earning a continuously compounded rate of interest r. Finally we want to consider a European call option on the stock with maturity T and strike K. The value of the call option at the end of the period is $C_u = \max(0, S_u - K)$ in the up state and $C_d = \max(0, S_d - K)$ in the down state.

We will assume

$$\frac{S_u}{S} > e^{rT} > \frac{S_d}{S} \ . \tag{1.1}$$

This condition means that the rate of return on the stock in the up state is greater than the risk-free rate, and the rate of return on the stock in the down state is less than the risk-free rate. If it were not true, there would be an arbitrage opportunity: if the rate of return on the stock were greater than the risk-free rate in both states, then one should buy an infinite amount of the stock on margin, and conversely if the rate of return on the stock were less than the risk-free rate in both states, then one should short an infinite amount of stock and put the proceeds in the risk-free asset. So what we are assuming is that there are no arbitrage opportunities in the market for the stock and risk-free asset.

The "delta" of the call option is $\delta = (C_u - C_d)/(S_u - S_d)$. Multiplying by $S_u - S_d$ gives us $\delta(S_u - S_d) = C_u - C_d$ and rearranging yields $\delta S_u - C_u = \delta S_d - C_d$, which is critical to what follows. Consider purchasing δ shares of the stock at date 0 and borrowing

$$\mathrm{e}^{-rT}(\delta S_u - C_u) = \mathrm{e}^{-rT}(\delta S_d - C_d)$$

dollars at date 0. Then you will owe

$$\delta S_u - C_u = \delta S_d - C_d$$

dollars at date T, and hence the value of the portfolio at date T in the up state will be

$$\text{Value of delta shares} - \text{Dollars owed} = \delta S_u - (\delta S_u - C_u) = C_u \ ,$$

and the value of the portfolio at date T in the down state will be

$$\text{Value of delta shares} - \text{Dollars owed} = \delta S_d - (\delta S_d - C_d) = C_d \ .$$

Thus, this portfolio of buying delta shares and borrowing money (i.e., buying delta shares on margin) "replicates" the call option. Consequently, the value C of the option at date 0 must be the date–0 cost of the portfolio; i.e.,

$$C = \text{Cost of delta shares} - \text{Dollars borrowed} = \delta S - \mathrm{e}^{-rT}(\delta S_u - C_u) \ . \quad (1.2)$$

Because the call option is equivalent to buying the stock on margin, it can be considered a levered investment in the stock.

We will now rewrite the option pricing formula (1.2) in terms of "state prices." By substituting for δ in (1.2), we can rearrange it as

$$C = \frac{S - \mathrm{e}^{-rT}S_d}{S_u - S_d} \times C_u + \frac{\mathrm{e}^{-rT}S_u - S}{S_u - S_d} \times C_d \ . \quad (1.3a)$$

A little algebra also shows that

$$S = \frac{S - \mathrm{e}^{-rT}S_d}{S_u - S_d} \times S_u + \frac{\mathrm{e}^{-rT}S_u - S}{S_u - S_d} \times S_d \ , \quad (1.3b)$$

and

$$1 = \frac{S - \mathrm{e}^{-rT} S_d}{S_u - S_d} \times \mathrm{e}^{rT} + \frac{\mathrm{e}^{-rT} S_u - S}{S_u - S_d} \times \mathrm{e}^{rT} . \tag{1.3c}$$

It is convenient to denote the factors appearing in these equations as

$$\pi_u = \frac{S - \mathrm{e}^{-rT} S_d}{S_u - S_d} \quad \text{and} \quad \pi_d = \frac{\mathrm{e}^{-rT} S_u - S}{S_u - S_d} . \tag{1.4}$$

The numbers π_u and π_d are called the "state prices," for reasons that will be explained below.

With these definitions, we can write (1.3a)–(1.3c) as

$$C = \pi_u C_u + \pi_d C_d , \tag{1.5a}$$
$$S = \pi_u S_u + \pi_d S_d , \tag{1.5b}$$
$$1 = \pi_u \mathrm{e}^{rT} + \pi_d \mathrm{e}^{rT} . \tag{1.5c}$$

These equations have the following interpretation: the value of a security today is its value in the up state times π_u plus its value in the down state times π_d. This applies to (1.5c) by considering an investment of \$1 today in the risk-free asset—it has value 1 today and will have value e^{rT} in both the up and down states at date T. Moreover, this same equation will hold for any other derivative asset. For example, if we considered a put option, then a delta–hedging argument analogous to that we just gave for the call option will lead to a formula for the value P of the put today which can be expressed as $P = \pi_u P_u + \pi_d P_d$ for the same π_u and π_d defined in (1.4).

In this model, we can think of any security as a portfolio of what are called "Arrow securities" (in recognition of the seminal work of Kenneth Arrow [1]). One of the Arrow securities pays \$1 at date T if the up state occurs and the other pays \$1 at date T if the down state occurs. For example, the stock is equivalent to a portfolio consisting of S_u units of the first Arrow security and S_d units of the second, because the stock is worth S_u dollars in the up state and S_d dollars in the down state. Equations (1.5a)–(1.5c) show that π_u is the price of the first Arrow security and π_d is the price of the second. For example, the right-hand side of (1.5b) is the value of the stock at date 0 viewed as a portfolio of Arrow securities when the Arrow securities have prices π_u and π_d. Because the stock clearly is such a portfolio, its price today must equal its value as that portfolio, which is what (1.5b) asserts.

As mentioned before, the prices π_u and π_d of the Arrow securities are called the "state prices," because they are the prices of receiving \$1 in the two states of the world. The state prices should be positive, because the payoff of each Arrow security is nonnegative in both states and positive in one. A little algebra shows that the conditions $\pi_u > 0$ and $\pi_d > 0$ are exactly equivalent to our "no-arbitrage" assumption (1.1). Thus, we conclude that **in the absence of arbitrage opportunities, there exist positive state prices such that the price of any security is the sum across the states of the world of its payoff multiplied by the state price.**

This conclusion generalizes to other models, including models in which the stock price takes a continuum of possible values. We will discuss more general models later in this chapter. It is a powerful result that tremendously simplifies derivative security pricing.

1.3 Probabilities and Numeraires

In this section, we will continue our analysis of the binomial example. To apply the statement about state prices appearing in boldface type above in the most convenient way, we will manipulate the state prices so we can interpret the sums on the right-hand sides of (1.5a)–(1.5c) in terms of expectations. The expectation (or "mean") of a random variable is of course its probability-weighted average value.

In general, there are different expectations that are useful. In this model, there are two that we can define: one corresponding to the risk-free asset and one corresponding to the stock. Many readers will have experience with the first in the form of "risk-neutral probabilities."

The risk-neutral probabilities are defined as $\pi_u e^{rT}$ for the up state and $\pi_d e^{rT}$ for the down state. Denoting these as p_u and p_d respectively, (1.5a)–(1.5c) can be written as

$$C = e^{-rT}[p_u C_u + p_d C_d] \,, \tag{1.6a}$$

$$S = e^{-rT}[p_u S_u + p_d S_d] \,, \tag{1.6b}$$

$$1 = p_u + p_d \,. \tag{1.6c}$$

The numbers p_u and p_d are both positive (because the state prices are positive under our no-arbitrage assumption) and (1.6c) states that they sum to one, so it is indeed sensible to consider them as probabilities. Equations (1.6a) and (1.6b) state that the value of a security today is its expected value at date T (the expectation taken with respect to the risk-neutral probabilities) discounted at the risk-free rate. Thus, these are "present value" formulas. Unlike the Capital Asset Pricing Model, for example, there is no risk premium in the discount rate. This is the calculation we would do to price assets under the actual probabilities if investors were risk neutral (or for zero-beta assets). So, we can act as if investors are risk neutral by adjusting the probabilities. Of course, we are not really assuming investors are risk neutral. We have simply embedded any risk premia in the probabilities.[5]

Equations (1.6a) and (1.6b) can be written in an equivalent form, which, though somewhat less intuitive, generalizes more readily. First, let's introduce some notation for the price of the risk-free asset. Considering an investment of \$1 today which grows to e^{rT} at date T, it is sensible to take the price

---

[5] This fundamental idea is due to Cox and Ross [20].

today to be $R - 1$ and the price in the up and down states at date T to be $R_u = R_d = e^{rT}$.[6] In terms of this notation, (1.6a)–(1.6c) can be written as:

$$\frac{C}{R} = p_u \frac{C_u}{R_u} + p_d \frac{C_d}{R_d} \; , \tag{1.7a}$$

$$\frac{S}{R} = p_u \frac{S_u}{R_u} + p_d \frac{S_d}{R_d} \; , \tag{1.7b}$$

$$1 = p_u + p_d \; . \tag{1.7c}$$

Each of equations (1.7a) and (1.7b) states that the price of a security today divided by the price of the risk-free asset equals the expected future value of the same ratio, when we take expectations using the risk-neutral probabilities. In other words, the mean of the date–T value of the ratio is equal to the ratio today. We will discuss the interpretation and significance of these equations further below. First, we consider the other type of expectation in this model, which is based on probabilities corresponding to the stock.

Note that the risk-neutral probabilities are the state prices multiplied by the gross return on the risk-free asset. Analogously, define numbers $q_u = \pi_u S_u / S$ and $q_d = \pi_d S_d / S$. Substituting for π_u and π_d in (1.5a)–(1.5c) and continuing to use the notation R for the price of the risk-free asset, we obtain

$$\frac{C}{S} = q_u \frac{C_u}{S_u} + q_d \frac{C_d}{S_d} \; , \tag{1.8a}$$

$$1 = q_u + q_d \; , \tag{1.8b}$$

$$\frac{R}{S} = q_u \frac{R_u}{S_u} + q_d \frac{R_d}{S_d} \; . \tag{1.8c}$$

Equation (1.8b) establishes that we can view the q's as probabilities (like the risk-neutral probabilities, they are positive because the state prices are positive). Equations (1.8a) and (1.8c) both state that the ratio of a security price to the price of the stock today equals the mean value of the same ratio at date T, when we compute expectations using the q's as probabilities.

Here is some useful terminology:

- An assignment of probabilities to events is called a *probability measure*, or simply a *measure* (because it "measures" the events, in a sense). Thus, we have described two different probability measures in this section.
- The ratio of one price to another is the value of the first (numerator) asset when we are using the second (denominator) asset as the *numeraire*. The term "numeraire" means a unit of measurement. For example, the ratio C/S is the value of the call when we use the stock as the unit of measurement: it is the number of shares of stock for which one call option can be exchanged (to see this, note that C/S shares is worth $C/S \times S = C$ dollars, so C/S shares is worth the same as one call.)

---

[6] All of the equations appearing below will also be true if instead we take $R = e^{-rT}$ and $R_u = R_d = 1$.

- A variable that changes randomly over time with the expected future value being always equal to the current value is called a *martingale*.

The right-hand sides of (1.7a)–(1.7b) and (1.8a) and (1.8c) are expectations under different *probability measures* (the p's or q's). The expected future (date–T) value equals the current (date–0) value, so the random variables (C/R and S/R or C/S and R/S) are *martingales*. The values C/R and S/R are the values of the call and stock *using the risk-free asset as numeraire*, and the values C/S and R/S are the values of the call and risk-free asset *using the stock as numeraire*. Thus, we will express (1.7a)–(1.7b) as "the call and stock are martingales when we use the risk-free asset as numeraire." Likewise, we will express (1.8a) and (1.8c) as "the call and risk-free asset are martingales when we use the stock as numeraire." It should be understood in both cases that "using an asset as numeraire" means that we also use the corresponding probability measure (i.e., the p's or q's). In general, our conclusion that assets can be priced in terms of positive state prices when there are no arbitrage opportunities can be rephrased as: **if there are no arbitrage opportunities, then for each (non-dividend-paying) asset, there exists a probability measure such that the ratio of any other (non-dividend-paying) asset price to the first (numeraire) asset price is a martingale.**[7]

For this exposition, it was convenient to first calculate the state prices and then calculate the various probabilities. However, that is not the most efficient way to proceed in most applications. In a typical application, we would view the prices of the stock and risk-free asset in the various states of the world as given, and we would be attempting to compute the value of the call option. Note that the sets of equations (1.5a)–(1.5c), (1.7a)–(1.7c), and (1.8a)–(1.8c) are all equivalent. In each case we would consider that there are three unknowns—the value C of the call option and either two state prices or two probabilities. In each case the state prices or probabilities can be computed from the last two equations in the set of three equations and then the call value C can be computed from the first equation in the set. All three sets of equations produce the same call value.

In fact, as we will see, it will not even be necessary to calculate the probabilities. The fact that ratios of non-dividend paying asset prices to the numeraire asset price are martingales will tell us enough about the probabilities to calculate derivative values without having to calculate the probabilities themselves.

We conclude this section with another reformulation of the pricing relations (1.5a)–(1.5c). This formulation will generalize more easily to pricing

---

[7] We have applied this statement to the risk-free asset, which pays dividends (interest). However, the price $R_u = R_d = e^{rT}$ includes the interest, so no interest has been withdrawn—the interest has been reinvested—prior to the maturity T of the option. This is what we mean by a "non-dividend-paying" asset. In general, we will apply the formulas developed in this and the following section to dividend-paying assets by considering the portfolios in which dividends are reinvested.

when there are a continuum of states, the subject of the next section. Let prob_u denote the actual probability of the up state and prob_d denote the probability of the down state. These probabilities are irrelevant for pricing derivatives in the binomial model, but we will use them to write the pricing relations (1.5a)–(1.5c) as expectations with respect to the actual probabilities. To do this, we can define

$$\phi_u = \frac{\pi_u}{\text{prob}_u},$$

$$\phi_d = \frac{\pi_d}{\text{prob}_d}.$$

Then (1.5a)–(1.5c) can be written as

$$C = \text{prob}_u\phi_u C_u + \text{prob}_d\phi_d C_d, \tag{1.9a}$$
$$S = \text{prob}_u\phi_u S_u + \text{prob}_d\phi_d S_d, \tag{1.9b}$$
$$R = \text{prob}_u\phi_u R_u + \text{prob}_d\phi_d R_d. \tag{1.9c}$$

The right-hand sides are expectations with respect to the actual probabilities. For example, the right-hand side of equation (1.9a) is the expectation of the random variable that equals $\phi_u C_u$ in the up state and $\phi_d C_d$ in the down state. The risk-neutral probabilities can be calculated from ϕ_u and ϕ_d as $p_u = \text{prob}_u\phi_u R_u/R$ and $p_d = \text{prob}_d\phi_d R_d/R$. Likewise, the probabilities using the stock as the numeraire can be calculated from ϕ_u and ϕ_d as $q_u = \text{prob}_u\phi_u S_u/S$ and $q_d = \text{prob}_d\phi_d S_d/S$. In the following section, we will assume (which can be shown to be true under some technical conditions) that relations such as (1.9a)–(1.9c) hold in a general (non-binomial) model given the absence of arbitrage opportunities. We will then show, using definitions analogous to the definitions of p_u, p_d, q_u, and q_d in this paragraph, that relations analogous to (1.7a)–(1.7c) and (1.8a)–(1.8c) hold.

1.4 Asset Pricing with a Continuum of States

In this section, we will define the concepts of state prices and probabilities corresponding to different numeraires in a more general framework than that of the preceding section. This leads to what we will call the "fundamental pricing equation," namely equation (1.17). There are really no new concepts in this section, only a bit more mathematics.

Consider a non-dividend-paying security having the random price $S(T)$ at date T. We call the contingencies that affect the price $S(T)$ the "states of the world." Our principle regarding state prices developed in the preceding section can in general be expressed as:[8] **if there are no arbitrage opportunities,**

---

[8] We have proven this in the binomial model, but we will not prove it in general. As is standard in the literature, we will simply adopt it as an assumption. A

there exists for each date T a positive random variable $\phi(T)$ such that the value at date 0 of a non-dividend-paying security with price S is

$$S(0) = E[\phi(T)S(T)] . \tag{1.10}$$

Here, $E[\phi(T)S(T)]$ denotes the expectation of the random variable $\phi(T)S(T)$. The random variable $\phi(T)$ is called the "state price density."[9] In a binomial model (or in any model with only a finite number of states of the world), the concept of an expectation is clear: it is just a weighted average of outcomes, the weights being the probabilities. In the binomial model, the right-hand side of equation (1.9b) is the same as the right-hand side of equation (1.10).[10]

To convert from state prices to probabilities corresponding to different numeraires, we follow the same procedure as at the end of the previous section: we multiply together (i) the probability of the state, (ii) the value of $\phi(T)$ in the state, and (iii) the gross return of the numeraire in the state. If there is a continuum of states, then the actual probability of any individual state will typically be zero, so this multiplication will produce a zero probability. However, we can nevertheless "add up" these probabilities to define the probability of any event A, an "event" being a set of states of the world. To do this, let 1_A denote the random variable that takes the value 1 when A is true and which is zero otherwise. Then the probability of A using S as the numeraire is defined as

$$E\left[1_A\phi(T)\frac{S(T)}{S(0)}\right] . \tag{1.11}$$

This makes sense as a probability because it is nonnegative and because, if A is the set of all states of the world, then its probability is $E[\phi(T)S(T)/S(0)]$, which equals one by virtue of (1.10). From the definition (1.11) of the probability of any event A, it can be shown that the expectation of any random variable X using S as the numeraire is

$$E\left[X\phi(T)\frac{S(T)}{S(0)}\right] . \tag{1.12}$$

The use of the symbol S to denote the price of the numeraire may be confusing, because S is usually used to denote a stock price. The numeraire

---

general proof is in fact difficult and requires a definition of "no arbitrage" that is considerably more complicated than the simple assumption (1.1) that is sufficient in the binomial model.

[9] The term "density" reflects the fact that in each state of the world $\phi(T)$ can be interpreted as the state price per unit of probability, just as the normal meaning of density is "mass per unit of volume."

[10] In general the expectation (or mean) of a random variable is an intuitive concept, and an intuitive understanding will be sufficient for this book, so I will not give a formal definition. It should be understood that we are assuming implicitly, whenever necessary, that the expectation exists (which is not always the case). In this regard, it is useful to note in passing that a product of two random variables XY has a finite mean whenever X and Y have finite variances.

here could be any non dividend-paying asset. For example, we can take $S(t) = e^{rt}$, the price of the risk-free asset. The definition of probabilities as

$$E[1_A\phi(T)e^{rT}] \tag{1.13}$$

will be called the "risk-neutral probability measure" or simply "risk-neutral measure" as before.

Different numeraires lead to different probability measures and hence to different expectations. To keep this straight, we will use the numeraire as a superscript on the expectation symbol: for example, E^S will denote expectation with respect to the probability measure that corresponds to S being the numeraire. Also, we will use the symbol $\text{prob}^S(A)$ to denote the probability of an event A when we use S as the numeraire. So, (1.11) and (1.12) will be written as

$$\text{prob}^S(A) = E\left[1_A\phi(T)\frac{S(T)}{S(0)}\right], \tag{1.14}$$

$$E^S[X] = E\left[X\phi(T)\frac{S(T)}{S(0)}\right]. \tag{1.15}$$

Our key result in the preceding section was that the ratio of the price of any non-dividend paying asset to the price of the numeraire asset has zero expected change when we use the probability measure corresponding to the numeraire. We will demonstrate the same result in this more general model. Recall that T denotes an arbitrary but fixed date at which we have defined the probabilities using S as the numeraire in (1.11). At each date $t < T$, let E_t^S denote the expectation given information at time t and using S as the numeraire (we will continue to write the expectation at date 0 without a subscript; i.e., E^S has the same meaning as E_0^S). Let Y denote the price of another non-dividend-paying asset. We will show that

$$\frac{Y(t)}{S(t)} = E_t^S\left[\frac{Y(T)}{S(T)}\right]. \tag{1.16}$$

Thus, the expected future (date-T) value of the ratio Y/S always equals the current (date-t) value when we use S as the numeraire. As discussed in the preceding section, the mathematical term for a random variable whose expected future value always equals its current value is "martingale." Thus, we can express equation (1.16) as stating that the ratio Y/S is a martingale when we compute expectations using the probability measure that corresponds to S being the numeraire.

The usefulness of equation (1.16) is that it gives us a formula for the asset price $Y(t)$ at any time t—and recall that this formula holds for every non-dividend paying asset. The formula is obtained from (1.16) by multiplying through by $S(t)$:

$$Y(t) = S(t)E_t^S\left[\frac{Y(T)}{S(T)}\right]. \tag{1.17}$$

We will call equation (1.17) **the fundamental pricing formula.** It is at the heart of modern pricing of derivative securities. It is a present value relation: the value at time t of the asset is the expectation of its value $Y(T)$ at time T "discounted" by the (possibly random) factor $S(t)/S(T)$. To emphasize that the numeraire can be any non-dividend-paying asset (and not necessarily a stock price, as the symbol S might suggest), we can write equation (1.17) in the equivalent form

$$Y(t) = \text{num}(t) E_t^{\text{num}} \left[\frac{Y(T)}{\text{num}(T)} \right] , \qquad (1.17')$$

where now $\text{num}(t)$ denotes the price of the (non-dividend-paying) numeraire asset at time t.

For example, letting $R(t)$ denote the value e^{rt} of the risk-free asset and using it as the numeraire, equation (1.17) becomes

$$Y(t) = e^{rt} E_t^R \left[\frac{Y(T)}{e^{rT}} \right] = e^{-r(T-t)} E_t^R [Y(T)] , \qquad (1.18)$$

which means that the value $Y(t)$ is the expected value of $Y(T)$ discounted at the risk-free rate for the remaining time $T - t$, when the expectation is computed under the risk-neutral probability measure.

We end this section with a proof of (1.16), a proof that the reader may skip if desired.[11]

Consider any time t and any event A that is distinguishable by time t. Consider the trading strategy of buying one share of the asset with price Y at time t when A has happened and financing this purchase by short selling $Y(t)/S(t)$ shares of the asset with price S. Each share of this asset that you short brings in $S(t)$ dollars, so shorting $Y(t)/S(t)$ shares brings in $Y(t)$ dollars, exactly enough to purchase the desired share of the first asset. Hold this portfolio until time T and then liquidate it. Liquidating it will generate

$$1_A \left(Y(T) - \frac{Y(t)}{S(t)} S(T) \right)$$

dollars. The multiplication by the random variable 1_A is because we only implement this strategy when A occurs (i.e., when $1_A = 1$). Consider the security that pays this number of dollars at time T. Because we obtained it with a trading strategy that required no investment at time t, its price at time 0 must be 0. We already observed that we can represent the price in terms of state prices, so we conclude that

$$E \left[\phi(T) 1_A \left(Y(T) - \frac{Y(t)}{S(t)} S(T) \right) \right] = 0 .$$

When we divide by $S(0)$, this will still equal zero. Factoring $S(T)$ outside the parentheses gives

---

[11] The proof is due to Harrison and Kreps [31]. See also Geman, El Karoui and Rochet [27]. We omit here technical assumptions regarding the existence of expectations.

$$E\left[1_A \frac{S(T)}{S(0)}\phi(T)\left(\frac{Y(T)}{S(T)} - \frac{Y(t)}{S(t)}\right)\right] = 0 \ .$$

We see from the formula (1.15) for expectations using S as the numeraire that we can write this as

$$E^S\left[1_A\left(\frac{Y(T)}{S(T)} - \frac{Y(t)}{S(t)}\right)\right] = 0 \ .$$

This is true for any event A distinguishable at time t, so the expectation of $Y(T)/S(T) - Y(t)/S(t)$ must be zero given any information at time t when we use S as the numeraire; i.e.,

$$E_t^S\left[\frac{Y(T)}{S(T)} - \frac{Y(t)}{S(t)}\right] = 0 \ ,$$

or, equivalently

$$E_t^S\left[\frac{Y(T)}{S(T)}\right] = \frac{Y(t)}{S(t)} \ .$$

1.5 Introduction to Option Pricing

A complete development of derivative pricing requires the continuous-time mathematics to be covered in the next chapter. However, we can present the basic ideas using the tools already developed. Consider the problem of pricing a European call option. Let T denote the maturity of the option and K its strike price, and let S denote the price of the underlying. We will assume for now that the underlying does not pay dividends, but we will make no assumptions about the distribution of its price $S(T)$ at the maturity of the option. Assume there is a risk-free asset with constant interest rate r.

Our convention will be that date 0 denotes the date at which we are attempting to value a derivative. The value of the option at maturity is $\max(0, S(T) - K)$. Consider a contract that pays $S(T)$ at date T when $S(T) \geq K$ and that pays zero when $S(T) < K$, and consider another contract that pays K at date T when $S(T) \geq K$ and zero when $S(T) < K$. In Chap. 3, we will call the first contract a "share digital" and the second contract a "digital." The call option is equivalent to a portfolio that is long the first contract and short the second, because the value of the call at maturity is $S(T) - K$ when $S(T) \geq K$ and it is zero otherwise. So, we can value the call if we can value the share digital and the digital. This "splitting up" of complex payoffs into simpler contracts is a key to analyzing many types of derivatives.

Pricing Share Digitals

Consider first the problem of valuing the share digital. Let $Y(t)$ denote its value at each date $t \leq T$. We seek to find $Y(0)$. Our fundamental pricing formula (1.17) tells us that

$$Y(0) = \text{num}(0)E^{\text{num}}\left[\frac{Y(T)}{\text{num}(T)}\right] \ ,$$

for any numeraire with price $\mathrm{num}(t)$. We want to choose the numeraire to simplify the calculation of the expectation. The expectation only involves the states of the world in which $S(T) \geq K$, because $Y(T) = 0$ when $S(T) < K$. In the states of the world in which $S(T) \geq K$, the value of the share digital is $S(T)$. The calculation of the expectation would be simplified if the value were a constant when it was nonzero, because, if you are to receive a constant amount in a certain event, your expected payoff is the constant times the probability of the event (e.g., the expected payoff of a gamble that pays \$1 when a fair die rolls a 6 is $1/6$). This suggests we should use the stock as the numeraire, because then we will have

$$\frac{Y(T)}{\mathrm{num}(T)} = \frac{S(T)}{S(T)} = 1$$

when $S(T) \geq K$, implying that

$$E^{\mathrm{num}} \left[\frac{Y(T)}{\mathrm{num}(T)} \right] = \mathrm{prob}^S \big(S(T) \geq K \big) \,,$$

where prob^S denotes the probability using S as the numeraire. This implies that the value of the share digital is

$$S(0) \times \mathrm{prob}^S \big(S(T) \geq K \big) \,.$$

The remaining question is obviously how to compute the probability. We will **not** use the formula (1.11) which expresses the probability in terms of an expectation involving state prices. To attempt to do so would simply raise the question of how to compute the state prices. Instead, we use the fundamental pricing formula again, this time replacing the derivative value Y with the value of the risk-free asset. This is exactly analogous to computing the "q probabilities" from (1.8b) and (1.8c) in Sects. 1.2. Recall that the fundamental formula holds for any non-dividend-paying asset, so it holds for $R(t) = \mathrm{e}^{rt}$, telling us that the ratio $R(t)/S(t)$ is a martingale when we use S as the numeraire. In a continuous-time model (at least until we introduce stochastic volatility) this will give us exactly the information we need to compute the distribution of $S(T)$ when we use S as the numeraire, and from the distribution of $S(T)$ we can easily compute $\mathrm{prob}^S \big(S(T) \geq K \big)$. This calculation will be covered in Chap. 3 for the Black-Scholes model.

Pricing Digitals

Now consider the problem of pricing the digital. We will change notation to let $Y(t)$ denote now the value of the digital at time t. Again we want to compute

$$Y(0) = \mathrm{num}(0) E^{\mathrm{num}} \left[\frac{Y(T)}{\mathrm{num}(T)} \right] \,,$$

and again this expectation only involves the states of the world in which $S(T) \geq K$. In these states of the world, the value of the digital is already a constant K, so we should take the numeraire to have a constant value at T, so that the ratio $Y(T)/\mathrm{num}(T)$ will be constant in the states in which $S(T) \geq K$. This means that we should take the numeraire to be the risk-free asset. For this numeraire, the pricing formula is

$$Y(0) = \mathrm{e}^{-rT} E^R[Y(T)] = \mathrm{e}^{-rT} K \times \mathrm{prob}^R(S(T) \geq K) \,,$$

so we need to compute the risk-neutral probability that $S(T) \geq K$. We will do this by using the fact that $S(t)/R(t) = \mathrm{e}^{-rt} S(t)$ is a martingale under the risk-neutral probability measure. This is analogous to computing the risk-neutral probabilities from (1.7b) and (1.7c) in Sects. 1.2. This calculation will also be covered in Chap. 3 for the Black-Scholes model.

Readers familiar with the Black-Scholes formula may already have surmised that, under the Black-Scholes assumptions,

$$\mathrm{prob}^S(S(T) \geq K) = \mathrm{N}(d_1) \quad \text{and} \quad \mathrm{prob}^R(S(T) \geq K) = \mathrm{N}(d_2) \,,$$

where N denotes the cumulative normal distribution function . The numbers d_1 and d_2 are different, and hence these are different probabilities, even though they are both probabilities of the option finishing in the money $(S(T) \geq K)$. They are different probabilities because they are computed under different numeraires.

A Remark

It seems worthwhile here to step back a bit from the calculations and try to offer some perspectives on the methods developed in this chapter. The change of numeraire technique probably seems mysterious. Even though one may agree that it works after following the steps in the chapter, there is probably a lingering question about why it works. The author's opinion is that it may be best to regard it simply as a "computational trick." Fundamentally it works because valuation is linear. Linearity simply means that the value of a cash flow $X = X_1 + X_2$ is the sum of the values of the cash flows X_1 and X_2 and the value of the cash flow aX is a times the value of X, for any constant a. This linearity is manifested in the statement that the value of a cash flow is the sum across states of the world of the state prices multiplied by the size of the cash flow in each state. The change of numeraire technique exploits the linearity to further simplify the valuation exercise. There are other ways the linearity can be used (for example, it produces solvable partial differential equations) but the particular trick we have developed in this chapter seems the most useful to the author (and to others, though perhaps not to everyone). After enough practice with it, it will seem as natural as other computational tricks one might have learned.

1.6 An Incomplete Markets Example

In this section, we consider a more difficult valuation problem than the binomial model and discuss the general implications of this example. We only need to make the problem slightly more difficult to see the issues. Consider a "trinomial" model, in which the asset price takes three possible values: $S_u > S_m > S_d$ ("m" for middle, medium, median, ...). We continue to make the "no arbitrage" assumption (1.1). State prices π_u, π_m and π_d must satisfy equations analogous to (1.5b)–(1.5c); specifically,

$$S = \pi_u S_u + \pi_m S_m + \pi_d S_d , \tag{1.19a}$$

$$1 = \pi_u e^{rT} + \pi_m e^{rT} + \pi_d e^{rT} . \tag{1.19b}$$

In the binomial case, these equations can be solved for π_u and π_d, as shown in (1.4). However, in the trinomial case, we have only two equations in three unknowns. Thus, there exist many solutions.

Given any particular solution (π_u, π_m, π_d) of (1.19), we can define the risk-neutral probabilities p_u, p_m and p_d as before—e.g., $p_u = \pi_u e^{rT}$. Likewise, we can define the probabilities using the stock as numeraire. Thus, we can value calls and puts and other derivative securities. However, the values we obtain will depend on the particular solution (π_u, π_m, π_d). There are many arbitrage-free values for a call option, one for each solution of (1.19).

The reason that there are many arbitrage-free values for a call (or put) is that a call cannot be replicated in a trinomial model using the stock and risk-free asset; we can say equivalently that there is no "delta hedge" for a call option. Recall that we first found the value of a call in the binomial model by finding the replicating portfolio and calculating its cost. A similar analysis is impossible in the trinomial model. To see this, consider a portfolio of a dollars invested in the risk free asset and b dollars invested in the stock. The value of the portfolio at date T will be $ae^{rT} + bS_x/S$, where $x \in \{u, m, d\}$. To replicate the call, we need a and b to satisfy

$$ae^{rT} + bS_u/S = \max(0, S_u - K) , \tag{1.20a}$$

$$ae^{rT} + bS_m/S = \max(0, S_m - K) , \tag{1.20b}$$

$$ae^{rT} + bS_d/S = \max(0, S_d - K) . \tag{1.20c}$$

These are three linear equations in the two unknowns a and b. For any strike price K between S_d and S_u, none of the equations is redundant, and the system has no solution. When there are state-contingent claims (such as the call option payoff) that cannot be replicated by trading in the marketed assets (the stock and risk-free asset in this case), one says that the market is "incomplete." Thus, the trinomial model is an example of an incomplete market.

To value derivative securities in this situation, we have to select some particular solution (π_u, π_m, π_d) of (1.19) and assume that the market uses that solution for valuation. Equivalently, we can assume the market uses a particular set of risk-neutral probabilities (p_u, p_m, p_d). This type of valuation is often called "equilibrium" valuation, as opposed to arbitrage valuation, because to give a foundation for our particular choice of risk-neutral probabilities, we would have to assume something about the preferences and endowments of investors and the production possibilities. We will encounter incomplete markets when we consider stochastic volatility in Chap. 4.

Problems

1.1. Create an Excel worksheet in which the user inputs S, S_d, S_u, K, r and T. Check that the no-arbitrage condition (1.1) is satisfied. Compute the value of a call option in each of the following ways:

(a) Compute the delta and use (1.2).
(b) Compute the state prices and use (1.5a).
(c) Compute the risk-neutral probabilities and use (1.6a).
(d) Compute the probabilities using the stock as numeraire and use (1.8a).

Verify that all of these methods produce the same answer.

1.2. In a binomial model, a put option is equivalent to δ_p shares of the stock, where $\delta_p = (P_u - P_d)/(S_u - S_d)$ (this will be negative, meaning a short position) and some money invested in the risk-free asset. Derive the amount of money x that should be invested in the risk-free asset to replicate the put option. The value of the put at date 0 must be $x + \delta_p S$.

1.3. Using the result of the previous exercise, repeat Problem 1.1 for a put option.

1.4. Here is a chance to apply option pricing theory to real life. Suppose you have a "significant other" who would marry you if you ask.

(a) What type of option do you have on marriage? Can you tell when it is in the money?
(b) Under what circumstances should you exercise this option early?
(c) What is the put option in a marriage contract called? (You shouldn't need a hint for this one, but, just in case, it is the name of a song made popular by Dolly Parton!).

Before anyone might be tempted to take this too literally, it should be pointed out that, in some "real option" settings, keeping one's options open has both advantages and disadvantages. Airbus' decision to build a new larger passenger plan can be seen as the early exercise of a call option, justified perhaps because by committing to do so it discouraged Boeing from launching a similar project, both companies presumably believing that the market is too small for both to enter. Thus, the exercise of a real option (commitment) can change the environment in ways that do not arise, or at least we assume not to arise, in financial markets.

2

Continuous-Time Models

This chapter has three objectives. The first is to introduce the concept of a Brownian motion. A Brownian motion is a random process (a variable that changes randomly over time) that evolves continuously in time and has the property that its change over any time period is normally distributed with mean zero and variance equal to the length of the time period. The "mean zero" feature means that a Brownian motion is a martingale. We will also give a different characterization (Levy's theorem) emphasizing the "quadratic variation" process, which is a property of the paths (how the variable evolves over time, in a given state of the world) of the process.

The second objective is to explain Itô's formula, which is the chain rule for stochastic calculus. In the Black-Scholes model, the stock price is assumed to satisfy

$$\frac{\mathrm{d}S}{S} = \mu \, \mathrm{d}t + \sigma \, \mathrm{d}B \, ,$$

where B is a Brownian motion. In the case that the stock pays no dividend, the rate of return is its price change $\mathrm{d}S$ divided by the initial price S, so the model states that the expected rate of return in each instant $\mathrm{d}t$ is $\mu \, \mathrm{d}t$ (of course, t denotes time, so $\mathrm{d}t$ is the change in time). The variance of the rate of return depends on σ. This model can be equivalently written in terms of the natural logarithm of S, which we will write as $\log S$. The above equation for the rate of return is equivalent to

$$\mathrm{d} \log S = \left(\mu - \frac{1}{2}\sigma^2 \right) \mathrm{d}t + \sigma \, \mathrm{d}B \, .$$

We will explain this equivalence and other similar calculations that are useful for pricing derivatives.

The third objective is to explain how, when we change numeraires, as described in the previous chapter, we can calculate the expectation in the fundamental pricing formula (1.17). The question is what effect does changing the numeraire (and hence the probability measure) have on the distribution of an asset price.

Everything in the remainder of the book is based on the mathematics presented in this chapter. For easy reference, the essential formulas have been highlighted in boxes.

2.1 Simulating a Brownian Motion

We begin with the fact that changes in the value of a Brownian motion are normally distributed with mean zero and variance equal to the length of the time period. Let $B(t)$ denote the value of a Brownian motion at time t. Then for any date $u > t$, given the information at time t, the random variable $B(u) - B(t)$ is normally distributed with mean zero and variance equal to $u - t$. Unless stated otherwise, our convention will be that a Brownian motion starts at $B(0) = 0$.

We can generate an approximate Brownian motion in Excel. To do so, we take a small time period Δt and define the value at the end of the period to be the value of the Brownian motion at the beginning plus a normally distributed variable with mean 0 and variance Δt. In the following procedure, the user is prompted to input the length T of the entire time period over which the Brownian motion is to be simulated and to input the number N of time periods of length Δt within the full interval $[0, T]$. The length Δt of each individual time period is then calculated as T/N. The quality of the approximation of this simulation to a true Brownian motion will be always be improved by increasing the number N. Plotting the output of the procedure creates a picture of what we call a "path" of the Brownian motion, which means that it shows the value taken at each time in one state of the world. Running the procedure again (for the same T and N) will create a different plot, which can be interpreted as the values of the Brownian motion in another state of the world. In other words, the path of the Brownian motion is itself random, depending in this approximation on the numbers produced by Excel's random number generating function.[1]

```
Sub Simulating_Brownian_Motion()
Dim T, dt, Sqrdt, BrownianMotion, i, N
T = InputBox("Enter the length of the time period (T)")
N = InputBox("Enter the number of periods (N)")
dt = T / N
Sqrdt = Sqr(dt)
```

---

[1] The generation of normally distributed random numbers in Excel is discussed in Appendix A. The function RandN() here is user-created (to simplify typing) to equal the function Application.NormSInv(Rnd()) supplied in VBA. The construction sqrtdt $* z$ scales the standard normal z so that its standard deviation is $\sqrt{\Delta t}$ and hence its variance is Δt, as desired. The subroutine creates two columns of data below the active cell in the Excel worksheet with headings "Time" and "Brownian Motion." To plot the path of the Brownian motion, select the two columns and insert an "XY (Scatter)" chart, with data points connected by lines.

```
ActiveCell.Value = "Time"
ActiveCell.Offset(0, 1) = "Brownian Motion"
ActiveCell.Offset(1, 0) = 0    ' beginning time
ActiveCell.Offset(1, 1) = 0    ' beginning value of Brownian motion
BrownianMotion = 0
For i = 1 To N
   ActiveCell.Offset(i + 1, 0) = i * dt              ' next time
   BrownianMotion = BrownianMotion + Sqrdt * RandN()
   ActiveCell.Offset(i + 1, 1) = BrownianMotion ' next value
Next i
End Sub
```

2.2 Quadratic Variation

If we take a large number N of time steps in the simulation of the preceding section, we will see the distinctive characteristic of a Brownian motion: it jiggles rapidly, moving up and down in a very erratic way. The name "Brownian motion" derives from the botanist Robert Brown's observations of the erratic behavior of particles suspended in a fluid. This has long been thought to be a reasonable model for the behavior of a stock price. The plot of other functions with which we may be familiar will be much smoother. This is captured in the concept of quadratic variation.

Consider a discrete partition

$$0 = t_0 < t_1 < t_2 < \cdots < t_N = T$$

of the time interval $[0, T]$. Let B be a Brownian motion and calculate the sum of squared changes

$$\sum_{i=1}^{N} [\Delta B(t_i)]^2 ,$$

where $\Delta B(t_i)$ denotes the change $B(t_i) - B(t_{i-1})$. If we consider finer partitions with the length of each time interval $t_i - t_{i-1}$ going to zero, the limit of the sum is called the "quadratic variation" of the process. For a Brownian motion, the quadratic variation over an interval $[0, T]$ is equal to T with probability one.

The functions with which we are normally familiar are continuously differentiable. If X is a continuously differentiable function of time (in each state of the world), then the quadratic variation of X will be zero. A simple example is a linear function: $X(t) = at$ for some constant a. Then, taking $t_i - t_{i-1} = \Delta t = T/N$ for each i, the sum of squared changes is

$$\sum_{i=1}^{N} [\Delta X(t_i)]^2 = \sum_{i=1}^{N} [a\,\Delta t]^2 = Na^2(\Delta t)^2 = Na^2 \left(\frac{T}{N}\right)^2 = \frac{a^2 T^2}{N} \to 0$$

as $N \to \infty$. Essentially the same argument shows that the quadratic variation of any continuously differentiable function is zero, because such a function is approximately linear at each point.

Thus, the jiggling of a Brownian motion, which leads to the nonzero quadratic variation, is quite unusual. To explain exactly how unusual it is, it is helpful to introduce the concept of "total variation," which is defined in the same way as quadratic variation but with the squared changes $[\Delta B(t_i)]^2$ replaced by the absolute value of the changes $|\Delta B(t_i)|$. If the quadratic variation of a continuous function is nonzero, then its total variation is necessarily infinite, so each path of a Brownian motion has infinite total variation (with probability one). It was mentioned above that, with a large number of time steps in the simulation of the preceding section, one could see the distinctive jiggling property of a Brownian motion. This is not quite right. Any plot drawn by a pencil (or a laser printer, for that matter) must have finite total variation, because the total variation is the total distance traveled by the pencil. Hence, no matter how many time steps one uses, one will never create a continuous plot with the nonzero quadratic variation (and infinite total variation) that a Brownian path has. Another way to understand this is to consider focusing on a small segment of a plot and viewing it with a magnifying glass. If the segment is small enough, and excluding the finite number of kinks that a pencil can draw in the plot of a function, it will look approximately like a straight line under the magnifying glass (with slope equal to the derivative of the function). However, if one could view a segment of a path of a true Brownian motion under a magnifying glass, it would look much the same as the entire picture does to the naked eye—no matter how small the segment, one would still see the characteristic jiggling.

One may well question why we should be interested in this curious mathematical object. The reason is that asset pricing inherently involves martingales (variables that evolve randomly over time in such a way that their expected changes are always zero), as our fundamental pricing equation (1.17) establishes. Furthermore, continuous processes (variables whose paths are continuous functions of time) are much more tractable mathematically than are processes that can jump at some instants. More importantly, it is possible in a mathematical model with continuous processes to define perfect hedges much more readily than it is in a model involving jump processes. So, we are led to a study of continuous martingales. An important fact is that any non-constant continuous martingale must have infinite total variation! So, the normal functions with which we are familiar are left behind once we enter the study of continuous martingales.

There remains perhaps the question of why we focus on Brownian motion within the world of continuous martingales. The answer here is that any continuous martingale is really just a transformation of a Brownian motion. This is a consequence of the following important fact, which is known as Levy's theorem:

> A continuous martingale is a Brownian motion if and only if its quadratic variation over each interval $[0, T]$ equals T.

Thus, among continuous martingales, a Brownian motion is defined by the condition that the quadratic variation over each interval $[0, T]$ is equal to T. This is really just a normalization. A different continuous martingale may have a different quadratic variation, but it can be converted to a Brownian motion just by deforming the time scale. Furthermore, many continuous martingales can be constructed as "stochastic integrals" with respect to a Brownian motion. We take up this topic in the next section.

2.3 Itô Processes

An Itô process is a variable X that changes over time as

$$dX(t) = \mu(t)\, dt + \sigma(t)\, dB(t) , \qquad (2.1)$$

where B is a Brownian motion, and μ and σ can also be random processes. Some regularity conditions are needed on μ and σ which we will omit, except for noting that $\mu(t)$ and $\sigma(t)$ should be known at time t. In particular, constant μ and σ are certainly acceptable. When we add the changes over time, we get

$$X(T) = X(0) + \int_0^T \mu(t)\, dt + \int_0^T \sigma(t)\, dB(t)$$

for any $T > 0$. There are other types of random processes, in particular, processes that can jump, but we will not consider them in this book.

We will not formally define the integral $\int_0^T \sigma(t)\, dB(t)$, but it should be understood as being approximately equal to a discrete sum of the form

$$\sum_{i=1}^N \sigma(t_{i-1})\, \Delta B(t_i) ,$$

where $0 = t_0 < \cdots t_N = T$ and the time periods $t_i - t_{i-1}$ are small. Given that we can simulate the changes $\Delta B(t_i)$ as random normals, we can approximately simulate the random variable $\int_0^T \sigma(t)\, dB(t)$ and hence we can approximately simulate $X(T)$.

An Itô process evolves continuously over time. We interpret $\mu(t)\, dt$ as the expected change in X in an instant dt. The quantity $\mu(t)$ is also called the "drift" of the process X at time t. The coefficient $\sigma(t)$ is called the "diffusion" coefficient of X at time t.

If μ and σ are constant, it is standard to refer to an Itô process X as a (μ, σ)–Brownian motion. Of course, it is not a martingale when $\mu \neq 0$. For example, when $\mu > 0$, X tends to increase over time. However, it has the jiggling property of a Brownian motion, scaled by the diffusion coefficient σ.

A very important fact is that an Itô process such as (2.1) can be a martingale only if $\mu = 0$. This should seem sensible, because $\mu \, dt$ is the expected change in X, and a process is a martingale only if its expected change is zero.[2] This observation plays a fundamental role in deriving asset pricing formulas, as we will begin to see in Sect. 2.9. Conversely, if $\mu = 0$ and

$$E\left[\int_0^T \sigma^2(t) \, dt\right] < \infty \qquad (2.2)$$

for each T, then the Itô process is a continuous martingale and the variance of its date–T value, calculated with the information available at date 0, is:

$$\text{var}[X(T)] = E\left[\int_0^T \sigma^2(t) \, dt\right].$$

Whether μ is zero or not, and independently of the assumption (2.2), the quadratic variation of the Itô process X is

$$\lim_{N \to \infty} \sum_{i=1}^N [\Delta X(t_i)]^2 = \int_0^T \sigma^2(t) \, dt \qquad (2.3)$$

with probability one. Thus we obtain (when $\mu = 0$ and (2.2) holds) a continuous martingale with a different quadratic variation than a Brownian motion via the diffusion function σ.

To "compute" the quadratic variation of an Itô process, we use the following simple and important rules (for the sake of brevity, we drop the "(t)" notation from $B(t)$ here and sometimes later):

$$(dt)^2 = 0 , \qquad (2.4a)$$
$$(dt)(dB) = 0 , \qquad (2.4b)$$
$$(dB)^2 = dt . \qquad (2.4c)$$

We apply these rules to "compute" the quadratic variation of X as follows:

[2] If the sources of uncertainty in the market can be modeled as Brownian motions, then in fact every martingale is an Itô process with $\mu = 0$. This is some justification for the assumption we will make in this book, when studying continuous-time models, that all martingales are Itô processes.

If $dX = \mu\,dt + \sigma\,dB$ for a Brownian motion B, then

$$\begin{aligned}
(dX)^2 &= (\mu\,dt + \sigma\,dB)^2 \\
&= \mu^2(dt)^2 + 2\mu\sigma(dt)(dB) + \sigma^2(dB)^2 \\
&= 0 + 0 + \sigma^2\,dt .
\end{aligned}$$

We integrate this from 0 to T to obtain the quadratic variation (2.3) over that time period:[3]

$$\int_0^T (dX(t))^2 = \int_0^T \sigma^2(t)\,dt . \tag{2.5}$$

2.4 Itô's Formula

First we recall some facts of the ordinary calculus. If $y = g(x)$ and $x = f(t)$ with f and g being continuously differentiable functions, then

$$\frac{dy}{dt} = \frac{dy}{dx} \times \frac{dx}{dt} = g'(x(t))f'(t) .$$

Over a time period $[0, T]$, this implies that

$$y(T) = y(0) + \int_0^T \frac{dy}{dt}\,dt = y(0) + \int_0^T g'(x(t))f'(t)\,dt .$$

Substituting $dx(t) = f'(t)\,dt$, we can also write this as

$$y(T) = y(0) + \int_0^T g'(x(t))\,dx(t) . \tag{2.6}$$

We can contrast (2.6) with a special case of Itô's formula for the calculus of Itô processes (the more general formula will be discussed in the next section). If B is a Brownian motion and $Y = g(B)$ for a twice-continuously differentiable function g, then

$$Y(T) = Y(0) + \int_0^T g'(B(t))\,dB(t) + \frac{1}{2}\int_0^T g''(B(t))\,dt . \tag{2.7}$$

---

[3] In a more formal mathematical presentation, one normally writes $d\langle X, X \rangle$ for what we are writing here as $(dX)^2$. This is the differential of the quadratic variation process, and the quadratic variation through date T is

$$\langle X, X \rangle(T) = \int_0^T d\langle X, X \rangle(t) = \int_0^T \sigma^2(t)\,dt .$$

Thus, relative to the ordinary calculus, Itô's formula has an "extra term" involving the second derivative g''. We can write (2.7) in differential form as

$$dY(t) = \frac{1}{2}g''(B(t))\,dt + g'(B(t))\,dB(t).$$

Thus, $Y = g(B)$ is an Itô process with drift $g''(B(t))/2$ and diffusion coefficient $g'(B(t))$.

To gain some intuition for the "extra term" in Itô's formula, we return to the ordinary calculus. Given dates $t < u$, the derivative defines a linear approximation of the change in y over this time period; i.e., setting $\Delta x = x(u) - x(t)$ and $\Delta y = y(u) - y(t)$, we have the approximation

$$\Delta y \approx g'(x(t))\,\Delta x \,.$$

A better approximation is given by the second-order Taylor series expansion

$$\Delta y \approx g'(x(t))\,\Delta x + \frac{1}{2}g''(x(t))\,[\Delta x]^2 \,.$$

An interpretation of (2.6) is that the linear approximation works perfectly for infinitesimal time periods dt, because we can compute the change in y over the time period $[0, T]$ by "summing up" the infinitesimal changes $g'(x(t))\,dx(t)$. In other words, the second-order term $\frac{1}{2}g''(x(t))\,[\Delta x]^2$ "vanishes" when we consider very small time periods.

The second-order Taylor series expansion in the case of $Y = g(B)$ is

$$\Delta Y \approx g'(B(t))\,\Delta B + \frac{1}{2}g''(B(t))\,[\Delta B]^2 \,.$$

For example, given a partition $0 = t_0 < t_1 < \cdots < t_N = T$ of the time interval $[0, T]$, we have, with the same notation we have used earlier,

$$Y(T) = Y(0) + \sum_{i=1}^{N} \Delta Y(t_i)$$

$$\approx Y(0) + \sum_{i=1}^{N} g'(B(t_{i-1}))\,\Delta B(t_i) + \frac{1}{2}\sum_{i=1}^{N} g''(B(t_{i-1}))\,[\Delta B(t_i)]^2 \,. \quad (2.8)$$

If we make the time intervals $t_i - t_{i-1}$ shorter, letting $N \to \infty$, we cannot expect that the "extra" term here will disappear, leading to the result (2.6) of the ordinary calculus, because we know that

$$\lim_{N \to \infty} \sum_{i=1}^{N} [\Delta B(t_i)]^2 = T \,,$$

whereas for the continuously differentiable function $x(t) = f(t)$, the same limit is zero. In fact it seems sensible to interpret the limit of $[\Delta B]^2$ as $(dB)^2 = dt$. This is perfectly consistent with Itô's formula: if we take the limit in (2.8), replacing the limit of $[\Delta B(t_i)]^2$ with $(dB)^2 = dt$, we obtain (2.7).

2.5 Multiple Itô Processes

Now consider two Itô processes

$$dX(t) = \mu_x(t)\,dt + \sigma_x(t)\,dB_x(t)\,, \tag{2.9a}$$
$$dY(t) = \mu_y(t)\,dt + \sigma_y(t)\,dB_y(t)\,, \tag{2.9b}$$

where B_x and B_y can be different Brownian motions. The relation between the two Brownian motions is determined by their covariance or correlation. Given dates $t < u$, we know that both changes $B_x(u) - B_x(t)$ and $B_y(u) - B_y(t)$ are normally distributed with mean 0 and variance equal to $u - t$. There will exist a (possibly random) process ρ such that the covariance of these two normally distributed random variables, given the information at date t, is

$$E_t\left[\int_t^u \rho(s)\,ds\right]\,.$$

The process ρ is called the correlation coefficient of the two Brownian motions, because when it is constant the correlation of the changes $B_x(u) - B_x(t)$ and $B_y(u) - B_y(t)$ is

$$\frac{\text{covariance}}{\text{product of standard deviations}} = \frac{\int_t^u \rho\,ds}{\sqrt{u-t}\sqrt{u-t}} = \frac{(u-t)\rho}{u-t} = \rho\,.$$

Moreover, given increasingly fine partitions $0 = t_0 < \cdots < t_N = T$ of an interval $[0, T]$ as before, we will have

$$\sum_{i=1}^N \Delta B_x(t_i) \times \Delta B_y(t_i) \to \int_0^T \rho(t)\,dt$$

as $N \to \infty$, with probability one.

We know that

$$\sum_{i=1}^N [\Delta X(t_i)]^2 \to \int_0^T \sigma_x^2(t)\,dt \quad \text{and} \quad \sum_{i=1}^N [\Delta Y(t_i)]^2 \to \int_0^T \sigma_y^2(t)\,dt\,. \tag{2.10}$$

Furthermore, it can be shown that the sum of products satisfies

$$\sum_{i=1}^N \Delta X(t_i) \times \Delta Y(t_i) \to \int_0^T \sigma_x(t)\sigma_y(t)\rho(t)\,dt\,. \tag{2.11}$$

By adding the rule
$$(dB_x)(dB_y) = \rho \, dt \tag{2.4d}$$
to the rules (2.4a)–(2.4c), we can "compute" the limit in (2.11) as

$$
\lim_{N \to \infty} \sum_{i=1}^{N} \Delta X(t_i) \times \Delta Y(t_i) = \int_0^T (dX)(dY)
$$

$$
= \int_0^T (\mu_x \, dt + \sigma_x \, dB_x)(\mu_y \, dt + \sigma_y \, dB_y)
$$

$$
= \int_0^T \sigma_x(t)\sigma_y(t)\rho(t) \, dt \,. \tag{2.12}
$$

The most general case of Itô's formula that we will need is for a function $Z(t) = g(t, X(t), Y(t))$ where X and Y are Itô processes as in (2.9). In this case, Itô's formula is[4]

$$
Z(T) = Z(0) + \int_0^T \frac{\partial g}{\partial t} \, dt + \int_0^T \frac{\partial g}{\partial x} \, dX(t) + \int_0^T \frac{\partial g}{\partial y} \, dY(t)
$$

$$
+ \frac{1}{2} \int_0^T \frac{\partial^2 g}{\partial x^2} \, (dX(t))^2 + \frac{1}{2} \int_0^T \frac{\partial^2 g}{\partial y^2} \, (dY(t))^2
$$

$$
+ \int_0^T \frac{\partial^2 g}{\partial x \partial y} \, (dX(t))(dY(t)) \,. \tag{2.13}
$$

In this equation, we apply the rules (2.4a)–(2.4d) to compute

$$(dX(t))^2 = \sigma_x^2(t) \, dt \,,$$
$$(dY(t))^2 = \sigma_y^2(t) \, dt \,,$$
$$(dX(t))(dY(t)) = \sigma_x(t)\sigma_y(t)\rho(t) \, dt \,.$$

Itô's formula (2.13) appears a bit simpler (and easier to remember) if we write it in "differential form." We have:

If $Z(t) = g(t, X(t), Y(t))$ where X and Y are Itô processes as in (2.9), then

$$
dZ = \frac{\partial g}{\partial t} \, dt + \frac{\partial g}{\partial x} \, dX + \frac{\partial g}{\partial y} \, dY + \frac{1}{2} \frac{\partial^2 g}{\partial x^2} \, (dX)^2 + \frac{1}{2} \frac{\partial^2 g}{\partial y^2} \, (dY)^2
$$

$$
+ \frac{\partial^2 g}{\partial x \partial y} \, (dX)(dY) \,. \tag{2.14}
$$

---

[4] We need to assume $g(t, x, y)$ is continuously differentiable in t and twice continuously differentiable in (x, y) for (2.13) and (2.14) to be valid. Note also that we are using a short-hand notation here. The partial derivatives of g will generally depend on t, $X(t)$ and $Y(t)$ just as g does.

2.6 Examples of Itô's Formula

The following are the applications of Itô's formula that will be used most frequently in the book. They follow from the boxed formula at the end of the previous section by taking $g(x, y) = xy$ or $g(x, y) = y/x$ or $g(x) = e^x$ or $g(x) = \log x$.

Products. If $Z = XY$, then $dZ = X\,dY + Y\,dX + (dX)(dY)$. We can write this as

$$\frac{dZ}{Z} = \frac{dX}{X} + \frac{dY}{Y} + \left(\frac{dX}{X}\right)\left(\frac{dY}{Y}\right). \qquad (2.15)$$

Ratios. If $Z = Y/X$, then

$$\frac{dZ}{Z} = \frac{dY}{Y} - \frac{dX}{X} - \left(\frac{dY}{Y}\right)\left(\frac{dX}{X}\right) + \left(\frac{dX}{X}\right)^2. \qquad (2.16)$$

Exponentials. If $Z = e^X$, then

$$\frac{dZ}{Z} = dX + \frac{(dX)^2}{2}. \qquad (2.17)$$

Logarithms. If $Z = \log X$, then

$$dZ = \frac{dX}{X} - \frac{1}{2}\left(\frac{dX}{X}\right)^2. \qquad (2.18)$$

Compounding/Discounting. Let

$$Y(t) = \exp\left(\int_0^t q(s)\,\mathrm{d}s\right)$$

for some (possibly random) process q and define $Z = XY$ for any Itô process X. The usual calculus gives us $\mathrm{d}Y(t) = q(t)Y(t)\,\mathrm{d}t$, and the product rule above implies

$$\frac{\mathrm{d}Z}{Z} = q\,\mathrm{d}t + \frac{\mathrm{d}X}{X} \,. \tag{2.19}$$

This is the same as in the usual calculus.

2.7 Reinvesting Dividends

Frequently, we will assume that the asset underlying a derivative security pays a "constant dividend yield," which we will denote by q. This means, for an asset with price $S(t)$, that the dividend "in an instant $\mathrm{d}t$" is $qS(t)\,\mathrm{d}t$. If the dividends are reinvested in new shares, the number of shares will grow exponentially at rate q. To see this, consider the portfolio starting with a single share of the asset and reinvesting dividends until some date T. Let $X(t)$ denote the number of shares resulting from this strategy at any time $t \leq T$. Then the dividend received at date t is $qS(t)X(t)\,\mathrm{d}t$, which can be used to purchase $qX(t)\,\mathrm{d}t$ new shares. This implies that $\mathrm{d}X(t) = qX(t)\,\mathrm{d}t$, or $\mathrm{d}X(t)/\mathrm{d}t = qX(t)$, and it is easy to check (and very well known) that this equation is solved by $X(t) = \mathrm{e}^{qt}X(0)$. In our case, with $X(0) = 1$, we have $X(t) = \mathrm{e}^{qt}$.

The dollar value of the trading strategy just described will be $X(t)S(t) = \mathrm{e}^{qt}S(t)$. Denote this by $V(t)$. This is the value of a non-dividend-paying portfolio, because all dividends are reinvested. From the Compounding/Discounting example in Sect. 2.6, we know that

$$\frac{\mathrm{d}V}{V} = q\,\mathrm{d}t + \frac{\mathrm{d}S}{S} \,. \tag{2.20}$$

This means that the rate of return on the portfolio is the dividend yield $q\,\mathrm{d}t$ plus the return $\mathrm{d}S/S$ due to capital gains.

2.8 Geometric Brownian Motion

Let

$$S(t) = S(0) \exp \left(\mu t - \sigma^2 t/2 + \sigma B(t) \right) \tag{2.21}$$

for constants μ and σ, where B is a Brownian motion. Using the product rule and the rule for exponentials, we obtain

$$\frac{dS}{S} = \mu \, dt + \sigma \, dB . \tag{2.22}$$

When we see an equation of the form (2.22), we should recognize (2.21) as the solution.

The process S is called a "geometric Brownian motion." In keeping with the discussion of Sect. 2.3, we interpret (2.22) as stating that $\mu \, dt$ is the expected rate of change of S and $\sigma^2 \, dt$ is the variance of the rate of change in an instant dt. We call μ the "drift" and σ the "volatility." The geometric Brownian motion will grow at the average rate of μ, in the sense that $E[S(t)] = e^{\mu t} S(0)$.

Taking the natural logarithm of (2.21) gives an equivalent form of the solution:

$$\log S(t) = \log S(0) + \left(\mu - \frac{1}{2}\sigma^2 \right) t + \sigma B(t) . \tag{2.23}$$

This shows that $\log S(t) - \log S(0)$ is a $(\mu - \sigma^2/2, \sigma)$–Brownian motion. Given information at time t, the logarithm of $S(u)$ for $u > t$ is normally distributed with mean $(u - t)(\mu - \sigma^2/2)$ and variance $(u - t)\sigma^2$. Because S is the exponential of its logarithm, S can never be negative. For this reason, a geometric Brownian motion is a better model for stock prices than is a Brownian motion.

The differential of (2.23) is

$$d \log S(t) = \left(\mu - \frac{1}{2}\sigma^2 \right) dt + \sigma \, dB(t) . \tag{2.24}$$

We conclude:

The equation
$$\frac{dS}{S} = \mu \, dt + \sigma \, dB$$
is equivalent to the equation
$$d \log S(t) = \left(\mu - \frac{1}{2}\sigma^2 \right) dt + \sigma \, dB(t) .$$
The solution of both equations is (2.21) or the equivalent formula (2.23).

Over a discrete time interval Δt, equation (2.24) implies that the change in the logarithm of S is

$$\Delta \log S = \left(\mu - \frac{1}{2}\sigma^2\right) \Delta t + \sigma \Delta B . \qquad (2.25)$$

If S is the price of a non-dividend-paying asset, then over the time period t_{i-1} to t_i, with $t_i - t_{i-1} = \Delta t$, we have

$$\Delta \log S = r_i \Delta t , \qquad (2.26)$$

where r_i is the continuously compounded annualized rate of return during the period Δt. This follows from the definition of the continuously compounded rate of return as the constant rate over the time period Δt that would cause S to grow (or fall) from $S(t_{i-1})$ to $S(t_i)$. To be precise, r_i is defined by

$$\frac{S(t_i)}{S(t_{i-1})} = e^{r_i \Delta t} ,$$

which is equivalent to (2.26). Thus, the geometric Brownian motion model (2.22) implies that the continuously compounded annualized rate of return over a period of length Δt is given by

$$r_i = \mu - \frac{1}{2}\sigma^2 + \frac{\sigma \Delta B}{\Delta t} .$$

This means that r_i is normally distributed with mean $\mu - \sigma^2/2$ and variance $\sigma^2/\Delta t$. Given historical data on the rates of return, the parameters μ and σ can be estimated by standard methods (see Chap. 4).

We can simulate a path of S by simulating the changes $\Delta \log S$. The random variable $\sigma \Delta B$ in (2.25) has a normal distribution with zero mean and variance equal to $\sigma^2 \Delta t$. We simulate it as $\sigma \sqrt{\Delta t}$ multiplied by a standard normal.

```
Sub Simulating_Geometric_Brownian_Motion()
Dim T, S, mu, sigma, dt, SigSqrdt, LogS, drift, i, N
T = InputBox("Enter the length of the time period (T)")
N = InputBox("Enter the number of periods (N)")
S = InputBox("Enter the initial stock price (S)")
mu = InputBox("Enter the expected rate of return (mu)")
sigma = InputBox("Enter the volatility (sigma)")
dt = T / N
SigSqrdt = sigma * Sqr(dt)
drift = (mu - 0.5 * sigma * sigma) * dt
LogS = Log(S)
ActiveCell.Value = "Time"
ActiveCell.Offset(0, 1) = "Stock Price"
ActiveCell.Offset(1, 0) = 0     ' beginning time
ActiveCell.Offset(1, 1) = S     ' beginning stock price
```

```
For i = 1 To N
  ActiveCell.Offset(i + 1, 0) = i * dt        ' next time
  LogS = LogS + SigSqrdt * RandN()
  ActiveCell.Offset(i + 1, 1) = Exp(LogS)     ' next stock price
Next i
End Sub
```

2.9 Numeraires and Probabilities

When we change probability measures, we cannot expect a process B that was a Brownian motion to remain a Brownian motion. The expected change in a Brownian motion must always be zero, but when we change probabilities, the expected change of B is likely to become nonzero. (Likewise, a martingale is unlikely to remain a martingale when we change probabilities.) However, the Brownian motion B will still be an Itô process under the new probability measure. In fact, every Itô process under one probability measure will still be an Itô process under the new probability measure, and the diffusion coefficient of the Itô process will be unaffected by the change in probabilities.[5] Changing probabilities only changes the drift of an Itô process.

In a sense, this should not be surprising. It was noted in Sect. 2.2 that a Brownian motion B can be defined as a continuous martingale with paths that jiggle in such a way that the quadratic variation over any interval $[0, T]$ is equal to T. Changing the probabilities will change the probabilities of the various paths (so it may affect the expected change in B) but it will not affect how each path jiggles. So, under the new probability measure, B should still be like a Brownian motion but it may have a nonzero drift. If we consider a general Itô process, the reasoning is the same. The diffusion coefficient σ determines how much each path jiggles, and this is unaffected by changing the probability measure. Furthermore, instantaneous covariances—the $(dX)(dY)$ terms—between Itô processes are unaffected by changing the probability measure. Only the drifts are affected.

As explained in Sect. 1.5, we need to know the distribution of the underlying under probability measures corresponding to different numeraires. Let S be the price of an asset that has a constant dividend yield q, and, as in Sect. 2.7, let $V(t) = e^{qt} S(t)$. This is the price of the portfolio in which all dividends are reinvested, and we have

$$\frac{dV}{V} = q\,dt + \frac{dS}{S} .$$

Let Y be the price of another another asset that does not pay dividends. Let $r(t)$ denote the instantaneous risk-free rate at date t and let $R(t) =$

---

[5] To be a little more precise, this is true provided sets of states of the world having zero probability continue to have zero probability when the probabilities are changed. Because of the way we change probability measures when we change numeraires (cf. (1.11)) this will always be true for us.

$\exp\left(\int_0^t r(s)\,\mathrm{d}s\right)$. Assume

$$\frac{\mathrm{d}S}{S} = \mu_s\,\mathrm{d}t + \sigma_s\,\mathrm{d}B_s\ ,$$

$$\frac{\mathrm{d}Y}{Y} = \mu_y\,\mathrm{d}t + \sigma_y\,\mathrm{d}B_y\ ,$$

where B_s and B_y are Brownian motions under the actual probability measure with correlation ρ, and where $\mu_s, \mu_y, \sigma_s, \sigma_y$ and ρ can be quite general random processes. We consider the dynamics of the asset price S under three different probability measures. In each case, we follow the same steps: (i) we note that the ratio of an asset price to the numeraire asset price must be a martingale, (ii) we use Itô's formula to calculate the drift of this ratio, and (iii) we use the fact that the drift of a martingale must be zero to compute the drift of $\mathrm{d}S/S$.

Risk-Neutral Probabilities

Under the risk-neutral measure, $Z(t)$ defined as

$$Z(t) = \frac{V(t)}{R(t)} = \exp\left(-\int_0^t r(s)\,\mathrm{d}s\right)V(t)$$

is a martingale. Using the compounding/discounting rule, we have

$$\frac{\mathrm{d}Z}{Z} = -r\,\mathrm{d}t + \frac{\mathrm{d}V}{V} = (q - r)\,\mathrm{d}t + \frac{\mathrm{d}S}{S}\ .$$

For Z to be a martingale, the drift ($\mathrm{d}t$ part) of $\mathrm{d}Z/Z$ must be zero. Therefore, the drift of $\mathrm{d}S/S$ must be $(r - q)\,\mathrm{d}t$ under the risk-neutral measure. Because the change of measure does not affect the volatility, this implies:

$$\frac{\mathrm{d}S}{S} = (r - q)\,\mathrm{d}t + \sigma_s\,\mathrm{d}B_s^*\ , \qquad (2.27)$$

where B_s^* is a Brownian motion under the risk-neutral measure.

Underlying as the Numeraire

When V is the numeraire, the process $Z(t)$ defined as

$$Z(t) = \frac{R(t)}{V(t)} = \frac{\exp\left(\int_0^t r(s)\,\mathrm{d}s\right)}{V(t)}$$

is a martingale. Using the rule for ratios, we have

$$\frac{dZ}{Z} = r\,dt - \frac{dV}{V} + \left(\frac{dV}{V}\right)^2 = (r - q + \sigma_s^2)\,dt - \frac{dS}{S}\,.$$

Because the drift of dZ/Z must be zero, this implies that the drift of dS/S is $(r - q + \sigma_s^2)\,dt$. We conclude that:

$$\frac{dS}{S} = (r - q + \sigma_s^2)\,dt + \sigma_s\,dB_s^*\,, \qquad (2.28)$$

where now B_s^* denotes a Brownian motion when $V(t) = e^{qt}S(t)$ is the numeraire.

Another Risky Asset as the Numeraire

When Y is the numeraire, $Z(t)$ defined as

$$Z(t) = \frac{V(t)}{Y(t)}$$

must be a martingale. Using again the rule for ratios, we have

$$\begin{aligned}
\frac{dZ}{Z} &= \frac{dV}{V} - \frac{dY}{Y} - \left(\frac{dV}{V}\right)\left(\frac{dY}{Y}\right) + \left(\frac{dY}{Y}\right)^2 \\
&= \frac{dV}{V} - \frac{dY}{Y} - \rho\sigma_s\sigma_y\,dt + \sigma_y^2\,dt \\
&= \frac{dS}{S} - \frac{dY}{Y} + (q - \rho\sigma_s\sigma_y\,dt + \sigma_y^2)\,dt\,.
\end{aligned}$$

We can apply our previous example to compute the dynamics of Y when Y is the numeraire. This shows that the drift of dY/Y is $(r + \sigma_y^2)\,dt$. Because the drift of dZ/Z must be zero, it follows that the drift of dS/S is $(r - q + \rho\sigma_s\sigma_y)\,dt$. We conclude that:

$$\frac{dS}{S} = (r - q + \rho\sigma_s\sigma_y)\,dt + \sigma_s\,dB_s^*\,, \qquad (2.29)$$

where B_s^* denotes a Brownian motion under the probability measure corresponding to the non-dividend-paying risky asset Y being the numeraire, and where ρ is the correlation of S and Y.

Notice that the formula (2.29), while more complicated, is also more general than the others. In fact, it includes the formulas (2.27) and (2.28) as special cases: (i) if Y is the price of the instantaneously risk-free asset, then $\sigma_y = 0$ and (2.29) simplifies to (2.27), and (ii) if $Y = V$, then $\sigma_y = \sigma_s$ and $\rho = 1$, so (2.29) simplifies to (2.28).

Further Discussion

It would be natural for one to ask at this point: "what is the Brownian motion B_s^* and where did it come from?" We have argued that once we know the drift, and the fact that the volatility does not change, we can immediately write down, for example,

$$\frac{\mathrm{d}S}{S} = (r - q)\,\mathrm{d}t + \sigma_s\,\mathrm{d}B_s^*$$

for a Brownian motion B_s^* under the risk-neutral measure. To answer this question, we will give here the definition of B_s^* under the risk-neutral measure. The definition shows that we are justified in writing down (2.27)–(2.29), but we will not repeat the definition each time we make a statement of this sort.

We showed that Z is a martingale under the risk-neutral measure, where Z satisfies

$$\frac{\mathrm{d}Z}{Z} = (q - r)\,\mathrm{d}t + \frac{\mathrm{d}S}{S} = (q - r + \mu_s)\,\mathrm{d}t + \sigma_s\,\mathrm{d}B_s \ . \tag{2.30}$$

Define $B_s^*(0) = 0$ and

$$\mathrm{d}B_s^* = \left(\frac{q - r + \mu_s}{\sigma_s}\right)\,\mathrm{d}t + \mathrm{d}B_s \ . \tag{2.31}$$

Then

$$\mathrm{d}B_s^* = \frac{1}{\sigma_s}\left(\frac{\mathrm{d}Z}{Z}\right)$$

and hence is a continuous martingale under the risk-neutral measure. We can compute its quadratic variation as

$$(\mathrm{d}B_s^*)^2 = \left(\frac{q - r + \mu_s}{\sigma_s}\right)^2 (\mathrm{d}t)^2 + 2\left(\frac{q - r + \mu_s}{\sigma_s}\right)(\mathrm{d}t)(\mathrm{d}B_s) + (\mathrm{d}B_s)^2 = \mathrm{d}t \ .$$

Therefore, by Levy's theorem (Sect. 2.2), B_s^* is a Brownian motion under the risk-neutral measure. From (2.30) and (2.31) we have

$$(q - r)\,\mathrm{d}t + \frac{\mathrm{d}S}{S} = \sigma_s\,\mathrm{d}B_s^* \quad \Longleftrightarrow \quad \frac{\mathrm{d}S}{S} = (r - q)\,\mathrm{d}t + \sigma_s\,\mathrm{d}B_s^* \ ,$$

as in (2.27).

2.10 Tail Probabilities of Geometric Brownian Motions

For each of the numeraires discussed in the previous section, we have

$$\mathrm{d}\log S = \alpha\,\mathrm{d}t + \sigma\,\mathrm{d}B \ , \tag{2.32}$$

for some α and σ, where B is a Brownian motion under the probability measure associated with the numeraire. Specifically, $\sigma = \sigma_s$, $B = B_s^*$, and

(1) for the risk-neutral measure, $\alpha = r - q - \sigma_s^2/2$,

(2) when $e^{qt} S(t)$ is the numeraire, $\alpha = r - q + \sigma_s^2/2$,

(3) when another risky asset price Y is the numeraire, $\alpha = r - q + \rho \sigma_s \sigma_y - \sigma_s^2/2$.

We will assume in this section that α and σ are constants. The essential calculation in pricing options, as we will see in the next chapter and in Chap. 8, is to compute $\mathrm{prob}(S(T) > K)$ and $\mathrm{prob}(S(T) < K)$ for a constant K (the strike price of an option), where prob denotes the probabilities at date 0 (the date we are pricing an option) associated with a particular numeraire.

Equation (2.32) gives us

$$\log S(T) = \log S(0) + \alpha T + \sigma B(T) .$$

Given this, we deduce

$$
\begin{aligned}
S(T) > K \quad &\Longleftrightarrow \quad \log S(T) > \log K \\
&\Longleftrightarrow \quad \sigma B(T) > \log K - \log S(0) - \alpha T \\
&\Longleftrightarrow \quad \frac{B(T)}{\sqrt{T}} > \frac{\log K - \log S(0) - \alpha T}{\sigma \sqrt{T}} \\
&\Longleftrightarrow \quad -\frac{B(T)}{\sqrt{T}} < \frac{\log S(0) - \log K + \alpha T}{\sigma \sqrt{T}} \\
&\Longleftrightarrow \quad -\frac{B(T)}{\sqrt{T}} < \frac{\log\left(\frac{S(0)}{K}\right) + \alpha T}{\sigma \sqrt{T}} .
\end{aligned}
\tag{2.33}
$$

The random variable on the left-hand side of (2.33) has the standard normal distribution—it is normally distributed with mean equal to zero and variance equal to one. As is customary, we will denote the probability that a standard normal is less than some number d as $\mathrm{N}(d)$. We conclude:

Assume $d \log S = \alpha \, dt + \sigma \, dB$, where B is a Brownian motion. Then, for any number K,

$$\mathrm{prob}(S(T) > K) = \mathrm{N}(d) , \tag{2.34}$$

where

$$d = \frac{\log\left(\frac{S(0)}{K}\right) + \alpha T}{\sigma \sqrt{T}} . \tag{2.35}$$

The probability $\mathrm{prob}(S(T) < K)$ can be calculated similarly, but the simplest way to derive it is to note that the events $S(T) > K$ and $S(T) < K$ are "complementary"—their probabilities sum to one (the event $S(T) = K$ having zero probability). Therefore $\mathrm{prob}(S(T) < K) = 1 - \mathrm{N}(d)$. This is the probability that a standard normal is greater than d, and by virtue of the symmetry of the standard normal distribution, it equals the probability that a standard normal is less than $-d$. Therefore, we have:

> Assume $\mathrm{d}\log S = \alpha\,\mathrm{d}t + \sigma\,\mathrm{d}B$, where B is a Brownian motion. Then, for any number K,
>
> $$\mathrm{prob}(S(T) < K) = \mathrm{N}(-d)\,, \tag{2.36}$$
>
> where d is defined in (2.35).

2.11 Volatilities

As mentioned in Sect. 2.8, when we encounter an equation of the form

$$\frac{\mathrm{d}S}{S} = \mu\,\mathrm{d}t + \sigma\,\mathrm{d}B$$

where B is a Brownian motion, we will say "σ is the volatility of S." For example, in the Black-Scholes model, the most important assumption is that the volatility of the underlying asset price is constant. We will occasionally need to compute the volatilities of products or ratios of random processes. These computations follow directly from Itô's formula.

Suppose

$$\frac{\mathrm{d}X}{X} = \mu_x\,\mathrm{d}t + \sigma_x\,\mathrm{d}B_x \qquad \text{and} \qquad \frac{\mathrm{d}Y}{Y} = \mu_y\,\mathrm{d}t + \sigma_y\,\mathrm{d}B_y\,,$$

where B_x and B_y are Brownian motions with correlation ρ, and μ_x, μ_y, σ_x, σ_y, and ρ may be quite general random processes.

Products

If $Z = XY$, then (2.15) gives us

$$\frac{\mathrm{d}Z}{Z} = (\mu_x + \mu_y + \rho\sigma_x\sigma_y)\,\mathrm{d}t + \sigma_x\,\mathrm{d}B_x + \sigma_y\,\mathrm{d}B_y\,. \tag{2.37}$$

The instantaneous variance of $\mathrm{d}Z/Z$ is calculated, using the rules for products of differentials, as

$$\left(\frac{\mathrm{d}Z}{Z}\right)^2 = (\sigma_x\,\mathrm{d}B_x + \sigma_y\,\mathrm{d}B_y)^2$$
$$= (\sigma_x^2 + \sigma_y^2 + 2\rho\sigma_x\sigma_y)\,\mathrm{d}t\,.$$

As will be explained below, the volatility is the square root of the instantaneous variance (dropping the $\mathrm{d}t$). This implies:

> The volatility of XY is
>
> $$\sqrt{\sigma_x^2 + \sigma_y^2 + 2\rho\sigma_x\sigma_y}\,. \tag{2.38}$$

Ratios

If $Z = Y/X$, then (2.16) gives us

$$\frac{dZ}{Z} = (\mu_y - \mu_x - \rho\sigma_x\sigma_y + \sigma_x^2)\, dt + \sigma_y\, dB_y - \sigma_x\, dB_x \,. \qquad (2.39)$$

The instantaneous variance of dZ/Z is therefore

$$\left(\frac{dZ}{Z}\right)^2 = (\sigma_y\, dB_y - \sigma_x\, dB_x)^2$$
$$= (\sigma_x^2 + \sigma_y^2 - 2\rho\sigma_x\sigma_y)\, dt \,.$$

This implies:

The volatility of Y/X is

$$\sqrt{\sigma_x^2 + \sigma_y^2 - 2\rho\sigma_x\sigma_y} \,. \qquad (2.40)$$

Further Discussion

To understand why taking the square root of $(dZ/Z)^2$ (dropping the dt) gives the volatility, consider for example the product case $Z = XY$. Define a random process B by $B(0) = 0$ and

$$dB = \frac{\sigma_x}{\sigma}\, dB_x + \frac{\sigma_y}{\sigma}\, dB_y \,, \qquad (2.41)$$

where σ is the volatility defined in (2.38). Then we can write (2.37) as

$$\frac{dZ}{Z} = (\mu_x + \mu_y + \rho\sigma_x\sigma_y)\, dt + \sigma\, dB \,. \qquad (2.42)$$

From the discussion in Sect. 2.3, we know that B is a continuous martingale. We can compute its quadratic variation from

$$(dB)^2 = \left(\frac{\sigma_x\, dB_x + \sigma_s\, dB_s}{\sigma}\right)^2$$
$$= \frac{(\sigma_x^2 + \sigma_s^2 + 2\rho\sigma_x\sigma_s)\, dt}{\sigma^2} \,,$$
$$= dt \,.$$

By Levy's theorem (see Sect. 2.2), any continuous martingale with this quadratic variation is necessarily a Brownian motion. Therefore, (2.42) shows that σ is the volatility of Z as defined at the beginning of the section.

Problems

2.1. Consider a discrete partition $0 = t_0 < t_1 < \cdots t_N = T$ of the time interval $[0, T]$ with $t_i - t_{i-1} = \Delta t = T/N$ for each i. Consider the function

$$X(t) = e^t .$$

Create a VBA subroutine, prompting the user to input T and N, which computes and prints $\sum_{i=1}^{N} [\Delta X(t_i)]^2$, where

$$\Delta X(t_i) = X(t_i) - X(t_{i-1}) = e^{t_i} - e^{t_{i-1}} .$$

Hint: The sum can be computed as follows.

```
sum = 0
For i = 1 To N
    DeltaX = Exp(i/N)-Exp((i-1)/N)
    sum = sum + DeltaX * DeltaX
Next i
```

2.2. Repeat the previous problem for the function $X(t) = t^3$. In both this and the previous problem, what happens to $\sum_{i=1}^{N} [\Delta X(t_i)]^2$ as $N \to \infty$?

2.3. Repeat the previous problem to compute $\sum_{i=1}^{N} [\Delta B(t_i)]^2$, where B is a simulated Brownian motion. For a given T, what happens to the sum as $N \to \infty$?

2.4. Repeat the previous problem, computing instead $\sum_{i=1}^{N} |\Delta B(t_i)|$ where $|\cdot|$ denotes the absolute value. What happens to this sum as $N \to \infty$?

2.5. Consider a discrete partition $0 = t_0 < t_1 < \cdots t_N = T$ of the time interval $[0, T]$ with $t_i - t_{i-1} = \Delta t = T/N$ for each i. Consider a geometric Brownian motion

$$\frac{\mathrm{d}Z}{Z} = \mu \, \mathrm{d}t + \sigma \, \mathrm{d}B .$$

An approximate path $\tilde{Z}(t)$ of the geometric Brownian motion can be simulated as

$$\Delta \tilde{Z}(t_i) = \tilde{Z}(t_{i-1}) [\mu \, \Delta t + \sigma \, \Delta B] . \tag{2.43}$$

The subroutine `Simulating_Geometric_Brownian_Motion` simulates a path Z of a geometric Brownian motion. Modify the subroutine to prompt the user to input T, N, σ, μ, and $Z(0)$ and to generate both a path $Z(t)$ and an approximate path $\tilde{Z}(t)$ according to (2.43), using the same ΔB for both paths and taking $\tilde{Z}(0) = Z(0)$. Plot both paths in the same figure. How well does the approximation work for large N? Warning: For N larger than about $100T$, the approximation will look perfect—you won't be able to tell that there are two plots in the figure.

3

Black-Scholes

In this chapter, we will study the value of European digital and share digital options and standard European puts and calls under the Black-Scholes assumptions. We will also explain how to calculate implied volatilities and the option "Greeks." The Black-Scholes assumptions are that the underlying asset pays a constant dividend yield q and has price S satisfying

$$\frac{\mathrm{d}S}{S} = \mu \, \mathrm{d}t + \sigma \, \mathrm{d}B \tag{3.1}$$

for a Brownian motion B. Here σ is assumed to be constant (though we will allow it to vary in a non-random way at the end of the chapter) and μ can be a quite general random process. It is also assumed that there is a constant continuously-compounded risk-free rate r.

Under these assumptions, we will complete the discussion of Sect. 1.5 to derive option pricing formulas. Recall that, to price a European call option, all that remains to be done is to calculate the probabilities of the option finishing in the money when we use the risk-free asset and the underlying asset as numeraires. We will do this using the results of Sect. 2.9. As in Sect. 1.5, we will approach the pricing of call and put options by first considering their basic building blocks: digitals and share digitals.

3.1 Digital Options

A digital (or "binary") option pays a fixed amount in a certain event and zero otherwise. Consider a digital that pays \$1 at date T if $S(T) > K$, where K is a number that is fixed by the contract. This means that the digital pays x dollars at date T where x is defined as

$$x = \begin{cases} 1 & \text{if } S(T) > K \,, \\ 0 & \text{otherwise} \,. \end{cases}$$

Using the risk-neutral pricing formula (1.18), the value of the digital at date 0 is $\mathrm{e}^{-rT} E^R[x]$. Note that

$$
\begin{aligned}
E^R[x] &= 1 \times \mathrm{prob}^R(x=1) + 0 \times \mathrm{prob}^R(x{=}0) \\
&= \mathrm{prob}^R(x=1) \\
&= \mathrm{prob}^R\big(S(T) > K\big) \,.
\end{aligned}
$$

So we need to calculate this probability of the digital finishing in the money.

In Sect. 2.9—see (2.27)—we learned that under the Black-Scholes assumption (3.1) we have

$$
\frac{\mathrm{d}S}{S} = (r - q)\,\mathrm{d}t + \sigma\,\mathrm{d}B^* \,,
$$

where B^* is a Brownian motion under the risk-neutral measure.[1] In Sect. 2.8, we observed that this is equivalent to

$$
\mathrm{d}\log S = \left(r - q - \frac{1}{2}\sigma^2\right)\,\mathrm{d}t + \sigma\,\mathrm{d}B^* \,.
$$

Now using the formulas (2.34)–(2.35), with $\alpha = r - q - \sigma^2/2$, we have $\mathrm{prob}^R\big(S(T) > K\big) = \mathrm{N}(d_2)$ where

$$
d_2 = \frac{\log\left(\frac{S(0)}{K}\right) + \left(r - q - \frac{1}{2}\sigma^2\right) T}{\sigma\sqrt{T}} \,. \tag{3.2}
$$

The notation d_2 is standard notation from the Black-Scholes formula, and we use it—rather than a simple d—to distinguish the number (3.2) from a similar number—to be called d_1 of course—that we will see in the next section. We conclude:

The value of a digital option that pays \$1 when $S(T) > K$ is $\mathrm{e}^{-rT}\,\mathrm{N}(d_2)$, where d_2 is defined in (3.2).

Consider now a digital that pays when the underlying asset price is low; i.e., consider a security that pays y dollars at date T where

$$
y = \begin{cases} 1 & \text{if } S(T) < K \,, \\ 0 & \text{otherwise} \,. \end{cases}
$$

Using risk-neutral pricing again, the value of this digital at date 0 is

---

[1] There is no other risky asset price Y in this model, so the subscripts we used in Sect. 2.9 on the volatility coefficients and on B and B^* to distinguish the Brownian motion driving S from the Brownian motion driving Y and to distinguish their volatilities are not needed here.

$$\mathrm{e}^{-rT}E^R[y] = \mathrm{e}^{-rT}\mathrm{prob}^R(y-1) = \mathrm{e}^{-rT}\mathrm{prob}^R\big(S(T) < K\big)\;.$$

From this fact and the formula (2.36), we conclude:

> The value of a digital option that pays \$1 when $S(T) < K$ is $\mathrm{e}^{-rT}\,\mathrm{N}(-d_2)$, where d_2 is defined in (3.2).

3.2 Share Digitals

Consider a derivative security that pays one share of the underlying asset at date T if $S(T) > K$ and pays zero otherwise. This is called a "share digital." As before, let

$$x = \begin{cases} 1 & \text{if } S(T) > K\;, \\ 0 & \text{otherwise}\;. \end{cases}$$

Then the payoff of the share digital at date T is $xS(T)$. Let $Y(t)$ denote the value of this claim for $0 \le t \le T$. We have $Y(T) = xS(T)$ and we want to find $Y(0)$.

From Sect. 2.7, we know that $V(t) = \mathrm{e}^{qt}S(t)$ is the price of a non-dividend-paying portfolio. From our fundamental pricing formula (1.17), using V as the numeraire, we have

$$Y(0) = S(0)E^V\left[\frac{Y(T)}{\mathrm{e}^{qT}S(T)}\right]$$
$$= \mathrm{e}^{-qT}S(0)E^V[x]\;.$$

As in the previous section, $E^V[x] = \mathrm{prob}^V(x = 1)$, so we need to compute this probability of the option finishing in the money.

We follow the same steps as in the previous section. From (2.28) we have

$$\frac{\mathrm{d}S}{S} = (r - q + \sigma^2)\,\mathrm{d}t + \sigma\,\mathrm{d}B^*,$$

where now B^* denotes a Brownian motion when V is the numeraire. This is equivalent to

$$\mathrm{d}\log S = \left(r - q + \frac{1}{2}\sigma^2\right)\,\mathrm{d}t + \sigma\,\mathrm{d}B^*\;. \tag{3.3}$$

Thus, from the formulas (2.34)–(2.35), with $\alpha = r - q + \sigma^2/2$, we have

$$\mathrm{prob}^V\big(S(T) > K\big) = \mathrm{N}(d_1)\;,$$

where

$$d_1 = \frac{\log\left(\frac{S(0)}{K}\right) + \left(r - q + \frac{1}{2}\sigma^2\right)T}{\sigma\sqrt{T}}.$$

(3.4)

This implies:

> The value of a share digital that pays one share when $S(T) > K$ is $e^{-qT}S(0)\,N(d_1)$, where d_1 is defined in (3.4).

Consider now a share digital that pays one share of the stock at date T if $S(T) < K$. Letting

$$y = \begin{cases} 1 & \text{if } S(T) < K\,, \\ 0 & \text{otherwise}\,, \end{cases}$$

the payoff of this option is $yS(T)$. Its value at date 0 is

$$e^{-qT}S(0)E^V[y] = e^{-qT}S(0) \times \text{prob}^V(y = 1)$$
$$= e^{-qT}S(0) \times \text{prob}^V(S(T) < K)\,,$$

and from the formula (2.36) we have

$$\text{prob}^V(S(T) < K) = N(-d_1)\,.$$

We conclude:

> The value of a share digital that pays one share when $S(T) < K$ is $e^{-qT}S(0)\,N(-d_1)$, where d_1 is defined in (3.4).

3.3 Puts and Calls

A European call option pays $S(T) - K$ at date T if $S(T) > K$ and 0 otherwise. Again letting

$$x = \begin{cases} 1 & \text{if } S(T) > K\,, \\ 0 & \text{otherwise}\,, \end{cases}$$

the payoff of the call can be written as $xS(T) - xK$. This is equivalent to one share digital minus K digitals, with the digitals paying in the event that $S(T) > K$. The share digital is worth $e^{-qT}S(0)\,N(d_1)$ at date 0 and each digital is worth $e^{-rT}N(d_2)$. Note that equations (3.2) and (3.4) for d_1 and d_2 imply $d_2 = d_1 - \sigma\sqrt{T}$. Therefore, combining the results of the previous two sections yields the Black-Scholes formula:

The value of a European call option at date 0 is

$$e^{-qT} S(0) \, N(d_1) - e^{-rT} K \, N(d_2) \,, \tag{3.5}$$

where d_1 is defined in (3.4) and $d_2 = d_1 - \sigma \sqrt{T}$.

A European put option pays $K - S(T)$ at date T if $S(T) < K$ and 0 otherwise. As before, let

$$y = \begin{cases} 1 & \text{if } S(T) < K \,, \\ 0 & \text{otherwise} \,. \end{cases}$$

The payoff of the put option is $yK - yS(T)$. This is equivalent to K digitals minus one share digital, all of the digitals paying when $S(T) < K$. Thus, we have:

The value of a European put option at date 0 is

$$e^{-rT} K \, N(-d_2) - e^{-qT} S(0) \, N(-d_1) \,, \tag{3.6}$$

where d_1 is defined in (3.4) and $d_2 = d_1 - \sigma \sqrt{T}$.

Again, this is the Black-Scholes formula.

The values of the European put and call satisfy put-call parity, and we can also find one from the other by[2]

$$e^{-rT} K + \text{Call Price} = e^{-qT} S(0) + \text{Put Price} \,. \tag{3.7}$$

3.4 Greeks

The derivatives (calculus derivatives, not financial derivatives!) of an option pricing formula with respect to the inputs are commonly called "Greeks." The most important Greek is the option "delta." This measures the sensitivity of the option value to changes in the value of the underlying asset. The following table shows the standard Greeks, with reference to the Black-Scholes pricing formula.

---

[2] The put-call parity relation follows from the fact that both the left and the right-hand sides are the prices of portfolios that have value $\max(S(T), K)$ at the maturity of the option. To see this for the left-hand side, note that $e^{-rT} K$ is sufficient cash to accumulate to K at date T, allowing exercise of the call when it is in the money and retention of the cash K otherwise. For the right-hand side, note that $e^{-qT} S(0)$ is enough cash to buy e^{-qT} shares of the stock at date 0 which, with reinvestment of dividends, will accumulate to one share at date T, enabling exercise of the put if it is in the money or retention of the share otherwise.

Table 3.1. Black-Scholes Greeks

Input	Input Symbol	Greek	Greek Symbol
Stock price	S	delta	δ
delta	δ	gamma	Γ
- Time to maturity	$-T$	theta	Θ
Volatility	σ	vega	$\mathcal{V}$
Interest rate	r	rho	ρ

The second line of the above shows δ as an input.[3] Of course, it is not an input but instead is calculated. Gamma, the derivative of δ, is the second derivative of the option price with respect to the underlying asset price. The reason for calculating Θ as the derivative with respect to $-T$ instead of T is that the time-to-maturity T decreasing ($-T$ increasing) is equivalent to time passing, so Θ measures the change in the option value when time passes.

We can calculate these from the Black-Scholes formula using the chain rule from differential calculus. The derivative of the normal distribution function N is the normal density function n defined as

$$\mathrm{n}(d) = \frac{1}{\sqrt{2\pi}}\mathrm{e}^{-d^2/2} \ .$$

One can easily verify directly that

$$\mathrm{e}^{-qT}S\,\mathrm{n}(d_1) = \mathrm{e}^{-rT}K\,\mathrm{n}(d_2) \ , \tag{3.8}$$

which simplifies the calculations for the Black-Scholes call option pricing formula. For this formula, the Greeks are as follows:

$$\delta = \mathrm{e}^{-qT}\,\mathrm{N}(d_1) + \mathrm{e}^{-qT}S\,\mathrm{n}(d_1)\frac{\partial d_1}{\partial S} - \mathrm{e}^{-rT}K\,\mathrm{n}(d_2)\frac{\partial d_2}{\partial S}$$

$$= \mathrm{e}^{-qT}\,\mathrm{N}(d_1) + \mathrm{e}^{-qT}S\,\mathrm{n}(d_1)\left(\frac{\partial d_1}{\partial S} - \frac{\partial d_2}{\partial S}\right)$$

$$= \mathrm{e}^{-qT}\,\mathrm{N}(d_1) \ ,$$

$$\Gamma = \mathrm{e}^{-qT}\,\mathrm{n}(d_1)\frac{\partial d_1}{\partial S} = \mathrm{e}^{-qT}\,\mathrm{n}(d_1)\frac{1}{S\sigma\sqrt{T}} \ ,$$

---

[3] The delta is frequently denoted by the upper case Δ, but we will use the lower case, reserving the upper case for discrete changes, e.g., Δt. One may have noticed also that the symbol for vega is a little different from the others; this reflects the fact that vega is not actually a Greek letter.

$$\Theta = -e^{-qT}S\,n(d_1)\frac{\partial d_1}{\partial T} + qe^{-qT}S\,N(d_1)$$

$$+ e^{-rT}K\,n(d_2)\frac{\partial d_2}{\partial T} - re^{-rT}K\,N(d_2)$$

$$= e^{-qT}S\,n(d_1)\left(\frac{\partial d_2}{\partial T} - \frac{\partial d_1}{\partial T}\right)$$

$$+ qe^{-qT}S\,N(d_1) - re^{-rT}K\,N(d_2)$$

$$= -e^{-qT}S\,n(d_1)\frac{\sigma}{2\sqrt{T}} + qe^{-qT}S\,N(d_1) - re^{-rT}K\,N(d_2)\,,$$

$$\mathcal{V} = e^{-qT}S\,n(d_1)\frac{\partial d_1}{\partial \sigma} - e^{-rT}K\,n(d_2)\frac{\partial d_2}{\partial \sigma}$$

$$= e^{-qT}S\,n(d_1)\left(\frac{\partial d_1}{\partial \sigma} - \frac{\partial d_2}{\partial \sigma}\right)$$

$$= e^{-qT}S\,n(d_1)\sqrt{T}\,,$$

$$\rho = e^{-qT}S\,n(d_1)\frac{\partial d_1}{\partial r} - e^{-rT}K\,n(d_2)\frac{\partial d_2}{\partial r} + Te^{-rT}K\,N(d_2)$$

$$= e^{-qT}S\,n(d_1)\left(\frac{\partial d_1}{\partial r} - \frac{\partial d_2}{\partial r}\right) + Te^{-rT}K\,N(d_2)$$

$$= Te^{-rT}K\,N(d_2)\,.$$

We can calculate the Greeks of a European put option from the call option Greeks and put-call parity:

$$\text{Put Price} = \text{Call Price} + e^{-rT}K - e^{-qT}S(0)\,.$$

For example, the delta of a put is the delta of a call (with the same strike and maturity) minus e^{-qT}, and the gamma of a put is the same as the gamma of the corresponding call.

3.5 Delta Hedging

The ability to create a fully hedged (risk-free) portfolio of the stock and an option is the essence of the arbitrage argument underlying the Black-Scholes formula, as we saw in Chap. 1 for the binomial model. For a call option, such a portfolio consists of delta shares of the underlying asset and a short call option, or a short position of delta shares of the underlying and a long call option. These portfolios have no instantaneous exposure to the price of the underlying. To create a perfect hedge, the portfolio must be adjusted continuously, because the delta changes when the price of the underlying changes and when time passes. In practice, any hedge will therefore be imperfect, even if the assumptions of the model are satisfied.

We first consider the continuous-time hedging argument. Consider a European call option with maturity T, and let $C(S,t)$ denote the value of the option at date $t < T$ when the stock price is S at date t. Consider a portfolio that is short one call option and long δ shares of the underlying asset and that has a (short) cash position equal to $C - \delta S$. This portfolio has zero value at date t.

The change in the value of the portfolio in an instant dt is

$$-dC + \delta\, dS + q\delta S\, dt + (C - \delta S)r\, dt \ . \tag{3.9}$$

The first term reflects the change in the value of the option, the second term is the capital gain or loss on δ shares of stock, the third term is the dividends received on δ shares of stock, and the fourth term is the interest expense on the short cash position.

On the other hand, we know from Itô's formula that

$$dC = \frac{\partial C}{\partial S}\, dS + \frac{\partial C}{\partial t}\, dt + \frac{1}{2}\frac{\partial^2 C}{\partial S^2}(dS)^2$$
$$= \delta\, dS + \Theta\, dt + \frac{1}{2}\Gamma\sigma^2 S^2\, dt \ . \tag{3.10}$$

Substituting (3.10) into (3.9) shows that the change in the value of the portfolio is

$$-\Theta\, dt - \frac{1}{2}\Gamma\sigma^2 S^2\, dt + q\delta S\, dt + (C - \delta S)r\, dt \ . \tag{3.11}$$

Several aspects of this are noteworthy. First, as noted earlier, the delta hedge (being long δ shares of the underlying) eliminates the exposure to changes in the price of the underlying—there is no dS term in (3.11). Second, Θ will be negative, because it captures the time decay in the option value; being short the option means the portfolio will profit from time decay at rate $-\Theta$. Third, this portfolio is "short gamma." We can also say it is "short convexity," the term "convexity" referring to the convex shape of the option value as a function of the price of the underlying, which translates mathematically to a positive second derivative (gamma). The volatility in the stock makes convexity valuable, and a portfolio that is short convexity will suffer losses. Finally, the portfolio is earning dividends but paying interest.

It is straightforward to check, from the definitions of Θ, Γ and δ in the preceding section, that the sum of the terms in (3.11) is zero. The time decay in the option value and dividends received on the shares of the underlying exactly offset the losses due to convexity and interest. Therefore, the delta hedge is a perfect hedge. The portfolio, which has a zero cost, neither earns nor loses money. This is true not only on average but for every possible change in the stock price.

To see how well this works with only discrete adjustments to the hedge, one can simulate the changes in S over time and sum the gains and losses over discrete rebalancing periods. One should input the actual (not risk-neutral) expected rate of return on the asset to compute the actual distribution of gains and losses. This is discussed further in Sect. 3.10.

3.6 Gamma Hedging

To attempt to improve the performance of a discretely rebalanced delta hedge, one can use another option to create a portfolio that is both delta and gamma neutral. Being delta neutral means hedged as in the previous section—the portfolio value has no exposure to changes in the underlying asset price. In other words, it means that the derivative of the portfolio value with respect to the price of the underlying (the portfolio delta) is zero. Being gamma neutral means that the delta of the portfolio has no exposure to changes in the underlying price, which is equivalent to the second derivative of the portfolio value with respect to the price of the underlying (the portfolio gamma) being zero. If the delta truly did not change, then there would be no need to rebalance continuously, and hence no hedging error introduced by only adjusting the portfolio at discrete times rather than continuously. However, there is certainly no guarantee that a discretely-rebalanced delta/gamma hedge will perform better than a discretely rebalanced delta hedge.

A delta/gamma hedge can be constructed as follows. Suppose we have written (shorted) a call option and we want to hedge both the delta and gamma using the underlying asset and another option, for example, another call option with a different strike. In practice, one would want to use a liquid option for this purpose, which typically means that the strike of the option will be near the current value of the underlying (i.e., the option used to hedge would be approximately at the money).

Let δ and Γ denote the delta and gamma of the written option and let δ' and Γ' denote the delta and gamma of the option used to hedge. Consider holding a shares of the stock and b units of the option used to hedge in conjunction with the short option. The delta of the stock is one ($dS/dS = 1$), so to obtain a zero portfolio delta we need

$$0 = -\delta + a + b\delta'. \tag{3.12}$$

The gamma of the stock is zero ($d^2S/dS^2 = d1/dS = 0$), so to obtain a zero portfolio gamma we need

$$0 = -\Gamma + b\Gamma'. \tag{3.13}$$

Equation (3.13) shows that we should hold enough of the second option to neutralize the gamma of the option we have shorted; i.e.,

$$b = \frac{\Gamma}{\Gamma'}$$

Equation (3.12) shows that we should use the stock to delta hedge the portfolio of options; i.e.,

$$a = \delta - \frac{\Gamma}{\Gamma'}\delta' \ .$$

3.7 Implied Volatilities

All of the inputs into the option pricing formulas are in theory observable, except for the volatility coefficient σ. We can estimate σ from historical data (see Chap. 4), or estimate it from the prices of other options. The latter method exploits the fact that there is a one-to-one relationship between the price given by the Black-Scholes formula and the σ that is input, so one can take the price as given and infer σ from the formula. The σ computed in this way is called the "implied volatility." The implied volatility from one option can be used to price another (perhaps non-traded or less actively traded) option. The calculation of implied volatilities is discussed in Sect. 3.10.

Even if we acknowledge that the model is not correct, the computation of implied volatilities is still useful for characterizing market prices, because we can quickly describe an option as "expensive" or "cheap" depending on whether its implied volatility is large or small. Somewhat paradoxically, it is less easy to see if an option is expensive or cheap by looking at its price, because one must consider the price in the context of the exercise price and maturity. To some extent, the implied volatility normalizes the price relative to the exercise price and maturity. Of course, it does not always pay to sell expensive options or buy cheap options, unless they are expensive or cheap relative to an accurate model!

3.8 Term Structure of Volatility

The option pricing formulas in this chapter are derived from the fact that the natural logarithm of the stock price at maturity is normally distributed with a certain mean (depending on the numeraire) and variance equal to $\sigma^2 T$. It is not actually necessary that the volatility be constant. The formulas are still valid if

$$\frac{dS(t)}{S(t)} = \mu(t)\,dt + \sigma(t)\,dB(t)$$

where $\sigma(t)$ is some non-random function of time (and again μ can be a quite general random process). In this case, the variance of $\log S(T)$ will be

$$\int_0^T \sigma^2(t)\,dt\,,\tag{3.14}$$

which is essentially the sum of the instantaneous variances $\sigma^2(t)\,dt$. In the d_1's and d_2's in the option pricing formulas, $\sigma^2 T$ should be replaced by (3.14). A convenient way of expressing this is as follows. Let σ_{avg} be the positive number such that

$$\sigma_{\text{avg}}^2 = \frac{1}{T}\int_0^T \sigma^2(t)\,dt\,.\tag{3.15}$$

Then we simply need to input σ_{avg} as sigma in our option pricing functions. We will call σ_{avg} the "average volatility," though note that it is not really the average of $\sigma(t)$ but instead is the square root of the average of $\sigma^2(t)$.

It is important to recognize that, throughout this chapter, date 0 means the date at which the option is being valued. It is not necessarily the date at which the option was first bought or sold. So σ_{avg} is the average (in a sense) volatility during the remaining lifetime of the option, which need not be the same as the average during the option's entire lifetime. It is this remaining volatility that is important for pricing and hedging. Moreover, it is a mistake at date 0 to use $\sigma(0)$ as the volatility to compute prices and hedges. Instead, prices and hedges should be based on σ_{avg}.

These considerations provide a way to address the following situation. If we compute implied volatilities for options with different maturities, we will normally get different numbers. For example, consider two at-the-money options with maturities T_1 and T_2 where $T_2 > T_1$. Denote the implied volatilities by $\hat{\sigma}_1$ and $\hat{\sigma}_2$. We want to interpret these as average volatilities for the time periods $[0, T_1]$ and $[0, T_2]$ respectively. This requires the existence of a function $\sigma(t)$ such that

$$\hat{\sigma}_1^2 = \frac{1}{T_1} \int_0^{T_1} \sigma^2(t) \, \mathrm{d}t \quad \text{and} \quad \hat{\sigma}_2^2 = \frac{1}{T_2} \int_0^{T_2} \sigma^2(t) \, \mathrm{d}t \,.$$

This would imply

$$\hat{\sigma}_2^2 T_2 - \hat{\sigma}_1^2 T_1 = \int_{T_1}^{T_2} \sigma^2(t) \, \mathrm{d}t \,,$$

which requires

$$\hat{\sigma}_2^2 T_2 - \hat{\sigma}_1^2 T_1 \geq 0 \,.$$

Equivalently,

$$\hat{\sigma}_2 \geq \sqrt{\frac{T_1}{T_2}} \hat{\sigma}_1 \,.$$

Provided this last inequality is satisfied, we can easily construct the function $\sigma(t)$ as

$$\sigma(t) = \begin{cases} \hat{\sigma}_1 & \text{for } t \leq T_1 \\ \sqrt{\frac{\hat{\sigma}_2^2 T_2 - \hat{\sigma}_1^2 T_1}{T_2 - T_1}} & \text{for } T_1 < t \leq T_2. \end{cases}$$

More generally, given a sequence of at-the-money options with maturities $T_1 < T_2 < \cdots T_N$ and implied volatilities $\hat{\sigma}_1, \ldots, \hat{\sigma}_N$, we define

$$\sigma(t) = \sqrt{\frac{\hat{\sigma}_{i+1}^2 T_{i+1} - \hat{\sigma}_i^2 T_i}{T_{i+1} - T_i}}$$

for $T_i < t \leq T_{i+1}$, provided the expression inside the square root symbol is positive. This $\sigma(t)$ is often called the "term structure of (implied) volatilities." Generally, we may expect $\sigma(t)$ to be a decreasing function of time t when the current market is especially volatile and to be an increasing function when the current market is especially quiet.

3.9 Smiles and Smirks

If we compute implied volatilities for options with the same maturity but different strikes, we will again obtain different implied volatilities for different options. If we plot implied volatility against the strike, the pattern one normally sees for equities and equity indices is the implied volatility declining as the strike increases until the strike is somewhere near the current value of the underlying (so the option is at the money). The implied volatility will then generally flatten out or increase slightly at higher strikes. The graph looks like a twisted smile (smirk). This pattern has been very pronounced in equity index option prices since the crash of 1987. In contrast to the term structure of implied volatilities, this "moneyness" structure of implied volatilities is simply inconsistent with the model. It suggests that the risk-neutral return distribution is not lognormal but instead exhibits a higher likelihood of extreme returns than the lognormal distribution (i.e., it has "fat tails") with the likelihood of extreme negative returns being higher than the likelihood of extreme positive returns (i.e., it is "skewed"). We will return to this subject in Sect. 4.6.

3.10 Calculations in VBA

The Black-Scholes call and put formulas and Greeks can easily be calculated in an Excel worksheet, using the standard functions Exp, Ln and the cumulative normal distribution function, which is provided in Excel as NormSDist. However, if these are to be used repeatedly, it is useful to create functions in VBA. In VBA, the cumulative normal distribution function is called Application.NormSDist. Also, the natural logarithm function in VBA is Log rather than Ln and the square root function in VBA is Sqr rather than Sqrt.

Black-Scholes Call and Put Formulas

The following function implements the Black-Scholes call pricing formula. For the sake of completeness, the function returns a value even when a volatility of zero is input, in which case the formula (3.5) is invalid (it involves division by zero in the calculation of d_1 and d_2). If the volatility is zero, then the stock is riskless and should appreciate at rate $r - q$. Moreover the option is riskless and its date–0 value should be the date–T value discounted at the risk-free rate. This implies that the call value at date 0 is[4]

---

[4] This result can be verified by a simple arbitrage argument. For example, if the call value were less than this formula, then put-call parity would show that the put price is negative, which is impossible. On the other hand, if the call price is greater than this formula (and hence positive), then put-call parity shows that the put price is positive, and it is impossible that both the put and call will finish in the money (so, given that they are riskless, only one should have a positive value).

$$\mathrm{e}^{-rT} \max\left(0, \mathrm{e}^{(r-q)T} S(0) - K\right) = \max\left(0, \mathrm{e}^{-qT} S(0) - \mathrm{e}^{-rT} K\right) .$$

```
Function Black_Scholes_Call(S, K, r, sigma, q, T)
'
' Inputs are S = initial stock price
'            K = strike price
'            r = risk-free rate
'            sigma = volatility
'            q = dividend yield
'            T = time to maturity
'
Dim d1, d2, N1, N2
If sigma = 0 Then
    Black_Scholes_Call = Application.Max(0,Exp(-q*T)*S-Exp(-r*T)*K)
Else
    d1 = (Log(S/K) + (r-q+0.5*sigma*sigma)*T) / (sigma*Sqr(T))
    d2 = d1 - sigma * Sqr(T)
    N1 = Application.NormSDist(d1)
    N2 = Application.NormSDist(d2)
    Black_Scholes_Call = Exp(-q*T)*S*N1 - Exp(-r*T)*K*N2
End If
End Function
```

It is useful to note that

```
                Black_Scholes_Call(S,K,r,sigma,q,T)
```

gives the same result as

```
            Black_Scholes_Call(exp(-q*T)*S,K,r,sigma,0,T).
```

In the latter formulation, we view the underlying asset as the portfolio which starts with e^{-qT} shares of the asset and reinvests dividends until date T. This portfolio has value $S(T)$ at date T, so a European call option on this non-dividend-paying portfolio is equivalent to a European call option on the stock. The initial value of the portfolio is $\mathrm{e}^{-qT} S(0)$, which is input as the asset price in the latter formulation.

The Black-Scholes formula for the value of a European put option can be implemented as follows.

```
Function Black_Scholes_Put(S, K, r, sigma, q, T)
'
' Inputs are S = initial stock price
'            K = strike price
'            r = risk-free rate
'            sigma = volatility
'            q = dividend yield
'            T = time to maturity
'
Dim d1, d2, N1, N2
```

```
If sigma = 0 Then
    Black_Scholes_Put = Application.Max(0,Exp(-r*T)*K-Exp(-q*T)*S)
Else
    d1 = (Log(S/K) + (r-q+0.5*sigma*sigma)*T) / (sigma*Sqr(T))
    d2 = d1 - sigma * Sqr(T)
    N1 = Application.NormSDist(-d1)
    N2 = Application.NormSDist(-d2)
    Black_Scholes_Put = Exp(-r*T)*K*N2 - Exp(-q*T)*S*N1
End If
End Function
```

Black-Scholes Greeks

The delta and gamma of a European call option can be computed with the following functions. The other Greeks are obviously calculated in a similar manner. Note that the constant $\pi = 3.14159...$ is provided in Excel as the "function" Pi() and can be accessed in Excel VBA as Application.Pi.

```
Function Black_Scholes_Call_Delta(S, K, r, sigma, q, T)
'
' Inputs are S = initial stock price
'            K = strike price
'            r = risk-free rate
'            sigma = volatility
'            q = dividend yield
'            T = time to maturity
'
Dim d1, d2, N1, N2
d1 = (Log(S/K) + (r-q+0.5*sigma*sigma)*T) / (sigma*Sqr(T))
d2 = d1 - sigma * Sqr(T)
N1 = Application.NormSDist(d1)
N2 = Application.NormSDist(d2)
Black_Scholes_Call_Delta = Exp(-q*T)*N1
End Function


Function Black_Scholes_Call_Gamma(S, K, r, sigma, q, T)
'
' Inputs are S = initial stock price
'            K = strike price
'            r = risk-free rate
'            sigma = volatility
'            q = dividend yield
'            T = time to maturity
'
Dim d1, d2, N1, N2, nd1
d1 = (Log(S/K) + (r-q+0.5*sigma*sigma)*T) / (sigma*Sqr(T))
d2 = d1 - sigma * Sqr(T)
N1 = Application.NormSDist(d1)
N2 = Application.NormSDist(d2)
```

```
nd1 = Exp(-d1 * d1 / 2) / Sqr(2 * Application.Pi)
Black_Scholes_Call_Gamma = Exp(-q*T)*nd1/(S*sigma*Sqr(T))
End Function
```

Implied Volatilities

We could find an implied volatility using the Solver tool, but then we would have to re-run Solver each time we changed one of the input values. We will need to solve similar problems on several occasions, so it seems worthwhile to program a Solver-like function in VBA. We will write this in such a way that it can easily be applied in other contexts. We will assume there is a single variable for which we want to solve, solving for multiple variables being more difficult.

Letting C denote the market price of a European call option, the implied volatility is sigma satisfying

$$\texttt{Black\_Scholes\_Call(S,K,r,sigma,q,T) - C = 0.}$$

The solution of this equation is called a "root" of the function

$$\texttt{Black\_Scholes\_Call(S,K,r,sigma,q,T) - C,}$$

and the problem of finding roots of functions is a standard numerical problem. Roots are found by what are essentially sophisticated trial-and-error methods. The simplest method is to start with upper and lower bounds for σ and repeatedly bisect the interval containing σ, each time finding a new upper or lower bound. The program below is a standard bisection routine.

For there to be a volatility that equates the market price to the Black-Scholes price, it is necessary for the call option price to satisfy the arbitrage bound[5] $C + \mathrm{e}^{-rT} K \geq \mathrm{e}^{-qT} S$. We check this condition at the beginning of the program and supply an error message if it is violated.

We need to input all of the inputs of Black_Scholes_Call other than σ, and we need to input the call option price. The following uses an error tolerance of 10^{-6}. Therefore, the value that is returned will equal the exact implied volatility to at least five decimal places. The bisection is begun with a lower bound of $\sigma = 0$. An iterative procedure is used to find an upper bound, starting with $\sigma = 100\%$.

---

[5] Note that by put-call parity—equation (3.7)—the difference between the left and right-hand sides of this inequality is the value of the put with the same strike and maturity as the call. Thus, the inequality is equivalent to the statement that the put value is nonnegative, which must be the case.

This same algorithm can be used to find a real number x such that $f(x) = 0$ for any (continuous) function f. The only changes necessary are in the right hand sides of the assignment statements for `flower`, `fupper`, and `fguess` and in finding lower and upper bounds (and obviously one would not check the arbitrage bound in general).[6] We will use this algorithm on several occasions to find roots of functions.

```
Function Black_Scholes_Call_Implied_Vol(S, K, r, q, T, CallPrice)
'
' Inputs are S = initial stock price
'            K = strike price
'            r = risk-free rate
'            q = dividend yield
'            T = time to maturity
'            CallPrice = call price
'
Dim tol, lower, flower, upper, fupper, guess, fguess
If CallPrice < Exp(-q * T) * S - Exp(-r * T) * K Then
    MsgBox ("Option price violates the arbitrage bound.")
    Exit Function
End If
tol = 10 ^ -6
lower = 0
flower = Black_Scholes_Call(S, K, r, lower, q, T) - CallPrice
upper = 1
fupper = Black_Scholes_Call(S, K, r, upper, q, T) - CallPrice
Do While fupper < 0         ' double upper until it is an upper bound
    upper = 2 * upper
    fupper = Black_Scholes_Call(S, K, r, upper, q, T) - CallPrice
Loop
guess = 0.5 * lower + 0.5 * upper
fguess = Black_Scholes_Call(S, K, r, guess, q, T) - CallPrice
Do While upper - lower > tol  ' until root is bracketed within tol
    If fupper * fguess < 0 Then  ' root is between guess and upper
        lower = guess                ' make guess the new lower bound
        flower = fguess
        guess = 0.5 * lower + 0.5 * upper  ' new guess = bi-section
        fguess = Black_Scholes_Call(S,K,r,guess,q,T) - CallPrice
    Else                          ' root is between lower and guess
        upper = guess                ' make guess the new upper bound
        fupper = fguess
        guess = 0.5 * lower + 0.5 * upper  ' new guess = bi-section
```

---

[6] The key to the function is checking each time whether the root is between the guess and the upper bound or between the guess and the lower bound. If fupper × fguess < 0, then there is a root between the guess and the upper bound. In this case, we define the new lower bound to be the old guess and define the new guess to be the midpoint of this new lower bound and the old upper bound. We do the opposite if we find the root is between the guess and the lower bound.

```
        fguess = Black_Scholes_Call(S,K,r,guess,q,T) - CallPrice
    End If
Loop
Black_Scholes_Call_Implied_Vol = guess
End Function
```

To compute an implied volatility from a put option price, one can first compute a corresponding call option price from put-call parity and then run the above program.

There are faster root-finding methods than bisection. These use other methods to update the guess than just halving the distance between the prior guess and the upper or lower bound. For example, one can use the vega (the derivative of the option formula with respect to σ) at the given guess for σ and replace the bisection with

```
            guess = guess - call/vega .
```

This amounts to approximating the Black-Scholes formula as being linear in σ and using the root of the approximation as the updated guess. This is the essence of the *Newton-Raphson* method. A similar idea that does not require the computation of vega is to keep track of the two most recent (`guess`, `call`) pairs and to approximate `vega` as:

```
    vega = (call - prior_call) / (guess - prior_guess) .
```

This is the essence of the *secant* method.

Discretely-Rebalanced Delta Hedges

To compute the real-world distribution of gains and losses from a discretely-rebalanced delta hedge, we input the expected rate of return μ. We consider adjusting the hedge at dates $0 = t_0 < t_1 < \cdots < t_N = T$, with $t_i - t_{i-1} = \Delta t = T/N$ for each i. The changes in the natural logarithm of the stock price between successive dates t_{i-1} and t_i are simulated as

$$\Delta \log S = \left(\mu - q - \frac{1}{2}\sigma^2\right) \Delta t + \sigma \, \Delta B \, ,$$

where ΔB is normally distributed with mean zero and variance Δt. The random variables ΔB are simulated as standard normals multiplied by $\sqrt{\Delta t}$. We begin with the portfolio that is short a call, long δ shares of the underlying, and short $\delta S - C$ in cash. After the stock price changes, say from S to S', we compute the new delta δ'. The cash flow from adjusting the hedge is $(\delta - \delta')S'$. Accumulation (or payment) of interest on the cash position is captured by the factor $e^{r\Delta t}$. Continuous payment of dividends is modelled similarly: the dividends earned during the period Δt is taken to be $\delta S \left(e^{q\Delta t} - 1\right)$. The cash position is adjusted due to interest, dividends, and the cash flow from adjusting the hedge. At date T, the value of the portfolio is the cash position less the intrinsic value of the option.

To describe the distribution of gains and losses, we compute percentiles of the distribution. You should see that the hedge becomes more nearly perfect as the number of periods N is increased. Note that this is true regardless of the μ that is input, which reaffirms the point that option values and hedges do not depend on the expected rate of return of the underlying. The percentile is calculated with the Excel Percentile function.[7]

```
Function Simulated_Delta_Hedge_Profit(S0,K,r,sigma,q,T,mu,M,N,pct)
'
' Inputs are S0 = initial stock price
'             K = strike price
'             r = risk-free rate
'             sigma = volatility
'             q = dividend yield
'             T = time to maturity
'             mu = expected rate of return
'             N = number of time periods
'             M = number of simulations
'             pct = percentile to be returned
'
Dim dt, SigSqrdt, drift, LogS0, Call0, Delta0, Cash0, Comp, Div
Dim S, LogS, Cash, NewS, Delta, NewDelta, HedgeValue, i, j
Dim Profit() As Double
ReDim Profit(M)
dt = T / N
SigSqrdt = sigma * Sqr(dt)
drift = (mu - q - 0.5 * sigma * sigma) * dt
Comp = Exp(r * dt)
Div = Exp(q * dt) - 1
LogS0 = Log(S0)                        ' store log of initial stock price
Call0 = Black_Scholes_Call(S0, K, r, sigma, q, T)
Delta0 = Black_Scholes_Call_Delta(S0, K, r, sigma, q, T)
Cash0 = Call0 - Delta0 * S0    ' initial cash position
For i = 0 To M
    LogS = LogS0                       ' initialize log of stock price
    Cash = Cash0                       ' initialize cash position
    S = S0                             ' initialize beginning stock price
    Delta = Delta0                     ' initialize beginning stock position
    For j = 1 To N - 1
        LogS = LogS + drift + SigSqrdt * RandN()    ' new log S
        NewS = Exp(LogS)                            ' new S
```

---

[7] If numsims = 11 and pct =0.1, the percentile function returns the second lowest element in the series. The logic is that 10% of the numbers, excluding the number returned, are below the number returned—i.e., 1 out of the other 10 are below—and 90% of the others are above. In particular, if pct = 0.5, the percentile function returns the median. When necessary, the function interpolates; for example, if numsims = 10 and pct=0.1, then the number returned is an interpolation between the lowest and second lowest numbers.

```
      NewDelta = Black_Scholes_Call_Delta(NewS,K,r,sigma,q,T-j*dt)
      Cash = Comp*Cash + Delta*S*Div - (NewDelta-Delta)*NewS
      S = NewS              ' update stock price
      Delta = NewDelta      ' update stock position
   Next j
   LogS = LogS+drift+SigSqrdt*RandN()   ' final log of stock price
   NewS = Exp(LogS)              ' final stock price
   HedgeValue = Comp*Cash + Delta*S*Div + Delta*NewS
   Profit(i) = HedgeValue - Application.Max(NewS-K,0)
 Next i
 Simulated_Delta_Hedge_Profit = Application.Percentile(Profit, pct)
 End Function
```

Problems

3.1. Create an Excel worksheet in which the user inputs K, r, σ, q and T. Compute the delta of a call option for stock prices $S = .01K$, $.02K$, $\ldots$, $1.99K$, $2K$ (i.e., $S = iK/100$ for $i = 1, \ldots 200$) and plot the delta against the stock price.

3.2. The delta of a digital option that pays \$1 when $S(T) > K$ is

$$\frac{e^{-rT} n(d_2)}{\sigma S\sqrt{T}}.$$

Repeat the previous problem for the delta of this digital. Given that in reality it is costly to trade (due to commissions, the bid-ask spread and possible adverse price impacts for large trades), do you see any problems with delta hedging a short digital near maturity if it is close to being at the money?

3.3. Repeat Prob. 3.1 for the gamma of a call option.

3.4. Use put-call parity to derive the Greeks of a put option, and write a VBA function that computes the value and Greeks.

3.5. Consider delta and gamma hedging a short call option, using the underlying and a put with the same strike and maturity as the call. Calculate the position in the underlying and the put that you should take, using the analysis in Sect. 3.6. Will you ever need to adjust this hedge? Relate your result to put-call parity.

3.6. The delta of a share digital that pays one share when $S(T) > K$ is

$$e^{-qT} N(d_1) + \frac{e^{-qT} n(d_1)}{\sigma\sqrt{T}}.$$

Repeat Prob. 3.1 for the delta of this share digital.

3.7. Create an Excel worksheet in which the user inputs K, r, q and T. Compute the value of an at-the-money call option ($S = K$) using the function `Black_Scholes_Call` for volatilities $\sigma = .01, .02, \ldots, 1.0$. Plot the call value against the volatility.

3.8. Repeat the previous problem for $S = 1.2K$ (an example of an in-the-money call option).

3.9. The file CBOEQuotes.txt (available at www.kerryback.net) contains price data for call options on the S&P 500 index. The options expired in February, 2003, and the prices were obtained on January 22, 2003. The first column lists various exercise prices. The second column gives the bid price and the third column the ask price. Import this data into an Excel worksheet and compute and plot the implied volatility against the exercise price using this data. Use the ask price as the market price for the option. The options have 30 days to maturity (so $T = 30/365$). At the time the quotes were downloaded, the S&P 500 was at 884.25. According to the CBOE, the dividend yield on the S&P 500 was 1.76%. Use 1.25% for the risk-free interest rate.

3.10. Attempt to repeat the previous problem using the bid price as the market price of the option. If this doesn't work, what is wrong? Does this indicate there is an arbitrage opportunity?

3.11. Suppose an investor invests in a portfolio with price S and constant dividend yield q. Assume the investor is charged a constant expense ratio α (which acts as a negative dividend) and at date T receives either his portfolio value or his initial investment, whichever is higher. This is similar to a popular type of variable annuity. Letting D denote the number of dollars invested in the contract, the contract pays

$$\max\left(D, \frac{De^{(q-\alpha)T}S(T)}{S(0)}\right) \tag{3.16}$$

at date T. We can rearrange the expression (3.16) as

$$\max\left(D, \frac{De^{(q-\alpha)T}S(T)}{S(0)}\right) = D + \max\left(0, \frac{De^{(q-\alpha)T}S(T)}{S(0)} - D\right)$$

$$= D + e^{-\alpha T}D\max\left(0, \frac{e^{qT}S(T)}{S(0)} - e^{\alpha T}\right) . \tag{3.17}$$

Thus, the contract payoff is equivalent to the amount invested plus a certain number of call options written on the gross holding period return $e^{qT}S(T)/S(0)$. Note that $Z(t) = e^{qt}S(t)/S(0)$ is the date–t value of the portfolio that starts with $1/S(0)$ units of the asset (i.e., with a $1 investment) and reinvests dividends. Thus, the call options are call options on a non-dividend paying portfolio with the same volatility as S and initial price of $1. This implies that the date–0 value of the contract to the investor is $e^{-rT}D$ plus

```
Exp(-alpha*T)*D*Black_Scholes_Call(1,Exp(alpha*T),r,sigma,0,T)
```

(a) Create a VBA function to compute the fair expense ratio; i.e., find α such that the date–0 value of the contract is equal to D. Hint: Modify the

```
Black_Scholes_Call_Implied_Vol
```

function. You can use $\alpha = 0$ as a lower bound. Because the value of the contract is decreasing as α increases, you can find an upper bound by iterating until the value of the contract is less than D.

(b) How does the fair expense ratio vary with the maturity T? Why?

3.12. Modify the function `Simulated_Delta_Hedge_Profit` to compute percentiles of gains and losses for an investor who writes a call option and constructs a delta and gamma hedge using the underlying asset and another call option. Include the exercise price of the call option used to hedge as an input, and assume it has the same time to maturity as the option that is written. Hint: In each period `j = 1 to N-1`, the updated cash position can be calculated as

```
Cash = exp(r*dt)*Cash + a*S*(exp(q*dt)-1) - (Newa-a)*NewS _
     - (Newb-b)*PriceHedge ,
```

where `a` denotes the number of shares of the stock held, `b` denotes the number of units held of the option that is used for hedging, and `PriceHedge` denotes the price of the option used for hedging (computed from the Black-Scholes formula each period). This expression embodies the interest earned (paid) on the cash position, the dividends received on the shares of stock and the cash inflows (outflows) from adjusting the hedge. At the final date `N`, the value of the hedge is

```
exp(r*dt)*Cash + a*S*(exp(q*dt)-1) + a*NewS _
    + b*Application.Max(NewS-KHedge,0) ,
```

and the value of the overall portfolio is the value of the hedge less

```
Application.Max(NewS-KWritten,0) ,
```

where `KHedge` denotes the strike price of the option used to hedge and `KWritten` denotes the strike of the option that was written.

4

Estimating and Modelling Volatility

Thus far, we have assumed that the volatility of the underlying asset is constant or varying in a non-random way during the lifetime of the derivative. In this chapter we will look at models that relax this assumption and allow the volatility to change randomly. This is very important, because there is plenty of evidence that volatilities do change over time in a random way.

In the first three sections, we will consider the problem of estimating the volatility. The discussion of estimation methods leads naturally into the discussion of modelling a changing volatility.

4.1 Statistics Review

We begin with a brief review of basic statistics. Given a random sample $\{x_1, \ldots, x_N\}$ of size N from a population with mean μ and variance σ^2, the best estimate of μ is of course the sample mean

$$\bar{x} = \frac{1}{N} \sum_{i=1}^{N} x_i .$$

The variance is the expected value of $(x - \mu)^2$, so an obvious estimate of the variance is the sample average of $(x_i - \mu)^2$, replacing μ with its estimate $\bar{x}$. This would be

$$\frac{1}{N} \sum_{i=1}^{N} (x_i - \bar{x})^2$$

However, because $\bar{x}$ is computed from the x_i, the x_i will deviate less on average from $\bar{x}$ than they do from the true mean μ. Hence the estimate proposed above will on average be less than σ^2. To eliminate this bias, it suffices just to scale the estimate up by a factor of $N/(N-1)$. This leads to the estimate

$$s^2 = \frac{1}{N-1} \sum_{i=1}^{N} (x_i - \bar{x})^2 ,$$

and the best estimate of σ is the square root

$$s = \sqrt{\frac{1}{N-1}\sum_{i=1}^{N}(x_i - \bar{x})^2}\,.$$

To calculate s^2, notice that

$$\sum_{i=1}^{N}(x_i - \bar{x})^2 = \sum_{i=1}^{N}(x_i^2 - 2x_i\bar{x} + \bar{x}^2)$$

$$= \sum_{i=1}^{N}x_i^2 - 2\bar{x}\sum_{i=1}^{N}x_i + \sum_{i=1}^{N}\bar{x}^2$$

$$= \sum_{i=1}^{N}x_i^2 - 2\bar{x}(N\bar{x}) + N\bar{x}^2$$

$$= \sum_{i=1}^{N}x_i^2 - N\bar{x}^2\,.$$

Therefore

$$s = \sqrt{\frac{1}{N-1}\left(\sum_{i=1}^{N}x_i^2 - N\bar{x}^2\right)}\,.$$

It is important to know how much variation there would be in $\bar{x}$ if one had access to multiple random samples. More variation means that an $\bar{x}$ computed from a single sample will be a less reliable estimate of μ. The variance of $\bar{x}$ in repeated samples is σ^2/N,[1] and our best estimate of this variance is s^2/N. The standard deviation of $\bar{x}$ in repeated samples, which is called the "standard error" of $\bar{x}$, is $\sigma/\sqrt{N}$, and we estimate this by $s/\sqrt{N}$, which equals

$$\sqrt{\frac{1}{N(N-1)}\left(\sum_{i=1}^{N}x_i^2 - N\bar{x}^2\right)}\,.$$

If the population from which x is sampled has a normal distribution, then a 95% confidence interval for μ will be $\bar{x}$ plus or minus 1.96 standard errors. Even if x does not have a normal distribution, by the Central Limit Theorem, $\bar{x}/\sqrt{N}$ will be approximately normally distributed if the sample size N is large enough, and plus or minus 1.96 standard errors will still be approximately a 95% confidence interval for μ.

---

[1] The variance of $\bar{x} = (1/N)(x_1 + \cdots + x_N)$ is, by independence of the x_i, equal to $(1/N)^2(\operatorname{var} x_1 + \cdots + \operatorname{var} x_N)$, and, because the x_i all have the same variance σ^2, this is equal to $(1/N)^2 \times N\sigma^2 = \sigma^2/N$.

4.2 Estimating a Constant Volatility and Mean

Consider an asset price that is a geometric Brownian motion under the actual probability measure:

$$\frac{\mathrm{d}S}{S} = \mu\,\mathrm{d}t + \sigma\,\mathrm{d}B\;,$$

where μ and σ are unknown constants and B is a Brownian motion. We can as usual write this in log form as

$$\mathrm{d}\log S = \left(\mu - \frac{1}{2}\sigma^2\right)\mathrm{d}t + \sigma\,\mathrm{d}B\;.$$

Over a discrete time period of length Δt, this implies

$$\Delta\log S = \left(\mu - \frac{1}{2}\sigma^2\right)\Delta t + \sigma\Delta B\;. \tag{4.1}$$

Suppose we have observed the asset price S at dates $0 = t_0 < t_1 < \cdots < t_N = T$, where $t_i - t_{i-1} = \Delta t$. If the asset pays dividends, we will take S to be the value of the portfolio in which the dividends are reinvested in new shares. Thus, in general, $S(t_i)/S(t_{i-1})$ denotes the gross return (one plus the rate of return) between dates t_{i-1} and t_i. This return is measured on a non-compounded and non-annualized basis. The annualized continuously-compounded rate of return is the rate r_i defined by

$$\frac{S(t_i)}{S(t_{i-1})} = \mathrm{e}^{r_i\Delta t}\;.$$

This implies that

$$r_i = \frac{\log S(t_i) - \log S(t_{i-1})}{\Delta t} = \mu - \frac{1}{2}\sigma^2 + \sigma\frac{B(t_i) - B(t_{i-1})}{\Delta t}\;. \tag{4.2}$$

Because $B(t_i) - B(t_{i-1})$ is normally distributed with mean zero and variance Δt, the sample $\{r_1, \ldots, r_N\}$ is a sample of independent random variables each of which is normally distributed with mean $\mu - \sigma^2/2$ and variance $\sigma^2/\Delta t$. We are focused on estimating σ^2, so it will simplify things to define

$$y_i = r_i\sqrt{\Delta t} = \frac{\log S(t_i) - \log S(t_{i-1})}{\sqrt{\Delta t}}\;. \tag{4.3}$$

The sample $\{y_1, \ldots, y_N\}$ is a sample of independent random variables each of which is normally distributed with mean $(\mu - \sigma^2/2)\sqrt{\Delta t}$ and variance σ^2. As was discussed in the previous section, the best estimate of the mean of y is the sample mean

$$\bar{y} = \frac{1}{N}\sum_{i=1}^{N} y_i\;,$$

and the best estimate of σ^2 is

$$\hat{\sigma}^2 = \frac{1}{N-1} \sum_{i=1}^{N} (y_i - \bar{y})^2 .$$

This means that we estimate μ as

$$\hat{\mu} = \frac{\bar{y}}{\sqrt{\Delta t}} + \frac{1}{2}\hat{\sigma}^2 = \bar{r} + \frac{1}{2}\hat{\sigma}^2 .$$

Let us digress for a moment to discuss the reliability of $\hat{\mu}$ as an estimate of μ. Notice that

$$\begin{aligned}
\bar{r} &= \frac{\sum_{i=1}^{N} \log S(t_i) - \log S(t_{i-1})}{N\Delta t} \\
&= \frac{\log S(T) - \log S(0)}{N\Delta t} \\
&= \frac{\log S(T) - \log S(0)}{T} .
\end{aligned} \tag{4.4}$$

Therefore the first component $\bar{r}$ of the estimate of μ depends only on the total change in S over the time period. Hence, the reliability of this component cannot depend on how frequently we observe S within the time period $[0, T]$. The standard deviation of $\bar{r}$ in repeated samples is the standard deviation of $[\log S(T) - \log S(0)]/T$, which is $\sigma/\sqrt{T}$. This is likely to be quite large. For example, with $\sigma = 0.3$ and ten years of data ($T = 10$), the standard deviation of $\bar{r}$ is 9.5%, which means that a 95% confidence interval will be a band of roughly 38%. Given that μ itself should be of the order of magnitude of 10%, such a wide confidence interval is useless for all practical purposes.

Fortunately, it is easier to estimate σ. We observed in the previous section that the $\hat{\sigma}^2$ defined above can be calculated as

$$\frac{1}{N-1} \sum_{i=1}^{N} y_i^2 - \frac{N\bar{y}^2}{N-1} . \tag{4.5}$$

From the definition (4.3) of y_i and equation (4.4), we have

$$\bar{y} = \frac{\sqrt{\Delta t}}{T} [\log S(T) - \log S(0)] .$$

Hence, the second term in (4.5) is

$$\frac{N}{N-1} \left(\frac{\Delta t}{T^2} \right) [\log S(T) - \log S(0)]^2 .$$

If we observe the stock price sufficiently frequently, so that Δt is very small, this term will be negligible. In this circumstance, $\hat{\sigma}^2$ is approximately

$$\frac{1}{N-1}\sum_{i=1}^{N} y_i^2 = \frac{1}{N-1}\sum_{i=1}^{N}\frac{[\log S(t_i) - \log S(t_{i-1})]^2}{\Delta t} \qquad (4.6)$$

$$= \frac{N}{N-1}\times\frac{1}{T}\times\sum_{i=1}^{N}[\log S(t_i) - \log S(t_{i-1})]^2 . \qquad (4.7)$$

If we observe S more and more frequently, letting $\Delta t \to 0$ and $N \to \infty$, the sum

$$\sum_{i=1}^{N}[\log S(t_i) - \log S(t_{i-1})]^2$$

will converge with probability one to $\sigma^2 T$, as explained in Sect. 2.2. This implies that $\hat\sigma^2$ will converge to σ^2. Thus, in theory, we can estimate σ^2 with any desired degree of precision by simply observing S sufficiently frequently. This is true no matter how short the overall time period $[0, T]$ may be.

In practice, this doesn't work out quite so well. If we observe minute-by-minute data, or we observe each transaction, much of the variation in the price S will be due to bouncing back and forth between the bid price and the ask price. This is not really what we want to estimate, and this source of variation will be much less important if we look at weekly or even daily data. So, there are practical limits to how frequently we should observe S. Nevertheless, it is still true that, if σ^2 were truly constant, we could estimate it with a very high degree of precision. In fact, we can estimate the volatility of a stock with enough precision to determine that it really isn't constant! The real problem that we face is to estimate and model a changing volatility.

4.3 Estimating a Changing Volatility

Without attempting yet to model how the volatility may change, we can say a few things about how we might estimate a changing volatility. In this and following sections, we will take the observation interval Δt to be fixed. We assume it is small (say, a day or a week) and focus on the estimate (4.7). Recall from Sect. 4.1 that the reason we are dividing by $N-1$ rather than N is that the sample standard deviation usually underestimates the actual standard deviation, because it uses the sample mean, which will be closer to the points x_i than will be the true mean. However, (4.7) does not employ the sample mean (it replaces it with zero), so there is no reason to make this correction. So, we take as our point of departure the estimate

$$\frac{1}{T}\sum_{i=1}^{N}[\log S(t_i) - \log S(t_{i-1})]^2 = \frac{1}{N}\sum_{i=1}^{N} y_i^2 .$$

An obvious response to the volatility changing over time is simply to avoid using data from the distant past. Such data is not likely to be informative about the current value of the volatility. What "distant" should mean in this context is not entirely clear, but, for example, we might want to use only the last 60 observations. If we are using daily data, this would mean that at the end of each day we would add that day's observation and drop the observation from 61 days past. This leads to a somewhat abruptly varying estimate. For example, a very large movement in the price on a particular day increases the volatility estimate for the next 60 days. On the 61st day, this observation would drop from the sample, leading to an abrupt drop in the estimate (presuming that there is not an equally large change in S on the 61st day). This seems unreasonable. An estimate in which the impact of each observation decays smoothly over time is more attractive.

We can construct such an estimate as

$$\hat{\sigma}^2_{i+1} = (1 - \lambda)y_i^2 + \lambda\hat{\sigma}_i^2 \tag{4.8}$$

for any constant $0 < \lambda < 1$. Here, $\hat{\sigma}^2_{i+1}$ denotes the estimate of the volatility from date t_i to date t_{i+1}. The estimate (4.8) is a weighted average of the estimate $\hat{\sigma}_i^2$ for the previous time period and the most recently observed squared change y_i^2. Following the same procedure, the next estimate will be

$$\begin{aligned}\hat{\sigma}^2_{i+2} &= (1 - \lambda)y_{i+1}^2 + \lambda\hat{\sigma}^2_{i+1} \\ &= (1 - \lambda)y_{i+1}^2 + \lambda(1 - \lambda)y_i^2 + \lambda^2\hat{\sigma}_i^2 \,.\end{aligned}$$

Likewise, the estimate at the following date will be

$$\hat{\sigma}^2_{i+3} = (1 - \lambda)y_{i+2}^2 + \lambda(1 - \lambda)y_{i+1}^2 + \lambda^2(1 - \lambda)^2 y_i^2 + \lambda^3\hat{\sigma}_i^2 \,.$$

This demonstrates the declining importance of the squared deviation y_i^2 for future estimates. At each date, y_i^2 enters with a weight that is lower by a factor of λ, compared to the previous date. If λ is small, the decay in the importance of each squared deviation will be fast. In fact, the formula (4.8) shows that, if λ is close to zero, the estimate $\hat{\sigma}^2_{i+1}$ is approximately equal to the squared deviation y_i^2—previous squared deviations are relatively unimportant. On the other hand, if λ is close to one, the decay will be slow; i.e., the importance of y_i^2 for the estimate $\hat{\sigma}^2_{i+2}$ will be nearly the same as for $\hat{\sigma}^2_{i+1}$, and nearly the same for $\hat{\sigma}^2_{i+3}$ as for $\hat{\sigma}^2_{i+2}$, etc. This will lead to a smooth (slowly varying) volatility estimate. The slowly varying nature of the estimate in this case is also clear from (4.8), because it shows that if λ is close to one, then $\hat{\sigma}^2_{i+1}$ will be approximately the same as $\hat{\sigma}_i^2$.

This method can also be used to estimate covariances, simply by replacing the squared deviations y_i^2 by the product of deviations for two different assets. And, of course, given covariance and variance estimates, we can construct estimates of correlations. To ensure that an estimated correlation is between -1 and $+1$, we will need to use the same λ to estimate each of the variances and the covariance. This is the method used by RiskMetrics.[2]

---

[2] See Mina and Xiao [53], available online at *www.riskmetrics.com*.

4.4 GARCH Models

We are going to adopt a subtle but important change of perspective now. Instead of considering (4.8) as simply an estimation procedure, we are going to assume that the actual volatility evolves according to (4.8), or a generalization thereof. We are also going to reintroduce the expected change in $\log S$, which we dropped in going from (4.5) to (4.7). Specifically, we return to (4.1), but we operate under the risk-neutral measure, so $\mu = r - q$, and we have

$$\log S(t_{i+1}) - \log S(t_i) = \left(r - q - \frac{1}{2}\sigma_{i+1}^2 \right) \Delta t + \sigma_{i+1}\Delta B . \qquad (4.9)$$

We assume the volatility σ_{i+1} between dates t_i and t_{i+1} is given by

$$\sigma_{i+1}^2 = a + by_i^2 + c\sigma_i^2 , \qquad (4.10)$$

for some constants $a > 0$, $b \geq 0$ and $c \geq 0$, with y_i now defined by

$$y_i = \frac{\log S(t_i) - \log S(t_{i-1}) - \left(r - q - \frac{1}{2}\sigma_i^2\right)\Delta t}{\sqrt{\Delta t}} .$$

From (4.9), applied to the period from t_{i-1} to t_i, this implies that y_i is normally distributed with mean zero and variance σ_i^2, and of course y_{i+1} has variance σ_{i+1}^2, etc. Under these assumptions, the random process $\log S$ is called a GARCH(1,1) process.[3] There are many varieties of GARCH processes that have been proposed in the literature, but we will only consider GARCH(1,1), which is the simplest.

We assume $b + c < 1$, in which case we can write the variance equation as a generalization of (4.8). Namely,

$$\sigma_{i+1}^2 = \kappa\theta + (1 - \kappa)\left[(1 - \lambda)y_i^2 + \lambda\sigma_i^2\right] , \qquad (4.11)$$

where $\lambda = c/(b + c)$, $\kappa = 1 - b - c$, and $\theta = a/(1 - b - c)$. Hence, σ_{i+1}^2 is a weighted average with weights κ and $1 - \kappa$, of two parts, one being the constant θ and the other being itself a weighted average of y_i^2 and σ_i^2. Whatever the variance might be at time t_i, the variance of y_j at any date t_j far into the future, computed without knowing the intervening $y_{i+1}, y_{i+2}, \ldots$, will be approximately the constant θ. The constant θ is called the "unconditional variance," whereas σ_i^2 is the "conditional variance" of y_i.

To understand the unconditional variance, it is useful to consider the variance forecasting equation. Specifically, we can calculate $E_{t_i}\left[\sigma_{i+n}^2\right]$, which is the estimate made at date t_i of the variance of y_{i+n}; i.e, we estimate the variance without having observed $y_{i+1}, \ldots, y_{i+n-1}$. Note that by definition $E_{t_i}[y_{i+1}^2] = \sigma_{i+1}^2$, so (4.11) implies

---

[3] GARCH is the acronym for "Generalized Autoregressive Conditional Heteroskedastic." "GARCH(1,1)" means that there is only one past y (no y_{i-1}, y_{i-2}, etc.) and one past σ (no σ_{i-1}, σ_{i-2}, etc.) in (4.10). See Bollerslev [7].

$$E_{t_i}\left[\sigma_{i+2}^2\right] = \kappa\theta + (1-\kappa)\left[(1-\lambda)E_{t_i}[y_{i+1}^2] + \lambda\sigma_{i+1}^2\right]$$
$$= \kappa\theta + (1-\kappa)\sigma_{i+1}^2 .$$

Likewise,

$$E_{t_{i+1}}\left[\sigma_{i+3}^2\right] = \kappa\theta + (1-\kappa)\sigma_{i+2}^2 ,$$

and taking the expectation at date t_i of both sides of this yields

$$E_{t_i}\left[\sigma_{i+3}^2\right] = E_{t_i}\left[E_{t_{i+1}}\left[\sigma_{i+3}^2\right]\right] = \kappa\theta + (1-\kappa)E_{t_i}\left[\sigma_{i+2}^2\right]$$
$$= \kappa\theta + (1-\kappa)\left[\kappa\theta + (1-\kappa)\sigma_{i+1}^2\right]$$
$$= \kappa\theta[1 + (1-\kappa)] + (1-\kappa)^2\sigma_{i+1}^2 .$$

This generalizes to

$$E_{t_i}\left[\sigma_{i+n}^2\right] = \kappa\theta\left[1 + (1-\kappa) + \cdots (1-\kappa)^{n-2}\right] + (1-\kappa)^{n-1}\sigma_{i+1}^2 .$$

Thus, there is decay at rate κ in the importance of the current volatility σ_{i+1}^2 for forecasting the future volatility. Furthermore, as $n \to \infty$, the geometric series

$$1 + (1-\kappa) + \cdots (1-\kappa)^{n-2}$$

converges to $1/\kappa$, so, as $n \to \infty$ we obtain

$$E_{t_i}\left[\sigma_{i+n}^2\right] \to \theta .$$

This means that our best estimate of the conditional variance, at some date far in the future, is approximately the unconditional variance θ.

The most interesting feature of the volatility equation is that large returns (in absolute value) lead to an increase in the variance and hence are likely to be followed by more large returns (whether positive or negative). This is the phenomenon of "volatility clustering," which is quite observable in actual markets. This feature also implies that the distribution of returns will be "fat tailed" (more technically, "leptokurtic"). This means that the probability of extreme returns is higher than under a normal distribution with the same standard deviation.[4] It is well documented that daily and weekly returns in most markets have this "fat-tailed" property.

We can simulate a path of an asset price that follows a GARCH process and the path of its volatility as follows. The following macro produces three columns of data (with headings), the first column being time, the second the asset price, and the third the volatility.

```
Sub Simulating_GARCH()
Dim S, sigma, r, q, dt, theta, kappa, lambda, LogS, Sqrdt
Dim a, b, c, y, i, N
```

---

[4] Conversely, the probability of returns very near the mean must also be higher than under a normal distribution with the same standard deviation—a fat-tailed distribution must also have a relatively narrow peak.

```
S = InputBox("Enter initial stock price")
sigma = InputBox("Enter initial volatility")
r = InputBox("Enter risk-free rate")
q = InputBox("Enter dividend yield")
dt = InputBox("Enter length of each time period (Delta t)")
N = InputBox("Enter number of time periods (N)")
theta = InputBox("Enter theta")
kappa = InputBox("Enter kappa")
lambda = InputBox("Enter lambda")
LogS = Log(S)
Sqrdt = Sqr(dt)
a = kappa * theta
b = (1 - kappa) * (1 - lambda)
c = (1 - kappa) * lambda
ActiveCell.Value = "Time"
ActiveCell.Offset(0, 1) = "Stock Price"
ActiveCell.Offset(0, 2) = "Volatility"
ActiveCell.Offset(1, 0) = 0              ' initial time
ActiveCell.Offset(1, 1) = S              ' initial stock price
ActiveCell.Offset(1, 2) = sigma          ' initial volatility
For i = 1 To N
    ActiveCell.Offset(i + 1, 0) = i * dt   ' next time
    y = sigma * RandN()
    LogS = LogS + (r - q - 0.5 * sigma * sigma) * dt + Sqrdt * y
    S = Exp(LogS)
    ActiveCell.Offset(i + 1, 1) = S          ' next stock price
    sigma = Sqr(a + b * y ^ 2 + c * sigma ^ 2)
    ActiveCell.Offset(i + 1, 2) = sigma      ' next volatility
Next i
End Sub
```

To price European options, we need to compute the usual probabilities $\text{prob}^S(S(T) > K)$ and $\text{prob}^R(S(T) > K)$. Heston and Nandi [35] provide a fast method for computing these probabilities in a GARCH (1,1) model.[5] Rather than developing this approach, we will show in Chap. 5 how to apply Monte-Carlo methods.

4.5 Stochastic Volatility Models

The volatility is stochastic (random) in a GARCH model, but it is determined by the changes in the stock price. In this section, in contrast, we will consider models in which the volatility depends on a second Brownian motion. The

---

[5] Actually, a slightly more general model is considered in [35], in which large negative returns lead to a greater increase in volatility than do large positive returns. This accommodates the empirically observed negative correlation between stock returns and volatility.

most popular model of this type is the model of Heston [34]. In this model, we have, as usual,

$$d\log S = \left(r - q - \frac{1}{2}\sigma^2\right) dt + \sigma \, dB_s \, , \qquad (4.12a)$$

where B_s is a Brownian motion under the risk-neutral measure but now σ is not a constant but instead evolves as $\sigma(t) = \sqrt{v(t)}$, where

$$dv(t) = \kappa\left[\theta - v(t)\right] dt + \gamma\sqrt{v(t)} \, dB_v \, , \qquad (4.12b)$$

where B_v is a Brownian motion under the risk-neutral measure having a constant correlation ρ with the Brownian motion B_s. In this equation, κ, θ and γ are positive constants. Given the empirical fact that negative return shocks have a bigger impact on future volatility than do positive shocks, one would expect the correlation ρ to be negative.

The term $\kappa(\theta - v)$ will be positive when $v < \theta$ and negative when $v > \theta$ and hence $\sigma^2 = v$ will tend to drift towards θ, which, as in the GARCH model, is the long-run or unconditional mean of σ^2. Thus, the volatility is said to "mean revert." The rate at which it drifts towards θ is obviously determined by the magnitude of κ, also as in the GARCH model.

The specification (4.12b) implies that the volatility of v approaches zero whenever v approaches zero. In this circumstance, one might expect the drift towards θ to dominate the volatility and keep v nonnegative, and this is indeed the case; thus, the definition $\sigma(t) = \sqrt{v(t)}$ is possible. Moreover, the parameter γ plays a role here that is similar to the role of $1 - \lambda$ in the GARCH model—the variance of the variance in the GARCH model (4.11) depends on the weight $1 - \lambda$ placed on the scaled return y_i, just as the variance of the variance in the stochastic volatility model (4.12b) depends on the weight γ placed on dB_v.

We could discretize (4.12) as:

$$\log S(t_{i+1}) = \log S(t_i) + \left(r - q - \frac{1}{2}\sigma(t_i)^2\right) \Delta t + \sqrt{v(t_i)}\,\Delta B_s, \qquad (4.13a)$$

$$v(t_{i+1}) = v(t_i) + \kappa\left[\theta - v(t_i)\right] \Delta t + \gamma\sqrt{v(t_i)}\,\Delta B_v \, . \qquad (4.13b)$$

However, even though in the continuous-time model (4.12) we always have $v(t) \geq 0$ and hence can define $\sigma(t) = \sqrt{v(t)}$, there is no guarantee that $v(t_{i+1})$ defined by (4.13b) will be nonnegative. A simple remedy is to define $v(t_{i+1})$ as the larger of zero and the right-hand side of (4.13b); thus, we will simulate the Heston model as (4.13a) and[6]

---

[6] There are better (but more complicated) ways to simulate the Heston model. An excellent discussion of ways to simulate the volatility process can be found in Glasserman [29]. Broadie and Kaya [16] present a method for simulating from the exact distribution of the asset price in the Heston model and related models.

$$v(t_{i+1}) = \max\left\{0, v(t_i) + \kappa\left[\theta - v(t_i)\right]\Delta t + \gamma\sqrt{v(t_i)}\,\Delta B_v\right\}. \qquad (4.13b')$$

A simple way to simulate the changes ΔB_s and ΔB_v in the two correlated Brownian motions is to generate two independent standard normals z_1 and z_2 and take

$$\Delta B_s = \sqrt{\Delta t}\,z \qquad \text{and} \qquad \Delta B_v = \sqrt{\Delta t}\,z^*,$$

where we define

$$z = z_1 \qquad \text{and} \qquad z^* = \rho z_1 + \sqrt{1 - \rho^2}\,z_2.$$

The random variable z^* is also a standard normal, and the correlation between z and z^* is ρ.

```
Sub Simulating_Stochastic_Volatility()
Dim S, sigma, r, q, dt, theta, kappa, Gamma, rho, LogS, var
Dim Sqrdt, Sqrrho, z1, Z2, Zstar, i, N
S = InputBox("Enter initial stock price")
sigma = InputBox("Enter initial volatility")
r = InputBox("Enter risk-free rate")
q = InputBox("Enter dividend yield")
dt = InputBox("Enter length of each time period (Delta t)")
N = InputBox("Enter number of time periods (N)")
theta = InputBox("Enter theta")
kappa = InputBox("Enter kappa")
Gamma = InputBox("Enter gamma")
rho = InputBox("Enter rho")
LogS = Log(S)
var = sigma * sigma
Sqrdt = Sqr(dt)
Sqrrho = Sqr(1 - rho * rho)
ActiveCell.Value = "Time"
ActiveCell.Offset(0, 1) = "Stock Price"
ActiveCell.Offset(0, 2) = "Volatility"
ActiveCell.Offset(1, 0) = 0               ' initial time
ActiveCell.Offset(1, 1) = S               ' initial stock price
ActiveCell.Offset(1, 2) = sigma           ' initial volatility
For i = 1 To N
   ActiveCell.Offset(i + 1, 0) = i * dt   ' next time
   z1 = RandN()
   LogS = LogS + (r-q-0.5*sigma*sigma)*dt + sigma*Sqrdt*z1
   S = Exp(LogS)
   ActiveCell.Offset(i + 1, 1) = S        ' next stock price
   Z2 = RandN()
   Zstar = rho * z1 + Sqrrho * Z2
   var = Application.Max(0, var+kappa*(theta-var)*dt _
         +Gamma*sigma*Sqrdt*Zstar)
   sigma = Sqr(var)
   ActiveCell.Offset(i + 1, 2) = sigma    ' next volatility
```

```
Next i
End Sub
```

To price European options, we again need to compute

$$\text{prob}^S(S(T) > K) \quad \text{and} \quad \text{prob}^R(S(T) > K).$$

The virtue of modelling volatility as in (4.12b) is that these probabilities can be computed quite efficiently, as shown by Heston [34].[7] There are many other ways in which one could model volatility, but the computations may be more difficult. For example, one could replace (4.12b) by

$$\sigma(t) = e^{v(t)} \quad \text{and} \quad dv(t) = \kappa(\theta - v(t))\,dt + \lambda\,dB^* . \qquad (4.12b')$$

This implies a lognormal volatility and is simpler to simulate than (4.12b)—because e^v is well defined even when v is negative—but it is easier to calculate the probabilities $\text{prob}^S(S(T) > K)$ and $\text{prob}^R(S(T) > K)$ if we assume (4.12b).

One way to implement the GARCH or stochastic volatility model is to imply both the initial volatility $\sigma(0)$ and the constants κ, θ and λ or κ, θ, γ and ρ from observed option prices. These four (or five) constants can be computed by forcing the model prices of four (or five) options to equal the observed market prices. Or, a larger set of prices can be used and the constants can be chosen to minimize the average squared error or some other measure of goodness-of-fit between the model and market prices.

4.6 Smiles and Smirks Again

As mentioned before, the GARCH and stochastic volatility models can generate "fat-tailed" distributions for the asset price $S(T)$. Thus, they can be more nearly consistent with the option smiles discussed in Sect. 3.9 than is the Black-Scholes model (though it appears that one must include jumps in asset prices as well as stochastic volatility in order to duplicate market prices with an option pricing formula). To understand the relation, let σ_{am} denote the implied volatility from an at-the-money call option, i.e., a call option with strike $K = S(0)$. The characteristic of a smile is that implied volatilities from options of the same maturity with strike prices significantly above and below $S(0)$ are higher than σ_{am}.

A strike price higher than $S(0)$ corresponds to an out-of-the money call option. The high implied volatility means that the market is pricing the right to buy at $K > S(0)$ above the Black-Scholes price computed from the volatility σ_{am}; thus, the market must attach a higher probability to stock prices $S(T) > S(0)$ than the volatility σ_{am} would suggest.

---

[7] Further discussion can be found in Epps [26].

A strike price lower than $S(0)$ corresponds to an in-the-money call option. The put option with the same strike is out of the money. The high implied volatility means that the market is pricing call options above the Black-Scholes price computed from the volatility σ_{am}. By put-call parity, the market must also be pricing put options above the Black-Scholes price computed from the volatility σ_{am}. The high prices for the rights to buy and sell at $K < S(0)$ means that the market must attach a higher probability to stock prices $S(T) < S(0)$ than the volatility σ_{am} would suggest. In particular, the high price for the right to sell at $K < S(0)$ means a high insurance premium for owners of the asset who seek to insure their positions, which is consistent with a market view that there is a significant probability of a large loss. This can be interpreted as a "crash premium." Indeed, the implied volatilities at strikes less than $S(0)$ are typically higher than the implied volatilities at strikes above $S(0)$ (giving the smile the appearance of a smirk, as discussed in Sect. 3.9), which is consistent with a larger probability of crashes than of booms (a fatter tail for low returns than for high).

4.7 Hedging and Market Completeness

The GARCH model is inherently a discrete-time model. If returns have a GARCH structure at one frequency (e.g., monthly), they will not have a GARCH structure at a different frequency (e.g., weekly). Hence, the return period (monthly, weekly, ...) is part of the specification of the model. One interpretation of the model is that the dates t_i at which the variance changes are the only dates at which investors can trade. Under this interpretation, it is impossible to perfectly hedge an option: the gross return $S(t_i)/S(t_{i-1})$ over the interval (t_{i-1}, t_i) is lognormally distributed, so no portfolio of the stock and riskless asset formed at t_{i-1} and held over the interval (t_{i-1}, t_i) can perfectly replicate the return of an option over the interval. As discussed in Sect. 1.6, we call a market in which some derivatives cannot be perfectly hedged an "incomplete market." Thus, the GARCH model is an example of an incomplete market, if investors can only trade at the frequency at which returns have a GARCH structure. However, it is unreasonable to assume that investors can only trade weekly or monthly or even daily.

Another interpretation of the GARCH model is that investors can trade continuously and the asset has a constant volatility within each period (t_{i-1}, t_i). Under this interpretation, the market is complete and options can be delta-hedged. The completeness is a result of the fact that the change $\sigma_{i+1} - \sigma_i$ in the volatility at date t_i (recall that σ_i is the volatility over the period (t_{i-1}, t_i) and σ_{i+1} is the volatility over the period (t_i, t_{i+1})) depends only on $\log S(t_i)$. Thus, the only random factor in the model that needs to be hedged is, as usual, the underlying asset price. However, this interpretation of the model is also a bit strange. Suppose for example that monthly returns are assumed to have a GARCH structure. Then the model states that the volatility in February will be higher if there is an unusually large return (in absolute value) in January. Suppose there is an unusually large return in the first half of January. Then, intuitively, one would expect the change in the volatility to occur in the second half of January rather than being delayed until February. However, the model specifies that the volatility is constant during each month, hence constant during January in this example.

The stochastic volatility model is more straightforward. The market is definitely incomplete. The value of a call option at date $t < T$, where T is the maturity of the option, will depend on the underlying asset price $S(t)$ and the volatility $\sigma(t)$. Denoting the value by $C(t, S(t), \sigma(t))$, we have from Itô's formula that

$$dC(t) = \text{something } dt + \frac{\partial C}{\partial S}\, dS(t) + \frac{\partial C}{\partial \sigma}\, d\sigma(t)\ .$$

A replicating portfolio must have the same dollar change at each date t. If we hold $\partial C/\partial S$ shares of the underlying asset, then the change in the value of the shares will be $(\partial C/\partial S)\, dS$. However, there is no way to match the $(\partial C/\partial \sigma)\, d\sigma$ term using the underlying asset and the riskless asset.

The significance of the market being incomplete is that the value of a derivative asset that cannot be replicated using traded assets (e.g., the underlying and riskless assets) is not uniquely determined by arbitrage considerations. As discussed in Sect. 1.6, one must use "equilibrium" pricing in this circumstance. That is what we have implicitly done in this chapter. By assuming particular dynamics for the volatility under the risk-neutral measure, we have implicitly selected a particular risk-neutral measure from the set of risk-neutral measures that are consistent with the absence of arbitrage.

Problems

4.1. The purpose of this exercise is to generate a fat-tailed distribution from a model that is simpler than the GARCH and stochastic volatility models but has somewhat the same flavor. The distribution will be a "mixture of normals." Create an Excel worksheet in which the user can input S, r, q, T, σ_1 and σ_2. Use these inputs to produce a column of 500 simulated $\log S(T)$. In each simulation, define $\log S(T)$ as

$$\log S(T) = \log S(0) + \left(r - q - \frac{1}{2}\sigma^2\right)T + \sigma\sqrt{T}z\ ,$$

where z is a standard normal, $\sigma = x\sigma_1 + (1-x)\sigma_2$, and x is a random variable that equals zero or one with equal probabilities. You can define z in each simulation as `NormSInv(Rand())` and x as `If(Rand()<0.5,1,0)`. Calculate the mean and standard deviation of the $\log S(T)$ and calculate the fraction that lie more than two standard deviations below the mean. If the $\log S(T)$

all came from a normal distribution with the same variance, then this fraction should equal N(−2) = 2.275%. If the fraction is higher, then the distribution is "fat tailed." (Of course, the actual fraction would differ from 2.275% in any particular case due to the randomness of the simulation, even if all of the log $S(T)$ came from a normal distribution with the same variance).

4.2. Create an Excel macro prompting the user to input the same inputs as in the Simulating_GARCH subroutine except for the initial volatility and θ. Simulate 500 paths of a GARCH process and output log $S(T)$ for each simulation (you don't need to output the entire paths as in the Simulating_GARCH macro). Take the initial volatility to be 0.3 and $\theta = 0.09$. Determine whether the distribution is fat-tailed by computing the fraction of the log $S(T)$ that lie two or more standard deviations below the mean, as in the previous exercise. For what values of κ and λ does the distribution appear to be especially fat-tailed?

4.3. Repeat Prob. 4.2 for the Heston stochastic volatility model, describing the values of κ, γ and ρ that appear to generate especially fat-tailed distributions.

Note

Excel provides some tools that are useful for exercises of this sort. If you load the Data Analysis add-in (click Tools/Add Ins), you can produce a histogram of the simulated data, which is useful for visually analyzing departures from normality. The Data Analysis add-in will also produce summary statistics, including the kurtosis and skewness of the data (without using the add-in, the kurtosis can be computed with the Excel function KURT and the skewness with the Excel function SKEW). The kurtosis of a random variable x is defined as $E[(x - \mu)^4]/\sigma^4$, where μ is the mean and σ is the standard deviation of x. The kurtosis of a normal distribution is 3. A kurtosis larger than 3 is "excess kurtosis," meaning the distribution is leptokurtic (fat tailed). Excel's KURT function actually computes excess kurtosis, so a positive value indicates a fat-tailed distribution. The skewness of x is defined as $E[(x - \mu)^3]/\sigma^3$. The skewness of a normal distribution is zero. Negative skewness indicates the distribution is skewed to the left, meaning the lower tail is fatter than the upper tail (crashes are more likely than booms). Positive skewness indicates the distribution is skewed to the right.

5

Introduction to Monte Carlo
and Binomial Models

In this chapter, we will introduce two principal numerical methods for valuing derivative securities: Monte Carlo and binomial models. We will consider two applications: valuing European options in the presence of stochastic volatility with Monte Carlo and valuing American options via binomial models. Additional applications of these methods will be presented in Chap. 9. Throughout the chapter, we will assume there is a constant risk-free rate. The last section, while quite important, could be skimmed on first reading—the rest of the book does not build upon it.

5.1 Introduction to Monte Carlo

According to our risk-neutral pricing formula (1.18), the value of a security paying an amount x at date T is

$$e^{-rT} E^R[x] .\tag{5.1}$$

To estimate this by Monte-Carlo means to simulate a sample of values for the random variable x and to estimate the expectation by averaging the sample values.[1] Of course, for this to work, the sample must be generated from a "population" having a distribution consistent with the risk-neutral probabilities.

The simplest example is valuing a European option under the Black-Scholes assumptions. Of course, for calls and puts, this is redundant, because we already have the Black-Scholes formulas. Nevertheless, we will describe how to do this for the sake of introducing the Monte Carlo method. In the case of a call option, the random variable x in (5.1) is $\max(0, S(T) - K)$. To simulate a sample of values for this random variable, we need to simulate the terminal

---

[1] Boyle [8] introduced Monte-Carlo methods for derivative valuation, including the variance-reduction methods of control variates and antithetic variates to be discussed in Chap. 9.

stock price $S(T)$. This is easy to do, because, under the Black-Scholes assumptions, the logarithm of $S(T)$ is normally distributed under the risk-neutral measure with mean $\log S(0) + \nu T$ and variance $\sigma^2 T$, where $\nu = r - q - \sigma^2/2$. Thus, we can simulate values for $\log S(T)$ as $\log S(0) + \nu T + \sigma\sqrt{T}z$, where z is a standard normal. We can average the simulated values of $\max(0, S(T) - K)$, or whatever the payoff of the derivative is, and then discount at the risk-free rate to compute the date–0 value of the derivative. This means that we generate some number M of standard normals z_i and estimate the option value as $e^{-rT}\bar{x}$, where $\bar{x}$ is the mean of

$$x_i = \max\left(0, e^{\log S(0) + \nu T + \sigma\sqrt{T}z_i} - K\right) .$$

To value options that are path-dependent we need to simulate the path of the underlying asset price. Path-dependent options are discussed in Chaps. 8 and 9.

There are two main drawbacks to Monte-Carlo methods. First, it is difficult (though not impossible) to value early-exercise features.[2] To value early exercise, we need to know the value at each date if not exercised, to compare to the intrinsic value. One could consider performing a simulation at each date to calculate the value if not exercised, but this value depends on the option to exercise early at later dates, which cannot be calculated without knowing the value of being able to exercise early at even later dates, etc. In contrast, the binomial model (and finite difference models discussed in Chap. 10) can easily handle early exercise but cannot easily handle path dependencies.

The second drawback of Monte Carlo methods is that they can be quite inefficient in terms of computation time (though, as will be explained in Chap. 9, they may be faster than alternative methods for derivatives written on multiple assets). As in statistics, the standard error of the estimate depends on the sample size. Specifically, we observed in Sect. 4.1 that, given a random sample $\{x_1, \ldots, x_M\}$ of size M from a population with mean μ and variance σ^2, the best estimate of μ is the sample mean $\bar{x}$, and the standard error of $\bar{x}$ (which means the standard deviation of $\bar{x}$ in repeated samples) is best estimated by

$$\sqrt{\frac{1}{M(M-1)}\left(\sum_{i=1}^{M} x_i^2 - M\bar{x}^2\right)} . \tag{5.2}$$

Recall that $\bar{x}$ plus or minus 1.96 standard errors is a 95% confidence interval for μ when the x_i are normally distributed. In the context of European option valuation, the expression (5.2) gives the standard error of the estimated option value at maturity, and multiplication of (5.2) by e^{-rT} gives the standard error of the estimated date–0 option value.

---

[2] Monte-Carlo methods for valuing early exercise include the stochastic mesh method of Broadie and Glasserman [15] and the regression method of Longstaff and Schwartz [48]. Glasserman [29] provides a good discussion of these methods and the relation between them.

To obtain an estimate with an acceptably small standard error may require a large sample size and hence a relatively large amount of computation time. The complexities of Monte Carlo methods arise from trying to reduce the required sample size. In Chap. 9, we will describe two such methods (antithetic variates and control variates). For those who want to engage in a more detailed study of Monte Carlo methods, the book of Glasserman [29] is highly recommended. Jäckel [39] is useful for more advanced readers, and Clewlow and Strickland [17] and Brandimarte [10] are useful references that include computer code.

5.2 Introduction to Binomial Models

As in the previous section, we will work with the dynamics of the logarithms of asset prices under the risk-neutral measure. Thus, our starting point is the equation

$$\mathrm{d}\log S = \left(r - q - \frac{\sigma^2}{2}\right)\mathrm{d}t + \sigma\,\mathrm{d}B\,, \qquad (5.3)$$

where B represents a Brownian motion under the risk-neutral measure.

In the binomial model, we assume that if the stock price is S at the beginning of the period, it will be either uS or dS at the end of the period, where the multiplicative factors u and d are constants to be determined. This means that the rate of return is $\Delta S/S = u-1$ in the "up" state and $\Delta S/S = d-1$ in the "down state." There are three parameters to the model: u, d, and the probability p of the up state (the probability of the down state being necessarily $1-p$). The following illustrates a three-period model.

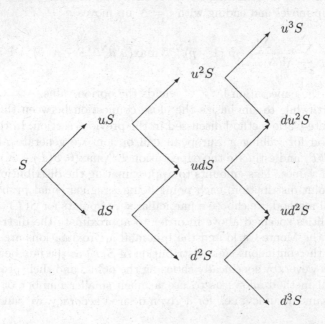

A tree constructed like this is "recombining" in the sense that the stock price after an up-down sequence is the same as after a down-up sequence. This is very important for reducing the computation time. For example, the number of nodes at the final date is $N+1$ in a recombining tree, where N is the number of periods, but it is 2^N for a non-recombining (sometimes called "bushy") tree. Hence, the computation time will increase linearly with N for a recombining tree but exponentially with N for a non-recombining tree. Unfortunately, this computational savings is generally not possible for path-dependent options, because the number of distinct paths through a tree (whether recombining or not) is again 2^N.

The value of a European derivative is of course the discounted expectation of its value at maturity, discounting at the risk-free rate and taking the expectation under the risk-neutral measure. The binomial tree allows us to approximate the expectation very easily. We simply sum over the nodes of the tree at the option maturity and weight each node by its binomial probability. In an N-period model, the probability of the top node is p^N, since the stock must go up each time to reach the top node. There are N paths reaching the second node from the top (since the period of the single down move could be any one of the N periods) and each such path has probability $p^{N-1}(1-p)$; therefore, the probability of reaching the second node from the top is $Np^{N-1}(1-p)$. More generally, the probability of going up i times and down $N-i$ times is

$$\frac{N!}{i!(N-i)!}p^i(1-p)^{N-i},$$

where as usual $x!$ denotes x factorial. Therefore, the expectation, for a European call option, is the following sum over the $N+1$ nodes at date N (starting with $i = 0$ up moves and ending with $i = N$ up moves):

$$\sum_{i=0}^{N}\frac{N!}{i!(N-i)!}p^i(1-p)^{N-i}\max(u^id^{N-i}S - K, 0). \tag{5.4}$$

Multiplying the expectation by e^{-rT} yields the option value.

It is worthwhile to emphasize the close connection between this method and the Monte-Carlo method discussed in the previous section. In the Monte-Carlo method for valuing a European call option, we generate M random values for $S(T)$ and estimate the expectation $E^R[\max(0, S(T) - K)]$ by averaging the M values. This amounts to approximating the distribution of $S(T)$ by an M–point distribution, each point being assigned equal probability. In the binomial method, we choose a particular set of points for $S(T)$ and assign the probabilities specified above in order to approximate the distribution of $S(T)$. Both the Monte-Carlo and the binomial approximations are known to converge to the continuous-time distribution of $S(T)$ as the number of points increases. However, by specifically choosing the points and their probabilities, the binomial method allows us to use a much smaller number of points to obtain the same accuracy; i.e., for a given desired accuracy, we can use many fewer periods N in the binomial model than we would need simulations M in the Monte-Carlo method. Thus, the binomial method will be much faster. Furthermore, as we will discuss in the next section, the binomial method is much better for pricing American options. On the other hand, as mentioned in the previous section, to value a path-dependent option in an N–period binomial tree would require the analysis of 2^N separate paths, so Monte Carlo may be faster for path-dependent options. Finally, as we will discuss in Sect. 9.2, Monte Carlo may be faster for options on multiple assets.

There is an important alternative method for calculating the sum (5.4), which is usually called "backward induction." We will describe it here and implement it in the next section to value American options. We begin at the last date, where there are $N + 1$ nodes. We calculate the option value at each of these nodes, storing the value at the bottom node as $C(0)$, the value at the next node up as $C(1)$, etc. This is illustrated in the diagram on the next page. Then we step back to the penultimate date. At each node at this date, we calculate the option value as the discounted expectation of its value at the last date. From each node, there are two nodes that can be reached at the next date, corresponding to a down move or an up move. So, the option value is calculated as

$$C = e^{-r\Delta t} p\, C_{\text{up}} + e^{-r\Delta t}(1 - p)C_{\text{down}} \,. \tag{5.5}$$

In terms of the vector notation shown in the diagram on the following page, the down move from node i is also node i and the up move is $i + 1$. So, we write over the elements of the C vector as

$$C(i) = e^{-r\Delta t} p\, C(i + 1) + e^{-r\Delta t}(1 - p)C(i) \,. \tag{5.6}$$

Discounting back through the tree like this, we reach date 0 and return the option value as $C(0)$. The virtue of this procedure is that it calculates a value for the option at each node in the tree, the value being the discounted expectation of the subsequent values attained by the option. This approach is essential for assessing the value of early exercise.

5.3 Binomial Models for American Options

Early exercise features are very simple to handle in a binomial framework. One only has to use the backward induction approach and check the optimality of early exercise at each node. Exercise is optimal when the intrinsic value of the option exceeds the discounted expected value of the option contingent on not exercising. When we back up in the tree, we check whether exercise is optimal, and, when it is, we replace the discounted expected value with the intrinsic value.

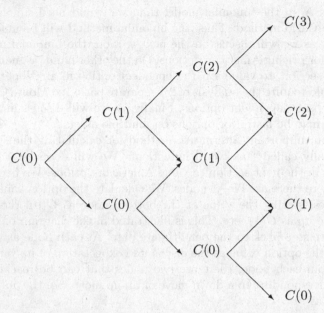

Early exercise is more important for puts than for calls (as discussed in Sect. 1.1, an American call on a non-dividend-paying stock should not be exercised early) so we will change our symbol for the option value from C to P. For a put option, we would calculate the value at each node at the end of the tree as described in the previous section:

$$P(i) = \max\left(0, K - u^i d^{N-i} S\right) , \tag{5.7}$$

for $i = 0, \ldots, N$. For a European put, we would also back up in the tree in accord with (5.6):

$$P(i) = e^{-r\Delta t} p\, P(i+1) + e^{-r\Delta t}(1-p)P(i) . \tag{5.8}$$

To accommodate early exercise, we simply need to assign to $P(i)$ the larger of this value and the value of early exercise. At node i at date n the stock price is $u^i d^{n-i} S$ and the intrinsic value of a put option is $\max(0, K - u^i d^{n-i} S)$. Therefore we replace (5.8) with

$$P(i) = \max\left(K - u^i d^{n-i} S,\ e^{-r\Delta t} p\, P(i+1) + e^{-r\Delta t}(1-p)P(i)\right) . \tag{5.9}$$

This will be explained in more detail in Sect. 5.8.

5.4 Binomial Parameters

Several different ways have been proposed for matching the binomial model to the continuous-time model. Consider an N–period binomial model for a time period of T years. This means that the length of each period is $\Delta t = T/N$. In the continuous-time model, over a discrete time period Δt, we have

$$\Delta \log S = \nu \Delta t + \sigma \Delta B ,$$

where $\nu = r - q - \sigma^2/2$ and B is a Brownian motion under the risk-neutral measure. The mean and variance, under the risk-neutral measure, of $\Delta \log S$ in the continuous-time model are

$$E^R[\Delta \log S] = \nu \Delta t ,$$
$$\text{var}^R[\Delta \log S] = \sigma^2 \Delta t ,$$

so

$$\frac{E^R[\Delta \log S]}{\Delta t} = \nu ,$$
$$\frac{\text{var}^R[\Delta \log S]}{\Delta t} = \sigma^2 .$$

In the binomial model, we have

$$\frac{E^R[\Delta \log S]}{\Delta t} = \frac{p \log u + (1-p) \log d}{\Delta t} ,$$
$$\frac{\text{var}^R[\Delta \log S]}{\Delta t} = \frac{p(1-p)(\log u - \log d)^2}{\Delta t} .$$

In order for the binomial model to converge in the appropriate sense to the continuous-time model as the number of periods $N \to \infty$ keeping the total amount of time T fixed (equivalently, as $\Delta t \to 0$), it is sufficient that

$$\frac{p \log u + (1-p) \log d}{\Delta t} \to \nu ,$$
$$\frac{p(1-p)(\log u - \log d)^2}{\Delta t} \to \sigma^2 .$$

The most popular model is probably that proposed by Cox, Ross and Rubinstein [21], who set $d = 1/u$ and

$$u = e^{\sigma \sqrt{\Delta t}} , \tag{5.10a}$$
$$p = \frac{e^{(r-q)\Delta t} - d}{u - d} . \tag{5.10b}$$

Another well-known model is that of Jarrow and Rudd [44], who take $p = 1/2$ and

$$u = \exp\left(\left(\left(r - q - \frac{1}{2}\sigma^2\right)\Delta t + \sigma\sqrt{\Delta t}\right)\right) , \tag{5.11a}$$
$$d = \exp\left(\left(\left(r - q - \frac{1}{2}\sigma^2\right)\Delta t - \sigma\sqrt{\Delta t}\right)\right) . \tag{5.11b}$$

Yet another method is proposed by Leisen and Reimer [46], and Jackson and Staunton [40] show that it is more efficient for approximating the Black-Scholes value of a European option than are the Cox-Ross-Rubinstein and Jarrow-Rudd trees.

For illustration, the Cox-Ross-Rubinstein tree will be implemented in Sect. 5.8. However, when we consider binomial models for multiple assets in Chap. 9, we will use the tree proposed by Trigeorgis [61], because it is the simplest to explain in that context. Trigeorgis proposes choosing p, u and d so that the mean and variance of $\Delta \log S$ in the binomial model match those in the continuous-time model exactly. This means that

$$\frac{p \log u + (1 - p) \log d}{\Delta t} = \nu \,,$$

$$\frac{p(1 - p)(\log u - \log d)^2}{\Delta t} = \sigma^2 \,.$$

These are two equations in the three unknowns, leaving one degree of freedom, so Trigeorgis takes $d = 1/u$, as do Cox, Ross and Rubinstein. As we will show in the next section, taking $d = 1/u$ simplifies the calculations of deltas and gammas. Solving these two equations yields[3]

$$\log u = \sqrt{\sigma^2 \Delta t + \nu^2 (\Delta t)^2} \,, \tag{5.13a}$$

$$p = \frac{1}{2} + \frac{\nu \Delta t}{2 \log u} \,. \tag{5.13b}$$

5.5 Binomial Greeks

To estimate Greeks in any valuation model, one can run the valuation program twice, for two different parameter values, and then estimate the Greek as the difference in value divided by the difference in parameters. For example, to estimate vega when the volatility of the underlying is σ, we could estimate the derivative value for a volatility of 0.99σ and for a volatility of 1.01σ.

---

[3] Notice that if we were to drop the $(\Delta t)^2$ term in (5.13a) (which we could do because it becomes increasingly negligible as $\Delta t \to 0$), then (5.13a) would be the same as (5.10a). The different choices of p in (5.10b) and (5.13b) can be understood as follows. Equation (5.10b) implies that the expected stock price $pS_u + (1-p)S_d$ equals $e^{(r-q)\Delta t}S$, so we have average growth at the rate $r - q$ as in the continuous-time model. On the other hand, (5.13b) implies that the expected *log* stock price $p \log S_u + (1-p) \log S_d$ equals $\log S + \nu \Delta t$, so the expected change in the logarithm is $\nu \Delta t$, also as in the continuous-time model. Thus, both match the binomial model to the continuous-time model, the Cox-Ross-Rubinstein method focusing on the expected return (equivalently, the expected change in the price of the underlying) and the Trigeorgis method focusing on the expected continuously-compounded return (the expected change in the logarithm of the price).

Denoting the former derivative value by C_d and the latter by C_u, the vega can be estimated by

$$\frac{C_u - C_d}{1.01\sigma - 0.99\sigma} = \frac{C_u - C_d}{0.02\sigma}.$$

We can in principle obtain a more precise estimate of the derivative by making a smaller change in the parameter (e.g., using 0.999σ and 1.001σ) but computer round-off errors limit how small a parameter change one should take in practice.

To estimate the gamma when the price of the underlying is S, we need to estimate the derivative value at two other prices for the underlying, which we will call S_u and S_d, with $S_u > S > S_d$. As just explained, the estimate of the delta (which we continue to denote by δ) would be

$$\delta = \frac{C_u - C_d}{S_u - S_d}, \tag{5.14}$$

where C_u denotes the derivative value when the underlying is equal to S_u and C_d denotes the derivative value when the underlying is equal to S_d. Letting C denote the derivative value when the underlying is equal to S, two other obvious estimates of the delta are

$$\delta_u = \frac{C_u - C}{S_u - S} \quad \text{and} \quad \delta_d = \frac{C - C_d}{S - S_d}.$$

The first of these should be understood as an estimate of the delta when the price of the underlying is at the midpoint of S_u and S, and the second is an estimate of the delta when the price of the underlying is at the midpoint of S_d and S. The distance between these midpoints is

$$\frac{S_u + S}{2} - \frac{S_d + S}{2} = \frac{S_u - S_d}{2},$$

so we obtain an estimate of Γ (the derivative of δ) as

$$\Gamma = \frac{\delta_u - \delta_d}{(S_u - S_d)/2}. \tag{5.15}$$

In a binomial model, it is possible to compute the most important Greeks, delta and gamma, more efficiently than by simply running the valuation program several times. Assume we have taken $d = 1/u$, so after an up and a down move (or a down and an up move) the stock price returns to its initial value S. After fixing the length $\Delta t = T/N$ of each time period, we redefine $N = N + 2$. This results in an $N + 2$ period tree covering a time period of length $T + 2\Delta t$. Now consider the tree starting two periods from the initial date. At the middle node shown below, the stock price is $udS = S$. Ignoring the top and bottom nodes and the branches that follow them, the result of adding two periods is that the tree starting from udS is an N–period tree for a time period of length T.

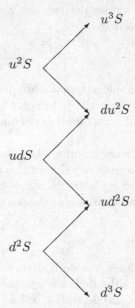

Hence, the derivative price calculated at the middle node will be the price we are trying to estimate. The derivative price at the top node will be the value of a derivative of maturity T when the initial price of the underlying is u^2S. Similarly, the derivative price at the bottom node will be the value of a derivative of maturity T when the initial price of the underlying is d^2S. Thus, when we back up in the tree to this date, we will have all of the information we need to return an estimate of the derivative value and to return estimates of the delta and gamma, taking $S_u = u^2S$ and $S_d = d^2S$ in equations (5.14) and (5.15). We are not interested in the tree to the left of what is shown above.

5.6 Monte Carlo Greeks I: Difference Ratios

As with binomial models, Greeks can be calculated by Monte Carlo by running the valuation program twice and computing a difference ratio, for example $(C_u - C_d)/(S_u - S_d)$ to estimate a delta. However, to minimize the error, and minimize the number of computations required, one should use the same set of random draws to estimate the derivative value for different values of the parameter. For path-independent options (e.g., European puts and calls) under the Black-Scholes assumptions, we only need to generate $S(T)$ and then we can compute $S_u(T)$ as $[S_u(0)/S(0)] \times S(T)$ and $S_d(T)$ as $[S_u(0)/S(0)] \times S(T)$. We can estimate standard errors for the Greeks in the same way that we estimate the standard error of the derivative value.

Actually, there is often a better method available that is just as simple. This is called "pathwise calculation." We will explain this in the next section.

Here we will describe how to estimate the delta and gamma of a derivative as sample means of difference ratios.

Consider initial prices for the underlying $S_u > S > S_d$. Denote the underlying price at the option maturity in a given simulation by $S_u(T)$ when the initial underlying price is S_u, by $S(T)$ when the initial underlying price is S, and by $S_d(T)$ when the initial underlying price is S_d. Under the Black-Scholes assumptions, the logarithm of the stock price at date T starting from the three initial prices S_d, S and S_u is

$$\log S_d(T) = \log S_d + \left(r - q - \frac{1}{2}\sigma^2 \right) T + \sigma B(T) \,,$$

$$\log S(T) = \log S + \left(r - q - \frac{1}{2}\sigma^2 \right) T + \sigma B(T) \,,$$

$$\log S_u(T) = \log S_u + \left(r - q - \frac{1}{2}\sigma^2 \right) T + \sigma B(T) \,,$$

so

$$\log S_d(T) = \log S(T) + \log S_d - \log S \implies S_d(T) = \left(\frac{S_d}{S} \right) S(T) \,,$$

and

$$\log S_u(T) = \log S(T) + \log S_u - \log S \implies S_u(T) = \left(\frac{S_u}{S} \right) S(T) \,.$$

Therefore, under the Black-Scholes assumptions, we only need to simulate $S(T)$ and then perform the multiplications indicated above to obtain $S_d(T)$ and $S_u(T)$.

Consider a particular simulation and let $C_d(T)$ denote the value of the derivative at maturity when the initial asset price is S_d, let $C(T)$ denote the value of the derivative at maturity when the initial asset price is S, and let $C_u(T)$ denote the value of the derivative at maturity when the initial asset price is S_u. For path-independent derivatives under the Black-Scholes assumptions, these can be computed directly from the simulation of $S(T)$ as just described. However, the following applies to general European derivatives under general assumptions about the underlying asset price (for example, it could follow a GARCH process).

The estimates C_d, C and C_u of the date–0 derivative values, for the different initial prices of the underlying, are the discounted sample means of the $C_d(T)$, $C(T)$ and $C_u(T)$. One way to estimate the delta is $(C_u - C_d)/(S_u - S_d)$. This is a difference of discounted sample means, multiplied by the reciprocal of $S_u - S_d$. Equivalently, it is the sample mean of the differences $C_u(T) - C_d(T)$, multiplied by $e^{-rT}/(S_u - S_d)$. As a sample mean, its standard error can be estimated as described in Chap. 4. The standard error is

$$\frac{e^{-rT}}{S_u - S_d} \sqrt{ \frac{1}{M(M-1)} \left(\sum_{i=1}^{M} [C_{ui}(T) - C_{di}(T)]^2 - M \left[\overline{C_u(T)} - \overline{C_d(T)} \right]^2 \right) } \,,$$

where the overline denotes the sample mean and where $C_{ui}(T)$ [respectively, $C_{di}(T)$] denotes the value of the derivative at maturity in simulation i when the initial asset price is S_u [respectively, S_d].

The corresponding Monte Carlo estimate of the gamma is also a sample mean. Simple algebra shows that the formula (5.15) is equivalent to

$$\Gamma = \frac{2}{(S_u - S)(S_u - S_d)}C_u - \frac{2}{(S_u - S)(S - S_d)}C + \frac{2}{(S - S_d)(S_u - S_d)}C_d .$$
(5.16)

Normally one would take $S_u = (1 + \alpha)S$ and $S_d = (1 - \alpha)S$ for some α (e.g., $\alpha = 0.01$). In this case (5.16) simplifies to

$$\Gamma = \frac{C_u - 2C + C_d}{\alpha^2 S^2} ,$$
(5.17)

and the standard error of the gamma is

$$\frac{e^{-rT}}{\alpha^2 S^2}\sqrt{\frac{1}{M(M-1)}}$$

$$\times \sqrt{\sum_{i=1}^{M}[C_{ui}(T) - 2C_i(T) + C_{di}(T)]^2 - M\left[\overline{C_u(T)} - 2\overline{C(T)} + \overline{C_d(T)}\right]^2} .$$

5.7 Monte Carlo Greeks II: Pathwise Estimates

We will examine the bias in the Monte Carlo delta estimate discussed in the preceding section and explain "pathwise" estimation of Greeks. By "biased," we mean that the expected value of an estimate is different from the true value. It is important to recognize that if a Monte Carlo estimate is biased, then, even if a large number of simulations is used and the standard error is nearly zero, the answer provided by the Monte Carlo method will be incorrect. For simplicity, consider a European call under the Black-Scholes assumptions.

The delta estimate we have considered is the discounted sample mean of

$$\frac{C_u(T) - C_d(T)}{S_u - S_d} .$$
(5.18)

This ratio takes on one of three values, depending on $S(T)$:

- If $S_u(T) \leq K$ then the option is out of the money in both the up and down cases; i.e.,

$$C_u(T) = C_d(T) = 0 ,$$

so the ratio (5.18) is zero.

- If $S_d(T) \geq K$ then the option is in the money in both the up and down cases; i.e.,

$$C_u(T) = S_u(T) - K = \left(\frac{S_u}{S}\right) S(T) - K \,,$$

$$C_d(T) = S_d(T) - K = \left(\frac{S_d}{S}\right) S(T) - K \,,$$

so the ratio (5.18) equals $S(T)/S$.

- If $S_u(T) > K > S_d(T)$, then the option is in the money in only the up case; i.e.,

$$C_u(T) = S_u(T) - K = \left(\frac{S_u}{S}\right) S(T) - K \,,$$

$$C_d(T) = 0 \,,$$

so the ratio (5.18) equals

$$\frac{\left(\frac{S_u}{S}\right) S(T) - K}{S_u - S_d} < \frac{S(T)}{S} \,.$$

The bias is induced by the third case above. We can see this as follows. We are trying to estimate

$$\frac{\partial}{\partial S} e^{-rT} E^R \big[\max(0, S(T) - K) \big] = e^{-rT} E^R \left[\frac{\partial}{\partial S} \max(0, S(T) - K) \right] \,. \quad (5.19)$$

The delta estimate $(C_u - C_d)/(S_u - S_d)$ replaces the mean E^R with the sample mean and replaces

$$\frac{\partial}{\partial S} \max(0, S(T) - K) \quad (5.20)$$

with the ratio (5.18). The derivative (5.20) takes on two possible values, depending on $S(T)$—we can ignore the case $S(T) = K$ because it occurs with zero probability:

- If $S(T) < K$, then $\max(0, S(T) - K) = 0$ and the derivative is zero.
- If $S(T) > K$, then $\max(0, S(T) - K) = S(T) - K$ and the derivative equals

$$\frac{\partial S(T)}{\partial S} = e^{(r-q-\sigma^2/2)T + \sigma B(T)} = \frac{S(T)}{S} \,.$$

Therefore, the true delta—the expectation (5.19)—equals[4]

---

[4] By changing numeraires, we can show that (5.21) equals $e^{-qT} E^V[x] = e^{-qT} \mathrm{N}(d_1)$, as we know from Chap. 3 is the delta of a European call in the Black-Scholes model (here, as in Chap. 3, $V(t) = e^{qt} S(t)$ denotes the value of the non-dividend-paying portfolio created from the stock).

$$e^{-rT} E^R \left[\frac{S(T)}{S} x \right] , \qquad (5.21)$$

where x is the random variable defined as

$$x = \begin{cases} 1 & \text{if } S(T) > K , \\ 0 & \text{otherwise} . \end{cases}$$

On the other hand, our analysis of the ratio (5.18) shows that the expected value of the delta estimate $(C_u - C_d)/(S_u - S_d)$ is

$$e^{-rT} E^R \left[\frac{S(T)}{S} y \right] + e^{-rT} E^R \left[\frac{S_u S(T) - SK}{S(S_u - S_d)} z \right] , \qquad (5.22)$$

where

$$y = \begin{cases} 1 & \text{if } S_d(T) > K , \\ 0 & \text{otherwise} . \end{cases}$$

and

$$z = \begin{cases} 1 & \text{if } S_u(T) > K > S_d(T) , \\ 0 & \text{otherwise} . \end{cases}$$

To contrast (5.21) and (5.22), note that if $y = 1$ then $x = 1$, so the term $E^R \left[\frac{S(T)}{S} y \right]$ in (5.22) is "part of" (5.21). However, there are two partially offsetting "errors" in (5.22): z sometimes equals one when x is zero, and when both z and x are one, then the factor multiplying z is smaller than the factor multiplying x. In any case, the expected value (5.22) is not the same as the true delta (5.21). As noted before, this implies that the delta estimate will be incorrect even if its standard error is zero. The bias can be made as small as one wishes by taking the magnitude $S_u - S_d$ of the perturbation to be small, but taking the perturbation to be very small will introduce unacceptable roundoff error.

The obvious way to estimate the delta in this situation is simply to compute the discounted sample average of $[S(T)/S]x$. This is called a "pathwise" estimate of the delta, because it only uses the sample paths of $S(t)$ rather than considering up and down perturbations. This method is due to Broadie and Glasserman [14]. Because the pathwise estimate is a sample average, its standard error can be computed in the usual way.

To compute pathwise estimates in other models and for other Greeks, we need the Greek to be an expectation as on the right-hand side of (5.19). Additional examples can be found in Glasserman [29] and Jäckel [39].

5.8 Calculations in VBA

Monte Carlo Valuation of a European Call

We will illustrate Monte Carlo by valuing a European call under the Black-Scholes assumptions. We will also estimate the delta by each of the methods described in Sects. 5.6 and 5.7. Of course, we know the call value and its delta from the Black-Scholes formulas, and they can be used to evaluate the accuracy of the Monte Carlo estimates.

In this circumstance, we only need to simulate the price of the underlying at the option maturity rather than the entire path of the price process. To estimate the option delta as a difference ratio $(C_u - C_d)/(S_u - S_d)$, we use the perturbations $S_u = 1.01S$ and $S_d = 0.99S$. The inputs are the same as for the Black-Scholes formula plus the sample size M (the number of stock prices to be simulated).

```
Function European_Call_MC(S, K, r, sigma, q, T, M)
'
' Inputs are S = initial stock price
'            K = strike price
'            r = risk-free rate
'            sigma = volatility
'            q = dividend yield
'            T = time to maturity
'            M = number of simulations
'
' This outputs the row vector (call value, delta 1, delta 2)
'
Dim LogS0,drift,SigSqrT,UpChange,DownChange,SumCall,SumCallChange
Dim SumPathwise,LogS,LogSd,LogSu,CallV,CallVu,CallVd,i,Delta1,Delta2
LogS0 = Log(S)
drift = (r - q - 0.5 * sigma * sigma) * T
SigSqrT = sigma * Sqr(T)
UpChange = Log(1.01)
DownChange = Log(0.99)
SumCall = 0
SumCallChange = 0
SumPathwise = 0
For i = 1 To M
    LogS = LogS0 + drift + SigSqrT * RandN()     ' log S(T)
    CallV = Application.Max(0, Exp(LogS) - K)     ' call value
    SumCall = SumCall + CallV                     ' sum call values
    LogSu = LogS + UpChange                       ' log Su(T)
    CallVu = Application.Max(0, Exp(LogSu) - K)   ' call value
    LogSd = LogS + DownChange                     ' Sd(T)
    CallVd = Application.Max(0, Exp(LogSd) - K)   ' call value
    SumCallChange = SumCallChange + CallVu - CallVd ' differences
    If Exp(LogS) > K Then
```

```
            SumPathwise = SumPathwise + Exp(LogS) / S    ' for pathwise
      End If
   Next i
   CallV = Exp(-r * T) * SumCall / M
   Delta1 = Exp(-r * T) * SumCallChange / (M * 0.02 * S)
   Delta2 = Exp(-r * T) * SumPathwise / M
   European_Call_MC = Array(CallV, Delta1, Delta2)
   End Function
```

Monte Carlo Valuation in a GARCH Model

For another example of Monte Carlo, we will value a European call option and estimate its standard error in a GARCH model. The underlying asset price is simulated as in Sect. 4.4. After each path of the underlying is simulated, we compute the date–T value of the option. We sum these as the simulations proceed in order to compute the average value. We also sum the squared date–T option values in order to compute the standard error of the estimate of the date–0 option value.

In addition to the inputs in the previous function, we input the number N of time periods in the interval $[0, T]$—implying a GARCH model for returns over time intervals of length $\Delta t = T/N$ under the risk-neutral measure—and the GARCH parameters κ, θ, and λ. To value a different type of European derivative, we would simply modify the statement

$$CallV = Application.Max(0, Exp(LogS) - K).$$

```
Function Eur_Call_GARCH_MC(S,K,r,sigma0,q,T,N,kappa,theta,lambda,M)
'
' Inputs are S = initial stock price
'            K = strike price
'            r = risk-free rate
'            sigma0 = initial volatility
'            q = dividend yield
'            T = time to maturity
'            N = number of time periods
'            kappa = GARCH parameter
'            theta = GARCH parameter
'            lambda = GARCH parameter
'            M = number of simulations
'
' This returns the row vector (call value, standard error).
'
Dim dt, Sqrdt, a, b, c, LogS0, SumCall, SumCallSq, LogS, sigma
Dim y, CallV, StdError, i, j
dt = T / N
Sqrdt = Sqr(dt)
a = kappa * theta               ' GARCH parameter
b = (1 - kappa) * lambda         ' GARCH parameter
```

```
c = (1 - kappa) * (1 - lambda)    ' GARCH parameter
LogS0 = Log(S)                    ' store log stock price
SumCall = 0                       ' initialize running total
SumCallSq = 0                     ' initialize running total
For i = 1 To M
    LogS = LogS0                  ' initialize log stock price
    sigma = sigma0               ' initialize volatility
    For j = 1 To N                ' generate path
        y = sigma * RandN()
        LogS = LogS + (r-q-0.5*sigma*sigma)*dt + Sqrdt*y
        sigma = Sqr(a + b * y ^ 2 + c * sigma ^ 2)  ' update vol
    Next j
    CallV = Application.Max(0, Exp(LogS) - K)        ' date-T value
    SumCall = SumCall + CallV                        ' update sum
    SumCallSq = SumCallSq + CallV * CallV            ' update sum
Next i
CallV = Exp(-r * T) * SumCall / M
StdError = Exp(-r * T) * Sqr((SumCallSq - SumCall * SumCall / M) / _
          (M * (M - 1)))
Eur_Call_GARCH_MC = Array(CallV, StdError)
End Function
```

Binomial Valuation of European Options

The binomial model for path-independent European options can be implemented as follows. We will use the Cox-Ross-Rubinstein parameters. To value a different type of European option in a binomial model, one would only have to change the formula

$$\texttt{Application.Max(S - K, 0)}$$

in the following. We first define the binomial parameters and some useful constants, denoting the probability p of an up move as pu and the probability $1 - p$ of a down move as pd.

```
Function European_Call_Binomial(S, K, r, sigma, q, T, N)
'
' Inputs are S = initial stock price
'            K = strike price
'            r = risk-free rate
'            sigma = volatility
'            q = dividend yield
'            T = time to maturity
'            N = number of time periods
'
Dim dt, u, d, pu, pd, u2, prob, CallV, i
dt = T / N                        ' length of time period
u = Exp(sigma * Sqr(dt))          ' size of up step
d = 1 / u                         ' size of down step
```

```
pu = (Exp((r - q) * dt) - d) / (u - d) ' probability of up step
pd = 1 - pu                            ' probability of down step
u2 = u * u
```

Now we calculate the stock price at the bottom node (the node corresponding to all down moves), the probability of reaching that node, and the first term in the sum (5.4).

```
S = S * d ^ N              ' stock price at bottom node at last date
prob = pd ^ N              ' probability of bottom node at last date
CallV = prob * Application.Max(S - K, 0)
```

To calculate the other N terms in the sum, we note that the stock price when there are i up moves is u^2 times the stock price with only $i-1$ up moves (because one more up move also means one fewer down move and adding an up and removing a down produces the factor $u/d = u^2$). Furthermore, the ratio of the probability of i up moves to $i - 1$ up moves is

$$\frac{p^i(1-p)^{N-i}N!/i!(N-i)!}{p^{i-1}(1-p)^{N-i+1}N!/(i-1)!(N-i+1)!} = \frac{(N-i+1)p}{(1-p)i} .$$

Therefore, as we increase the index i in computing the sum (5.4), we need to multiply the previous stock price by u^2 and multiply the previous probability by $(N - i + 1)p/[(1 - p)i]$. We add the result to CallV each time and, at the end, discount by e^{-rT}.

```
For i = 1 To N
    S = S * u2
    prob = prob * (pu / pd) * (N - i + 1) / i
    CallV = CallV + prob * Application.Max(S - K, 0)
Next i
European_Call_Binomial = Exp(-r * T) * CallV
End Function
```

Binomial Valuation of American Options

We will consider an American put. It may also be optimal to exercise an American call early, if there is a positive dividend yield, and the same procedure can be used for American calls. We begin as in the previous subsection by defining the binomial parameters, some useful constants, and the stock price at the bottom node at the last date.[5] We also compute the put value $P(0)$ at the bottom node at the last date.

---

[5] Note that the variable name $S0$ is assigned to the initial stock price. The variable S is modified as we step up across the nodes at each date. When we back up to the previous date, the initial stock price is still available in the variable $S0$.

```
Function American_Put_Binomial(S0, K, r, sigma, q, T, N)
'
' Inputs are S0 = initial stock price
'            K = strike price
'            r = risk-free rate
'            sigma = volatility
'            q = dividend yield
'            T = time to maturity
'            N = number of time periods
'
Dim dt, u, d, pu, dpu, dpd, u2, S, i, j
Dim PutV() As Double
ReDim PutV(N)
dt = T / N                              ' length of time period
u = Exp(sigma * Sqr(dt))                ' size of up step
d = 1 / u                               ' size of down step
pu = (Exp((r - q) * dt) - d) / (u - d)  ' probability of up step
dpu = Exp(-r * dt) * pu                 ' discount x up prob
dpd = Exp(-r * dt) * (1 - pu)           ' discount x down prob
u2 = u * u
S = S0 * d ^ N                          ' bottom stock price
PutV(0) = Application.Max(K - S, 0)     ' bottom put value
```

Now we loop over the other nodes at the last date, increasing the stock price by a factor of u^2 each time as before, and defining the put value as its intrinsic value at maturity.

```
For j = 1 To N
    S = S * u2
    PutV(j) = Application.Max(K - S, 0)
Next j
```

Now we do the backward induction. Note that a "period" is the time period between successive dates. In a one-period model, there are two dates (the beginning and end) and in general there are $N + 1$ dates in an N–period model. We index the dates as $i = 0, \ldots, N$. Since we are backing up in the tree, we step backwards from $i = N - 1$ to $i = 0$. At each date we start by defining the stock price at the bottom node. At date i there have been i past periods, so the bottom node corresponds to i down moves. The put value at each node is computed as the larger of the discounted expected value and the value of immediate exercise (the intrinsic value). Having already dealt with the bottom node ($j = 0$) we loop over the nodes $j = 1, \ldots, i$ at each date i, increasing the stock price by a factor of u^2 each time. When we have backed up to date 0, we return the put value $P(0)$, the value at the bottom node, which is the only node at date 0.

```
For i = N - 1 To 0 Step -1               ' back up in time to date 0
    S = S0 * d ^ i                       ' bottom stock price
    PutV(0) = Application.Max(K - S, dpd * PutV(0) + dpu * PutV(1))
```

```
        For j = 1 To i                        ' step up over nodes
            S = S * u2
            PutV(j) = Application.Max(K-S, dpd*PutV(j)+dpu*PutV(j+1))
        Next j
    Next i
    American_Put_Binomial = PutV(0)
    End Function
```

Binomial Estimation of Delta and Gamma '

We add two periods to the model and then stop the backward induction at
date $n = 2$, as described in Sect. 5.5.

```
Function American_Put_Binomial_DG(S0, K, r, sigma, q, T, N)
'
' Inputs are S0 = initial stock price
'            K = strike price
'            r = risk-free rate
'            sigma = volatility
'            q = dividend yield
'            T = time to maturity
'            N = number of time periods
'
' This returns the row vector (put value, delta, gamma).
'
Dim dt, u, d, pu, dpu, dpd, u2, S, Su, Sd, Deltau, Deltad, dist
Dim i, j, NewN, Delta, Gamma
Dim PutV() As Double
ReDim PutV(N + 2)
dt = T / N                              ' length of time period
NewN = N + 2                            ' now we add 2 periods
u = Exp(sigma * Sqr(dt))               ' size of up step
d = 1 / u                              ' size of down step
pu = (Exp((r - q) * dt) - d) / (u - d) ' probability of up step
dpu = Exp(-r * dt) * pu                 ' discount x up prob
dpd = Exp(-r * dt) * (1 - pu)           ' discount x down prob
u2 = u * u
S = S0 * d ^ NewN                       ' bottom stock price
PutV(0) = Application.Max(K - S, 0)     ' bottom put value
For j = 1 To NewN                       ' step up over nodes
    S = S * u2
    PutV(j) = Application.Max(K - S, 0)
Next j
For i = NewN - 1 To 2 Step -1           ' back up in time
    S = S0 * d ^ i                      ' bottom stock price
    PutV(0) = Application.Max(K - S, dpd * PutV(0) + dpu * PutV(1))
    For j = 1 To i                      ' step up over nodes
        S = S * u2
        PutV(j) = Application.Max(K-S, dpd*PutV(j)+dpu*PutV(j+1))
```

```
     Next j
  Next i
  Su = S0 * u2                          ' higher stock price
  Sd = S0 / u2                          ' lower stock price
  Deltau = (PutV(2) - PutV(1)) / (Su - S0) ' midpoint delta
  Deltad = (PutV(1) - PutV(0)) / (S0 - Sd) ' midpoint delta
  dist = S0 * (u2 - d * d)         ' dist between Su and Sd
  Delta = (PutV(2) - PutV(0)) / dist
  Gamma = 2 * (Deltau - Deltad) / dist
  American_Put_Binomial_DG = Array(PutV(1), Delta, Gamma)
  End Function
```

Problems

5.1. Consider an at-the-money European call option on a non-dividend-paying stock with six months to maturity. Take the initial stock price to be $50, the interest rate to be 5% and $\sigma =30\%$. Compute the value in a binomial model with $N = 10, 11, \ldots, 20$ and plot the values against N. Is convergence monotone?

5.2. Consider the same option as in the previous problem. Roughly what value of N is needed to get penny accuracy? (To evaluate the accuracy, compare the price to the price given by the Black-Scholes formula.)

5.3. The "early exercise premium" is the difference between the value of an American option and the value of a European option with the same parameters. Compute the early exercise premium for an American put and various values for the interest rate, exercise price, and stock parameters. Under what circumstances is the early exercise premium relatively large?

5.4. Create an Excel worksheet in which the user inputs S, r, σ, q, T, N, κ, θ, and λ. Use the function `European_Call_GARCH_MC` to compute call option prices for exercise prices $K = 0.6S$, $0.7S$, $0,8S$, $0.9S$, S, $1.1S$, $1.2S$, $1.3S$ and $1.4S$, taking $M = 500$ in each case. For each computed price, use the function `Black_Scholes_Call_Implied_Vol` to compute an implied Black-Scholes volatility. Plot the implied volatilities against the exercise prices.

5.5. Create a VBA function using Monte Carlo to estimate the value of a European call option in the Heston stochastic volatility model. The inputs should be the initial stock price S, the strike price K, the risk-free rate r, the initial volatility σ, the dividend yield q, the time to maturity T, the number of time periods N, the parameters κ, θ, γ, and ρ, and the number of simulations M. Return the estimated option value and its standard error.

5.6. Modify the VBA function in the previous exercise to also return the estimated delta of the option and the standard error of the delta.

5.7. Repeat Prob. 5.4 using the Heston model (the function developed in Prob. 5.5) to compute the call option prices, allowing the user to input γ and ρ (instead of λ).

Part II

Advanced Option Pricing

6

Foreign Exchange

We will see in this chapter how to apply the Black-Scholes formulas to value currency options and options on foreign assets. We will also discuss currency forwards and futures, quanto forwards, and return swaps.

For concreteness, we will call one currency the "domestic" currency and the other the "foreign" currency. Let $X(t)$ denote the exchange rate at time t measured in units of the domestic currency per unit of the foreign currency. Exchange rates can be confusing, because we can look at them from the perspective of either currency, so it may help to keep in mind that $X(t)$ here means the price of a unit of the foreign currency, just as we might consider the price of a stock. When we speak of the "cost" or "value" of something without specifying the currency, it should be understood to be the domestic currency that we mean. If S is the price of a foreign asset, denominated in units of the foreign currency, we can convert it into a domestic asset price simply by multiplying by the exchange rate: $X(t)S(t)$ is the price of the asset, denominated in the domestic currency. For example, if the domestic currency is dollars and the foreign currency is yen, then S is in units of yen and X is units of dollars per unit of yen, so XS is in units of dollars.

Throughout the chapter, we will maintain assumptions similar to the Black-Scholes assumptions. There is a foreign asset with price S in the foreign currency. It has a constant dividend yield q and a constant volatility σ_s. The exchange rate has a constant volatility σ_x and a constant correlation ρ with the foreign asset. There is a domestic risk-free asset with constant interest rate r and a foreign risk-free asset with constant interest rate r_f. The term "risk free" means of course that they are risk-free in their respective currencies. For example, an investment in the foreign risk-free asset is not risk free to a domestic investor, because of exchange rate risk.

6.1 Currency Options

A European call option on the exchange rate X pays $\max(0, X(T) - K)$ at its maturity T, where K is the strike price (in domestic currency). The underlying asset should be regarded as the foreign risk-free asset, the domestic price of which fluctuates with the exchange rate. An investment in the foreign risk-free asset grows via reinvestment of interest at rate r_f, just as the number of shares held of a stock grows via reinvestment of dividends at rate q, if q is its constant dividend yield. In particular, the cost at date 0 of obtaining one unit of foreign currency at date T is the cost at date 0 of $e^{-r_f T}$ units of foreign currency, which is $e^{-r_f T} X(0)$. Thus, the exchange rate is analogous to a stock price, with the foreign risk-free rate being its dividend yield. This means we can apply the Black-Scholes formulas to value currency calls and puts:

Calls and puts on foreign currency can be valued by the Black-Scholes formulas with inputs $X(0) =$ initial asset price, $r =$ risk-free rate, $\sigma_x =$ volatility, and $r_f =$ dividend yield.

6.2 Options on Foreign Assets Struck in Foreign Currency

An option on a foreign asset, with the strike price defined in the foreign currency, can be priced with the Black-Scholes formula, assuming the volatility and dividend yields of the asset are constant and that the (foreign) interest rate is constant. This must be true, because we did not need to specify the currency (dollars, yen, etc.) when deriving the Black-Scholes formula. The value given by the Black-Scholes formula is in the same currency as the asset. To obtain a value in domestic currency for an option on a foreign asset, we simply multiply the Black-Scholes formula by the current exchange rate.

6.3 Options on Foreign Assets Struck in Domestic Currency

A call option with domestic strike price K on the foreign asset with price S pays

$$\max(X(T)S(T) - K, 0)$$

at its maturity T. The underlying price $X(T)S(T)$ is the value in domestic currency of the portfolio that with starts with e^{-qT} units of the asset and reinvests dividends until date T. Thus, we can use the Black-Scholes formula to value it, taking the initial asset price to be e $^{qT}X(0)S(0)$ and the dividend rate to be zero (or taking the initial asset price to be $X(0)S(0)$ and the dividend rate to be q). The volatility that should be input into the Black-Scholes formula is the volatility of the domestic currency price $e^{-q(T-t)}X(t)S(t)$, which is the same as the volatility of $X(t)S(t)$. According to the formula (2.38), the volatility of the domestic currency price XS is

$$\sigma = \sqrt{\sigma_x^2 + \sigma_s^2 + 2\rho\sigma_x\sigma_s} \,. \tag{6.1}$$

We conclude:

Calls and puts on foreign assets struck in domestic currency can be valued by the Black-Scholes formulas with inputs $X(0)S(0)$ = initial price, r = risk-free rate, (6.1) = volatility, q = dividend yield.

6.4 Currency Forwards and Futures

Consider a forward contract maturing at some date T on one unit of foreign currency. In keeping with our convention for options, we will always assume (without loss of generality) that a forward contract is written on a single unit of currency. Let $F(t)$ denote the forward price (in domestic currency) at date $t \leq T$. This means that someone who purchases (goes long) the contract at date t will receive a unit of foreign currency, worth $X(T)$, at date T and must pay $F(t)$ at date T. The value of the long contract at date T is therefore $X(T) - F(t)$. The value at date T of a short contract initiated at date t is the opposite: $F(t) - X(T)$. Naturally, the forward price $F(t)$ is called the "forward exchange rate."

The deepest market for currency is the inter-bank forward market, but futures contracts are also traded on exchanges. The difference between forwards and futures is that futures are "marked to market" daily. Thus, there are daily cash flows with a futures contract, whereas the only cash flows on a forward contract occur at the maturity of the forward. In both cases, there is no cash flow at the time the contract is bought/sold, so its market value is zero. In Sect. 7.7 we will discuss futures contracts further. In particular, we will show, assuming continuous marking to market, that if there is a constant (domestic) risk-free rate—or, more generally, if there is an instantaneous risk-free rate that changes over time in a non-random way—then futures prices must equal forward prices in the absence of arbitrage opportunities. Thus, our assumptions in this chapter imply that currency futures prices should equal currency forward prices. We will consider currency forwards in the remainder of this section.

A forward contract on a traded asset can always be created synthetically simply by buying the asset and holding it, using borrowed money to finance the purchase and to finance any storage costs, assuming the storage costs can be estimated in advance. If the asset pays dividends or generates other positive cash flows, then we do not need to purchase the entire amount covered by the forward contract, because we can accumulate additional amounts of the asset by reinvesting the dividends. There are no storage costs on currency and its dividend yield is equal to the foreign risk-free rate. A forward contract on one unit of foreign currency maturing at date T can be created synthetically at date 0 by buying $e^{-r_f T}$ units of foreign currency and borrowing the cost $e^{-r_f T} X(0)$ at the domestic risk-free rate. This will lead to ownership of one unit of foreign currency at date T and a liability, including interest, of $e^{(r-r_f)T} X(0)$ at date T. Thus, the forward price at date 0 must be $F(0) = e^{(r-r_f)T} X(0)$; otherwise, one could arbitrage by buying the forward and "selling" the synthetic forward, or vice versa. More generally,

---

The forward exchange rate at date t, for a contract maturing at T, must be

$$F(t) = e^{(r-r_f)(T-t)} X(t) . \qquad (6.2)$$

The relation (6.2) is called "covered interest parity." The name stems from the fact that an investment in one of the risk-free assets (foreign or domestic) financed by borrowing in the other, with the currency risk hedged ("covered") by a forward contract, is certain to generate zero value (otherwise, it would be an arbitrage opportunity).[1]

Suppose that one has made a commitment to pay a certain amount of foreign currency (perhaps to a foreign manufacturer) at some date in the future. The exchange rate risk that this commitment entails can obviously be hedged by buying the currency forward. However, one can also create a synthetic forward, by buying currency today and investing it in the foreign risk-free asset. The cash outflow can be incurred today, or it can be deferred by borrowing the cost of the currency at the domestic risk-free rate. In the latter case, we have created a true synthetic forward. In either case, we would call this a "money market hedge" because we have utilized the foreign money market (risk-free asset) to create the hedge.

Later in this chapter we will construct replicating strategies for various contracts using the foreign risk-free asset and the domestic risk-free asset.

---

[1] A relation analogous to covered interest parity holds for any forward contract if the underlying asset has a constant dividend yield and storage costs that are a constant proportion of the value of the units stored. For commodities, the term "dividend yield" must be interpreted in a broad sense, and is usually called "convenience yield," because ownership of the physical asset may produce abnormal profits during temporary shortages, an advantage that is not obtained by owning a forward contract on the asset, just as dividends are not received by the owner of a forward contract. Thus, one must consider the "convenience" of owning the physical asset as an advantage analogous to dividends.

One can interpret these replicating strategies as money market hedges or synthetic currency forwards. In practice, it will often be more convenient to use actual forwards rather than using the foreign risk-free asset. Using actual currency forwards produces an equivalent (given that we are not considering transaction costs) replicating strategy. Here is, in abstract, the way we convert from money market hedges to hedges using forwards. As we have discussed,

Long Currency Forward = Long Synthetic Currency Forward

= Long Foreign Risk-Free Asset

+ Short Domestic Risk-Free Asset .

Subtracting a short position is the same as adding a long position, so we can rearrange this as

Long Currency Forward + Long Domestic Risk-Free Asset

= Long Foreign Risk-Free Asset .

Thus, an investment in the foreign risk-free asset can be replaced in any replicating strategy by long currency forwards and an investment in the domestic risk-free asset.

To be more precise about the sizes of the investments, consider replacing a money market hedge with a forward hedge at some date t prior to the maturity of the forward and analyze the replacement per unit of the money market hedge (per unit of foreign currency invested in the foreign risk-free asset). One unit of foreign currency invested in the foreign risk-free asset at date t will grow to $e^{r_f(T-t)}$ units by date T. Thus, the corresponding forward contract should be on $e^{r_f(T-t)}$ units of currency. The value at date t of both sides of the above equation should be the same, and the value of a forward contract at the date of initiation is zero, so the investment in the domestic risk-free asset should be the domestic currency equivalent of one unit of foreign currency, which is the exchange rate $X(t)$. Thus, we have

---

$e^{r_f(T-t)}$ Long Currency Forwards

$+ X(t)$ Long in the Domestic Risk-Free Asset

= 1 Unit of Foreign Currency Long in the Foreign Risk-Free Asset .

$$(6.3)$$

To check this, consider holding the portfolios until date T. As explained in the first paragraph of this section, the currency forwards will have value $e^{r_f(T-t)}[X(T) - F(t)]$, which by covered interest parity is $e^{r_f(T-t)}X(T) - e^{r(T-t)}X(t)$. When we include the long position in the domestic risk-free asset with accumulated interest, the value at date T of the portfolio on the left-hand side of (6.3) is $e^{r_f(T-t)}X(T)$. On the other, the right-hand side of (6.3) with accumulated interest will consist of $e^{r_f(T-t)}$ units of foreign currency, also worth $e^{r_f(T-t)}X(T)$.

6.5 Quantos

A "quanto" is a derivative written on a foreign asset the value of which is converted to domestic currency at a fixed exchange rate. In other words, the contract pays in the domestic currency and the exchange rate is part of the contract. Such contracts are very useful for investors who want to bet on foreign assets but do not want exposure to exchange rate risk. Such an investor could simply buy the foreign asset and hedge the currency risk by selling currency futures or forwards, but doing so is a bit tricky because the amount of currency that needs to be hedged depends on how well the foreign asset does. Thus, quantos can be desirable contracts. Of course, when an investor purchases a quanto, the problem of hedging the exchange rate risk has simply been transferred to the seller. In this and the following section, we will see how to value and how to replicate a contract that pays the price of a foreign asset at some future date T with the price translated into the domestic currency at a fixed exchange rate. The replicating strategy is the strategy that would be followed by the seller (or by an investor who wants to create a synthetic on his own). Specifically, in this section we will determine the value at date 0 (in domestic currency) of a contract that pays $\bar{X}S(T)$ at date T, where $\bar{X}$ is a fixed exchange rate. Later in the chapter, we will consider quanto forwards and quanto options.

In addition to being practically useful, this contract is an excellent example for demonstrating the methodology of pricing and hedging. The best way to proceed in problems of this general type is to first value the contract and then calculate the replicating strategy.[2] As discussed in Sect. 1.5, valuation is simplified by choosing a numeraire that will cancel the randomness in the contract payoff. Our numeraire must be a non-dividend-paying (domestic) asset price, so we can choose $Z(t) = X(t)e^{qt}S(t)$ to be the numeraire asset price. This is the value in the domestic currency of a strategy that is long one unit of the foreign asset at date 0 and which reinvests the dividends of the asset into new shares. As we will see immediately, using it as numeraire introduces randomness into the payoff through the exchange rate, and that poses some complications. Applying our fundamental pricing formula (1.17), the value of the contract is

---

[2] We did the same thing in Chap. 3: we first derived the Black-Scholes formula and then found the replicating strategy (delta hedge) by differentiating the formula.

$$Z(0)E^Z \left[\frac{\bar{X}S(T)}{Z(T)} \right] = e^{-qT} X(0)S(0)E^Z \left[\frac{\bar{X}S(T)}{X(T)S(T)} \right]$$

$$= e^{-qT} \bar{X} S(0) E^Z \left[\frac{X(0)}{X(T)} \right] . \tag{6.4}$$

Now we need to evaluate $E^Z[X(0)/X(T)]$, which is the expected growth of $1/X$ when Z is used as the numeraire. We will show that

$$E^Z \left[\frac{X(0)}{X(T)} \right] = \exp \left\{ (r_f - r - \rho\sigma_x\sigma_s)T \right\} . \tag{6.5}$$

This implies:

The value at date 0 of a contract that pays $\bar{X}S(T)$ at date T, where $\bar{X}$ is a fixed exchange rate and S is the foreign price of an asset with a constant dividend yield q, is

$$\exp \left\{ (r_f - r - q - \rho\sigma_x\sigma_s)T \right\} \bar{X}S(0) . \tag{6.6}$$

We will now prove (6.5). The assumption that S and X have constant volatilities and correlation means that

$$\frac{dX}{X} = \mu_x \, dt + \sigma_x \, dB_x ,$$

$$\frac{dS}{S} = \mu_s \, dt + \sigma_s \, dB_s ,$$

for some (possibly random) μ_x and μ_s, where B_s and B_x are Brownian motions with correlation equal to ρ. From Itô's formula, we have

$$\frac{dZ}{Z} = q \, dt + \frac{d(XS)}{XS}$$

$$= (q + \mu_x + \mu_s + \rho\sigma_x\sigma_s) \, dt + \sigma_x \, dB_x + \sigma_s \, dB_s$$

$$= (q + \mu_x + \mu_s + \rho\sigma_x\sigma_s) \, dt + \sigma \left(\frac{\sigma_x}{\sigma} \, dB_x + \frac{\sigma_s}{\sigma} \, dB_s \right)$$

$$= (q + \mu_x + \mu_s + \rho\sigma_x\sigma_s) \, dt + \sigma \, dB ,$$

where we define σ in (6.1) and B by $B(0) = 0$ and

$$dB = \frac{\sigma_x}{\sigma} \, dB_x + \frac{\sigma_s}{\sigma} \, dB_s .$$

As discussed in Sect. 2.11, B is a Brownian motion and σ is the volatility of Z. Notice that the correlation of X and Z is

$$(dB)(dB_x) = \left(\frac{\sigma_x}{\sigma} \, dB_x + \frac{\sigma_s}{\sigma} \, dB_s \right) (dB_x)$$

$$= \frac{\sigma_x + \rho\sigma_s}{\sigma} \, dt .$$

Now we use (2.29) in Sect. 2.9 which gives the drift of an asset when another risky asset is used as the numeraire. We use Z as the numeraire and X as the other asset, regarding X as the domestic price of an asset with dividend yield r_f as before. Therefore, we substitute r_f for q in (2.29), substitute σ for the volatility of the numeraire asset price, substitute σ_x for the volatility of the other asset, and substitute $(\sigma_x + \rho\sigma_s)/\sigma$ for their correlation. This yields

$$\frac{\mathrm{d}X}{X} = \left(r - r_f + \sigma_x^2 + \rho\sigma_x\sigma_s\right)\,\mathrm{d}t + \sigma_x\,\mathrm{d}B_x^*\,,$$

where B_x^* is a Brownian motion when Z is the numeraire. Now we apply Itô's formula for ratios to obtain

$$\frac{\mathrm{d}(1/X)}{1/X} = -\frac{\mathrm{d}X}{X} + \left(\frac{\mathrm{d}X}{X}\right)^2$$
$$= (r_f - r - \rho\sigma_x\sigma_s)\,\mathrm{d}t + \sigma_x\,\mathrm{d}B_x^*\,.$$

This implies that $1/X$ is a geometric Brownian motion with growth rate $r_f - r - \rho\sigma_x\sigma_s$, from which (6.5) follows.

6.6 Replicating Quantos

The assets we will use to replicate the payoff $\bar{X}S(T)$ are the foreign asset with price S, the foreign risk-free asset, and the domestic risk-free asset. At the end of this section, we will explain how to replace the foreign risk-free asset with currency forwards, as discussed in Sect. 6.4. Before beginning the calculations, we can make the following intuitive observations:

- The payoff $\bar{X}S(T)$ has exposure to the foreign asset price S, so the replicating portfolio must be long the foreign asset.
- The payoff $\bar{X}S(T)$ has no exposure to the exchange rate, so the replicating portfolio cannot have any exposure to the exchange rate either. Thus, the long position in the foreign risky asset must be offset by an equal short position in the foreign risk-free asset.
- As a result of the previous observation, the value of the replicating portfolio, displayed in (6.6), will equal the investment in the domestic risk-free asset.

Consequently, our real task is to compute the number of shares of the foreign asset that should be held, the remainder of the replicating portfolio being thereby determined.

The value of the replicating portfolio at any date $t \leq T$ must be the value at date t of receiving the payoff $\bar{X}S(T)$ at date T. We have calculated this value at date 0, and, clearly, the formula (6.6) applies to general dates t, when we replace the time T to maturity by $T - t$ and the asset price $S(0)$ at the date of valuation by $S(t)$. That is, the value of the portfolio at any date $t \leq T$ must be $V(t)$ defined as

$$V(t) = \exp\{(r_f - r \quad q - \rho\upsilon_x\upsilon_s)(T - t)\} \bar{X} S(t) .$$ (6.7)

As just noted, we will need to invest this amount in the domestic risk-free asset at date t. What remains to be done is to calculate the size of the long position in the foreign risky asset and the offsetting short position in the foreign risk-free asset.

From Itô's formula, we have

$$\frac{dV}{V} = -(r_f - r - q - \rho\sigma_x\sigma_s)\, dt + \frac{dS}{S} .$$ (6.8)

Equivalently,

$$dV = (r + q - r_f + \rho\sigma_x\sigma_s)V\, dt + V\frac{dS}{S} .$$ (6.9)

On the other hand, consider a strategy that invests $a(t)$ units of the domestic currency in the foreign asset, $b(t)$ units of the domestic currency in the foreign risk-free asset, and $c(t)$ units of the domestic currency in the domestic risk-free asset. Let $W = a + b + c$ denote the value of this portfolio. The return on the foreign asset, per unit of domestic currency invested, is

$$\frac{d(Xe^{qt}S)}{Xe^{qt}S} = q\, dt + \frac{dX}{X} + \frac{dS}{S} + \left(\frac{dX}{X}\right)\left(\frac{dS}{S}\right)$$

$$= (q + \rho\sigma_x\sigma_s)\, dt + \frac{dX}{X} + \frac{dS}{S} .$$ (6.10)

Similarly, the rate of return on the foreign risk-free asset is

$$\frac{d(e^{r_f t}X)}{e^{r_f t}X} = r_f\, dt + \frac{dX}{X} ,$$ (6.11)

and of course the rate of return on the domestic risk-free asset is $r\, dt$. Therefore, the change in the value of the portfolio will be

$$dW = a\left[(q + \rho\sigma_x\sigma_s)\, dt + \frac{dX}{X} + \frac{dS}{S}\right] + b\left[r_f\, dt + \frac{dX}{X}\right] + cr\, dt$$

$$= (aq + a\rho\sigma_x\sigma_s + br_f + cr)\, dt + (a + b)\frac{dX}{X} + a\frac{dS}{S} .$$ (6.12)

The change (6.12) of the portfolio value will match the change (6.9) of V if and only if

$$a = V, \qquad b = -V, \qquad c = V .$$ (6.13)

This implies:

The strategy that replicates the payoff $\bar{X}S(T)$ at date T is to invest $V(t)$ units of domestic currency in the foreign asset, where $V(t)$ is defined in (6.7). This will purchase

$$\frac{V(t)}{X(t)S(t)} = \frac{\bar{X}}{X(t)} \exp\{(r_f - r - q - \rho\sigma_x\sigma_s)(T-t)\} \qquad (6.14)$$

shares of the foreign asset. This position is financed entirely by borrowing at the foreign risk-free rate. On the other hand, the same amount $V(t)$ of the domestic currency is invested in the domestic risk-free asset.

From our analysis at the end of Sect. 6.4, we know that the foreign risk-free asset in this replicating strategy can be replaced by currency forwards. The strategy here involves borrowing at the foreign risk-free rate, so we should replace "long" by "short" in (6.3). Borrowing $V(t)$ units of domestic currency means borrowing $V(t)/X(t)$ units of the foreign currency. Therefore, (6.3) gives us:

An equivalent strategy for replicating the payoff $\bar{X}S(T)$ at date T is to invest $V(t)$ units of domestic currency in the foreign asset and to be short $e^{r_f(T-t)}V(t)/X(t)$ currency forwards at date t.

At the beginning of the previous section, we noted that an investor who wants to bet on a foreign asset but does not want the exchange rate exposure could simply buy the asset and sell the currency forward. This shows how much of the asset he should buy and how much currency he should sell forward.

It is important to note that this strategy involves continuously buying and selling forwards, just as it involves continuously trading the foreign asset. Buying at date t a forward contract sold at date $s < t$ cancels the delivery obligation on the contract sold at s and leaves a cash flow of $F(s) - F(t)$ to be paid/received at the maturity date T. Therefore, the strategy accumulates a liability or asset, depending on the direction the forward price moves, to be received at T. On the other hand, maintaining an investment of $V(t)$ in the foreign asset will generate cash flows as the asset is sold or purchased over time. As (6.14) shows, whether it is sold or purchased depends on the direction the exchange rate moves. These cash flows should be invested or borrowed at the domestic risk-free rate. Thus, there is a liability or asset to be received at date T that is not shown in the boxed statement immediately above, and there is an investment or liability in the domestic risk-free asset that is not shown. It can be demonstrated that these cancel each other: if profits are made from trading forwards, then they (more precisely, their present value) will be consumed by the cost of buying the foreign asset, and vice versa. Hedging with forwards (and with futures) is considered in more detail in Sect. 7.10.

6.7 Quanto Forwards

In this section, we consider a contract similar to that of the previous section, except that it is a pure forward, meaning that the payment for the contract occurs at date T. We maintain all of the assumptions of the previous section. The payment at date T is in domestic currency, and we define the quanto forward price in units of domestic currency. Specifically, a long quanto forward contract, initiated at date t and maturing at date T and initiated at the forward price $F^*(t)$ will pay

$$\bar{X}S(T) - F^*(t)$$

at date T. The forward price $F^*(t)$ should be the price that makes this contract have a value of 0 at date t.

We already know how to replicate the underlying payoff $\bar{X}S(T)$ of the forward contract at the cost $V(t)$ defined in (6.7). Thus, the synthetic quanto forward is to purchase the replicating strategy and to borrow the cost $V(t)$ at the domestic risk-free rate. This leads to the liability $e^{r(T-t)}V(t)$ at date T. Therefore, we have:

The quanto forward price is

$$F^*(t) = e^{r(T-t)}V(t) = \exp\left\{(r_f - q - \rho\sigma_x\sigma_s)(T-t)\right\}\bar{X}S(t) \,. \qquad (6.15)$$

Notice that borrowing V in domestic currency to finance the replicating strategy of the previous section — i.e., the domestic currency investments described in (6.13) — means eliminating the domestic risk-free investment $c = V$ required in the previous section. The replicating strategy for the quanto forward is simply to invest V in the foreign asset and to finance the investment entirely by borrowing at the foreign risk-free rate. As in the previous section, borrowing at the foreign risk-free rate can be replaced by borrowing at the domestic risk-free rate and selling currency forwards.

6.8 Quanto Options

Consider now a European call option on a foreign asset, with strike K set in the domestic currency and the value of the foreign asset being converted to domestic currency at a fixed exchange rate $\bar{X}$. This is called a "quanto call." We maintain all of the assumptions of the previous two sections.

The value of the quanto call at maturity is $\max(0, \bar{X}S(T) - K)$. To value this, we make use of what we learned in Sect. 6.5. Namely, the portfolio with value V defined in (6.7) replicates the payoff $\bar{X}S(T)$: in each state of the world, $V(T) = \bar{X}S(T)$. Therefore, the quanto call is equivalent to a standard

European call on the portfolio with domestic currency price V. The value is therefore given by the Black-Scholes formula. From the formula (6.8) for the dynamics of V, we see that the volatility of V is the same as that of S; therefore, we should input σ_s as the volatility in the Black-Scholes formula. Furthermore, the portfolio V is non-dividend-paying (it is the value of a claim to $\bar{X}S(T)$ at date T with no interim cash flows), so the dividend rate in the Black-Scholes formula should be zero. Thus, we have:

The value of a quanto call is

$$V(0)\,\mathrm{N}(d_1) - \mathrm{e}^{-rT}K\,\mathrm{N}(d_2)$$
$$= \exp\left\{(r_f - r - q - \rho\sigma_x\sigma_s)T\right\}\bar{X}S(0)\,\mathrm{N}(d_1) - \mathrm{e}^{-rT}K\,\mathrm{N}(d_2)\,, \quad (6.16)$$

where

$$d_1 = \frac{\log\left(\frac{V(0)}{K}\right) + \left(r + \frac{1}{2}\sigma_s^2\right)T}{\sigma_s\sqrt{T}}$$

$$= \frac{\log\left(\frac{\bar{X}S(0)}{K}\right) + \left(r_f - q - \rho\sigma_x\sigma_s + \frac{1}{2}\sigma_s^2\right)T}{\sigma_s\sqrt{T}}\,, \quad (6.17a)$$

$$d_2 = d_1 - \sigma_s\sqrt{T}\,. \quad (6.17b)$$

Likewise, the value of a quanto put is given by the Black-Scholes formula:

$$\mathrm{e}^{-rT}K\,\mathrm{N}(-d_2) - V(0)\,\mathrm{N}(-d_1)\,.$$

Notice that this is simply the Black-Scholes option formula with inputs $V(0) =$ initial asset price, $K =$ exercise price, $r =$ interest rate, $\sigma_s =$ volatility, $0 =$ dividend yield, and $T =$ time to maturity.

We can hedge a written quanto call the same way we hedge a written ordinary call: we buy delta shares of the underlying and borrow the difference between the cost of the delta shares and the option value. However, for the quanto call, the underlying should be regarded as the portfolio with value V described in Sect. 6.5. This portfolio consists of investing $V(0)$ units of domestic currency in the foreign asset, borrowing the same amount at the foreign risk-free rate, and investing $V(0)$ units of domestic currency in the domestic risk-free asset. The delta of the call is $\mathrm{N}(d_1)$, so the hedge consists of investing $\mathrm{N}(d_1)V(0)$ units of domestic currency in the foreign asset, borrowing the same amount at the foreign risk-free rate, and investing $\mathrm{N}(d_1)V(0)$ in the domestic risk-free asset. The difference between the cost of this portfolio and the value of the option is

$$\mathrm{N}(d_1)V(0) - [V(0)\,\mathrm{N}(d_1) - \mathrm{e}^{-rT}K\,\mathrm{N}(d_2)] = \mathrm{e}^{-rT}K\,\mathrm{N}(d_2)\,.$$

This amount is to be borrowed at the domestic risk-free rate. Thus, the net investment in the domestic risk-free asset is

$$\mathrm{N}(d_1)V(0) - \mathrm{e}^{-rT}K\,\mathrm{N}(d_2)\,,$$

which is just the value of the option. To summarize:

> To delta-hedge a written quanto call, one should invest $\mathrm{N}(d_1)V(0)$ units of domestic currency in the foreign asset, borrow the same amount at the foreign risk-free rate, and invest the value of the option in the domestic risk-free asset.

As in Sect. 6.6, borrowing $\mathrm{N}(d_1)V(0)$ units of domestic currency at the foreign risk-free rate can be replaced by borrowing the same amount at the domestic risk-free rate and selling $\mathrm{e}^{r_f T}\mathrm{N}(d_1)V(0)/X(0)$ currency forwards. This results in:

> An equivalent delta hedge for a written quanto call is to invest $\mathrm{N}(d_1)V(0)$ units of domestic currency in the foreign asset, sell $\mathrm{e}^{r_f T}\mathrm{N}(d_1)V(0)/X(0)$ currency forward contracts at the market forward price $F(0)$, and borrow $\mathrm{e}^{-rT}K\,\mathrm{N}(d_2)$ at the domestic risk-free rate.

6.9 Return Swaps

There are many types and applications of return swaps, but here is one important example that involves the concepts discussed in this chapter. Suppose an investor wants to receive at date T the difference in the rates of return of two assets that are denominated in different currencies. The return will be calculated on a given "notional principal." For example, an investor may want to receive at the end of a year the Nikkei rate of return minus the rate of return on the S&P 500, calculated on a $1 million notional principal. If the Nikkei earns 15% over the year and the S&P earns 10%, then the payment to the investor is 5% of $1 million. If the reverse happens—the Nikkei earns 10% and the S&P earns 15%—then the investor must pay 5% of $1 million to the counterparty.

To model this, let S_f denote the price of a foreign asset and S_d the price of a domestic asset. Assume they have constant dividend yields q_f and q_d. If the returns are calculated excluding dividends, as is likely to be the case, then the payment to the investor is

$$\left(\frac{S_f(T) - S_f(0)}{S_f(0)} - \frac{S_d(T) - S_d(0)}{S_d(0)} \right) A = \left(\frac{S_f(T)}{S_f(0)} - \frac{S_d(T)}{S_d(0)} \right) A\,,$$

where A denotes the notional principal. Of course, the investor may want the reverse swap, and we consider this particular case only for concreteness.

The swap may have nonzero market value at date 0, which means that some payment will have to be made upfront. To eliminate this, we can add a "swap spread" into the contract, affecting the cash flow at date T. This is a constant number a (which may be positive or negative), and including it changes the payment to the investor to

$$\left(a + \frac{S_f(T)}{S_f(0)} - \frac{S_d(T)}{S_d(0)} \right) A . \tag{6.18}$$

The question we will address here is: what is the "fair" swap spread; i.e., for what number a does the cash flow (6.18) have zero market value at date 0?

If the value is zero, then it is zero for any notional principal A, so we can conveniently take $A = 1$. The cash flow consists of three pieces, all of which are to be received/paid at date T: the constant a, the gross return on the foreign asset, and the gross return on the domestic asset. The value at date 0 of receiving a units of domestic currency is obviously $e^{-rT} a$. As we have observed several times before, the value at date 0 of receiving $S_d(T)$ units of domestic currency at date T is $e^{-q_d T} S_d(0)$, because this is the cost of enough shares to accumulate to one share at date T via reinvestment of dividends. Therefore, the value at date 0 of receiving $S_d(T)/S_d(0)$ at date T is $e^{-q_d T} S_d(0)/S_d(0) = e^{-q_d T}$.

What remains is to calculate the value of receiving $S_f(T)/S_f(0)$ units of domestic currency at date T. We can do this by interpreting $1/S_f(0)$ as the fixed exchange rate $\bar{X}$ in the definition of a quanto.[3] We need to assume as before that the foreign asset price S_f and the exchange rate have constant volatilities and a constant correlation. Denoting the volatilities by σ_s and σ_x and the correlation by ρ as before, equation (6.6) shows that the value of receiving $\bar{X} S_f(T) = S_f(T)/S_f(0)$ units of domestic currency at date T is

$$\exp \left\{ (r_f - r - q_f - \rho \sigma_x \sigma_s) T \right\} .$$

Adding up the pieces, the value at date 0 of the cash flow (6.18) (with $A = 1$) is

$$e^{-rT} a + \exp \left\{ (r_f - r - q_f - \rho \sigma_x \sigma_s) T \right\} - e^{-q_d T} ,$$

so we conclude:

The fair swap spread, which equates the value at date 0 of receiving the cash flow (6.18) at date T to zero, is

$$a = \exp \left\{ (r - q_d) T \right\} - \exp \left\{ (r_f - q_f - \rho \sigma_x \sigma_s) T \right\} . \tag{6.19}$$

---

[3] To make sense of the units, note that the cash flow of $S_f(T)/S_f(0)$ units of domestic currency can be calculated as $S_f(T)$ units of foreign currency times $1/S_f(0)$ units of domestic currency per unit of foreign currency. Therefore, the units of $1/S_f(0)$ can be taken to be the units of an exchange rate.

6.10 Uncovered Interest Parity in the Risk-Neutral Probabilities

When we use numerical methods to value American and path-dependent options, as in Chap. 5, we typically focus on the dynamics of asset prices under the risk-neutral measure. To apply these results to currency options or options on foreign assets, we need to know the dynamics of the exchange rate under the risk-neutral measure. Because we can view the exchange rate as the domestic price of an asset with the foreign risk-free rate r_f being its dividend yield, we have already calculated these dynamics in equation (2.27) of Sect. 2.9. The result is:

The exchange rate X must satisfy

$$\frac{\mathrm{d}X}{X} = (r - r_f)\,\mathrm{d}t + \sigma_x\,\mathrm{d}B_x^*\,, \qquad (6.20)$$

where B_x^* is a Brownian motion under the risk-neutral measure.

This equation has an interesting interpretation in terms of "uncovered interest parity," which is the theory that differences in interest rates across currencies will be offset on average by appreciation/depreciation of the currencies. In other words, it is the theory that the strategy of borrowing in low-interest-rate currencies to invest in high-interest-rate currencies will not earn money on average because of depreciation of the high-interest-rate currency relative to the low-interest-rate currency. It is well known that this theory is not always true in reality. However, equation (6.20) shows that it is true when we calculate expectations using the risk-neutral measure.

To see the interpretation of equation (6.20) as uncovered interest parity, suppose that the foreign interest rate r_f is lower than the domestic rate r. Then one may be tempted to borrow at the foreign rate and invest at the domestic rate. This would create a short position in the foreign currency. Equation (6.20) states that the exchange rate is expected (under the risk-neutral measure) to appreciate at the rate $r-r_f$; thus, repayment of the foreign currency will be more expensive in terms of domestic currency, offsetting the interest rate differential.

Problems

6.1. Create an Excel worksheet to compare the values of call options on foreign assets that are (i) struck in foreign currency or (ii) struck in domestic currency. Prompt the user to input $X(0)$, $S(0)$, K, r, r_f, σ_x, σ_s, ρ, q and T. Take the strike price of the option struck in foreign currency to be K and take the strike price of the option struck in domestic currency to be $X(0)K$ (so K is interpreted as an amount in foreign currency). You should be able to confirm, for example, that if $r = r_f$ and $\rho \geq 0$ then the option struck in domestic currency is more valuable.

6.2. Repeat the preceding problem comparing (i) call options struck in foreign currency, versus (ii) quanto call options. Use the same inputs as in the preceding problem and take the fixed exchange rate in the quanto to be $\bar{X} = X(0)$. You should be able to confirm, for example, that if $r = r_f$ and $\rho \geq 0$ then the option struck in foreign currency is more valuable.

6.3. Create an Excel worksheet in which the user inputs r and r_f and the exchange rate. Compute the forward exchange rate at maturities $T = 0.1, 0.2, \ldots, 2.0$ and plot the forward rate against the maturity in a scatter plot. A market is said to be in "contango" if this curve is upward sloping and to be in "backwardation" if this curve is downward sloping. For currencies, what determines whether the market is in contango or in backwardation?

6.4. Create a VBA subroutine to simulate a path of the exchange rate and the forward exchange rate under the risk-neutral measure, prompting the user to input $X(0)$, r, r_f, σ_x, and the maturity T of the forward contract.

6.5. Create a VBA subroutine to simulate a path of the exchange rate under the actual probability measure, prompting the user to input $X(0)$, σ_x, and the expected rate of growth μ of the exchange rate under the actual probability measure. Prompt the user also to input $S(0)$, r, r_f, σ_s, q, ρ, a fixed exchange rate $\bar{X}$, a maturity T, and a number of periods N. Calculate the gain/loss from the portfolio that promises to pay $\bar{X}S(T)$ at date T and uses a discretely rebalanced hedge, rebalancing at dates $t_1, \ldots t_N = T$, where $t_i - t_{i-1} = T/N$, similar to the calculation in the function `Simulated_Delta_Hedge_Profit`. Use the money-market hedge, which means investing $V(0)$ at date 0, holding the number of shares of the foreign asset shown in (6.14) at each date t_i, and having a short position in the foreign risk-free asset of the same value at each date t_i. Cash flows generated at each date from buying/selling the foreign asset and lending/borrowing at the foreign risk-free rate should be withdrawn/deposited in the domestic risk-free asset. Note: Because of discrete rebalancing, this is not a perfect hedge, and the investment in the domestic risk-free asset will not always equal $V(t)$.

6.6. Repeat the previous exercise using the forward contract hedge discussed in Sect. 6.6. The cash flows generated from trading forwards cannot be withdrawn/deposited in the domestic risk-free asset, because they do not materialize until the maturity of the forward. You will have to create a variable to keep track of the net asset/liability and include it in the valuation at date T.

6.7. Derive the money-market hedge and the forward contract hedge for a written quanto put.

6.8. Suppose a customer has contracted with you for a return swap in which the customer will receive the cash flow (6.18) for some number a, where $A = 1$. How can you hedge this?

7

Forward, Futures, and Exchange Options

In this chapter, we will derive three important generalizations of the Black-Scholes formula. We will derive them from the Black-Scholes formula, which shows that all of the formulas are equivalent. We will start with Margrabe's [50] formula for an option to exchange one asset for another. Standard calls and puts are special cases, involving the exchange of cash for an asset or an asset for cash. From Margrabe's formula, we will derive Black's [3] formulas for options on forward and futures contracts. Then, from Black's formulas, we will derive Merton's [51] formulas for calls and puts in the absence of a constant risk-free rate.

Unless explicitly stated otherwise, we will not assume in this chapter the existence of a risk-free asset (or even an instantaneously risk-free asset as described in Sect. 1.1). This implies that the market is incomplete and there are many risk-neutral measures. Nevertheless, we can price exchange options, forward and futures options, and stock options by arbitrage. Understanding this issue is not essential for deriving the formulas in this chapter—as mentioned, they will all be derived from the Black-Scholes formula—but the issue is nonetheless important. It is discussed in the final section of the chapter.

Naturally, all of the option-pricing formulas discussed in this chapter are quite similar. The similarity can be seen from the Black-Scholes formula for a call option, which we can write as follows (replacing d_1 by x and d_2 by y):

$$e^{-qT} S(0) \, N(x) - e^{-rT} K \, N(y) \,, \tag{7.1a}$$

where

$$x = \frac{\log\left(\frac{S(0)}{K}\right) + \left(r - q + \frac{1}{2}\sigma^2\right) T}{\sigma\sqrt{T}} \tag{7.1b}$$

$$y = x - \sigma\sqrt{T} \,. \tag{7.1c}$$

Note that $e^{-qT} S(0)$ is the present value at date 0 of the stock that would be acquired if the option is exercised, because it is the cost that one must pay at date 0 to have one share of the stock at date T with no withdrawal of dividends in the interim. Obviously, $e^{-rT}K$ is the present value of the cash that is paid if the option is exercised. Moreover, x is equal to

$$\frac{\log\left(\frac{e^{-qT}S(0)}{e^{-rT}K}\right) + \frac{1}{2}\sigma^2 T}{\sigma\sqrt{T}},$$

and the logarithm in the numerator is of the ratio of present values. All of the option pricing formulas in this chapter have the same form: the present value of the asset to be acquired multiplied by $\mathrm{N}(x)$ minus the present value of the asset to be delivered multiplied by $\mathrm{N}(y)$. Moreover in each case x is the logarithm of the ratio of present values plus one-half $\sigma^2 T$ all divided by $\sigma\sqrt{T}$, and in each case y is defined by (7.1c). Notice that the Black-Scholes put option formula has this structure also. The Black-Scholes put option formula is

$$e^{-rT}K\,\mathrm{N}(x) - e^{-qT}S(0)\,\mathrm{N}(y), \tag{7.2a}$$

where

$$x = -d_2$$

$$= -\frac{\log\left(\frac{S(0)}{K}\right) + \left(r - q - \frac{1}{2}\sigma^2\right)T}{\sigma\sqrt{T}}$$

$$= \frac{\log\left(\frac{e^{-rT}K}{e^{-qT}S(0)}\right) + \frac{1}{2}\sigma^2 T}{\sigma\sqrt{T}}, \tag{7.2b}$$

$$y = -d_1$$

$$= -\frac{\log\left(\frac{S(0)}{K}\right) + \left(r - q + \frac{1}{2}\sigma^2\right)T}{\sigma\sqrt{T}}$$

$$= x - \sigma\sqrt{T}. \tag{7.2c}$$

This similarity is discussed further in Sect. 7.5, where the pricing formulas are implemented in VBA.

7.1 Margrabe's Formula

Consider two assets with prices S_1 and S_2 and a European option to exchange asset 2 for asset 1 at date T. The value of the option at maturity is

$$\max(0, S_1(T) - S_2(T)).$$

Note that there is no real difference between a put and a call: the exchange option can be viewed as a call on the first asset with random strike $S_2(T)$ or as a put on the second asset with random strike $S_1(T)$.

Assume the assets pay constant dividend yields q_i and assume the prices satisfy

$$\frac{\mathrm{d}S_i}{S_i} = \mu_i \, \mathrm{d}t + \sigma_i \, \mathrm{d}B_i$$

where each B_i is a Brownian motion under the actual probability measure. As before, the drifts μ_i can be quite general random processes. We also allow the volatilities σ_i and the correlation ρ of the Brownian motions to be random processes; however, we make the assumption that σ defined as

$$\sigma = \sqrt{\sigma_1^2 + \sigma_2^2 - 2\rho\sigma_1\sigma_2} \tag{7.3}$$

is a constant. As shown in (2.40), σ is the volatility of S_1/S_2 (and also S_2/S_1). So, the assumption we are making is that the volatility of the ratio of the asset prices is constant. In Sect. 7.9, we will relax this assumption to allow σ to be time-varying (though still non-random).

The following is the formula of Margrabe [50]:

The value of a European option to exchange two assets at date T is

$$e^{-q_1 T} S_1(0) \, \mathrm{N}(d_1) - e^{-q_2 T} S_2(0) \, \mathrm{N}(d_2) \,, \tag{7.4a}$$

where

$$d_1 = \frac{\log\left(\frac{S_1(0)}{S_2(0)}\right) + \left(q_2 - q_1 + \frac{1}{2}\sigma^2\right) T}{\sigma\sqrt{T}} \,, \tag{7.4b}$$

$$d_2 = d_1 - \sigma\sqrt{T} \,. \tag{7.4c}$$

Margrabe's derivation is a very simple argument based on the Black-Scholes formula. We noted in Chap. 6 that the Black-Scholes formula does not depend on the currency—if the underlying asset and risk-free asset are dollar denominated, the formula gives the dollar value of an option; if they are yen denominated, the formula gives the yen value of an option, etc. So we can take the "currency" to be units of the second asset; i.e., we will use the second asset as numeraire. With this numeraire, the value of the first asset is S_1/S_2. The value of the exchange option at maturity is

$$\max(0, S_1(T) - S_2(T)) = S_2(T) \max\left(0, \frac{S_1(T)}{S_2(T)} - 1\right) \,.$$

This is the value in the natural currency (e.g., dollars). The value using the second asset as numeraire is obtained by dividing by $S_2(T)$, so it is

$$\max\left(0, \frac{S_1(T)}{S_2(T)} - 1\right) \,.$$

This is the value of a standard call option, the underlying being the first asset measured in units of the second. We can apply the Black-Scholes formula to obtain the value of the option (in units of the second asset) at date 0. Multiplying this value by $S_2(0)$ will give the option value in the natural currency.

The risk-free rate when the second asset is the numeraire is the dividend yield on the second asset q_2. To see this, note that the price of the second asset is always equal to one; moreover, an investment in the second asset will accumulate at the rate q_2 via reinvestment of dividends. Therefore, q_2 is a risk-free rate of return.

The dividend yield on the first asset remains q_1. To see this, note that the dividend paid in the natural currency is $q_1 S_1(t) \, \mathrm{d}t$ in an instant $\mathrm{d}t$ and the value of this dividend using the second asset as numeraire is $[q_1 S_1(t)/S_2(t)] \, \mathrm{d}t$, which is the fraction $q_1 \, \mathrm{d}t$ of the value $S_1(t)/S_2(t)$ of the first asset using the second asset as numeraire.

The volatility of the first asset using the second as numeraire is the volatility of the ratio $S_1(t)/S_2(t)$, which is σ defined in (7.3). Applying the Black-Scholes formula with these inputs yields Margrabe's formula directly.[1]

7.2 Black's Formula

Black [3] gives formulas for the values of options on futures contracts when interest rates are deterministic (i.e., non-random). It is well known (and we will establish this in Sect. 7.7) that, when interest rates are deterministic, futures prices should equal forward prices, so Black's formulas also yield formulas for the values of options on forward contracts when interest rates are deterministic. However, the formulas for options on forwards are valid more generally (even when interest rates vary randomly) and now a mention of Black's formulas is more likely to be referring to the formulas for options on forwards, instead of the formulas for options on futures. In any case, we will start with the formulas for options on forwards and then in Sect. 7.8 derive the formulas for options on futures when interest rates are deterministic.

We consider a forward contract that matures at some date T' and a call or put option on the forward that matures at $T \leq T'$. The meaning of a call option on a forward is that exercise of the call creates a long position in the forward contract with forward price equal to the strike price of the option. The long forward contract means that the investor will receive the underlying asset at T' and pay the forward price (the strike of the option) at T'. Thus, the strike price is not paid at the date of exercise but instead is paid when the underlying asset is delivered. Symmetrically, the exercise of a put creates a short position in the forward contract with forward price equal to the strike of the put, which means that the exerciser must deliver the underlying at T' and will receive the strike price at T'.

---

[1] Of course, it is possible to give a direct proof, without relying on the Black-Scholes formula. A sketch is given in Sect. 7.11.

We will denote the market forward price by $F(t)$. We assume the forward price satisfies

$$\frac{\mathrm{d}F}{F} = \mu\,\mathrm{d}t + \sigma\,\mathrm{d}B\,, \tag{7.5}$$

where B is a Brownian motion. As before, μ can be a quite general random process. We will assume in this section that the volatility σ is a constant and generalize to a time-varying (but non-random) volatility in Sect. 7.9. In Sect. 7.3, we will discuss the relations of the forward price and its volatility to the price and volatility of the underlying.

Black's formulas are particularly useful when interest rates are assumed to be random, as we will see in Part III of the book when we study fixed income derivatives. Therefore, we do not assume here that there is a constant risk-free rate. Instead we will assume that there is a "discount bond" that pays \$1 at date T'. It is called a "discount bond" because its price is the appropriate discount factor for computing the present value of nonrandom cash flows at date T'. Such a bond is also called a "zero coupon" bond because it does not pay any cash flows until T', when it pays its face value (which we take simply for convenience to be \$1). We will let $P(t, T')$ denote the price of the bond at date t.[2]

Black's formulas are:

The values at date 0 of European options with strike K and maturity T on a forward contract with maturity T' are

$$\text{Call Price} = P(0,T')F(0)\,\mathrm{N}(d_1) - P(0,T')K\,\mathrm{N}(d_2)\,, \tag{7.6a}$$
$$\text{Put Price} = P(0,T')K\,\mathrm{N}(-d_2) - P(0,T')F(0)\,\mathrm{N}(-d_1)\,, \tag{7.6b}$$

where

$$d_1 = \frac{\log\left(\frac{F(0)}{K}\right) + \frac{1}{2}\sigma^2 T}{\sigma\sqrt{T}}\,, \tag{7.6c}$$
$$d_2 = d_1 - \sigma\sqrt{T}\,. \tag{7.6d}$$

Black's formulas are a simple consequence of Margrabe's formula. To see this, we first need to describe the value of an option on a forward at the maturity date T of the option. Consider a call option. Exercise of the call results in a long forward position with forward price K. The value of the long forward is given by its market price $F(T)$, but we must keep in mind that the forward price is not paid until the underlying is delivered at date T'. So

---

[2] In this section we could drop the T' in $P(t, T')$ and simply write $P(t)$, because we only consider one maturity date, but we will use the same notation when discussing multiple maturities in Part III.

suppose that you exercise the call and then sell a forward contract at the market forward price $F(T)$. The delivery/receipt obligations of the long and short forwards cancel, leaving you with the obligation to pay K dollars at date T' and with an asset of $F(T)$ dollars to be received at date T'. The value of the net cash flow at date T is $P(T, T')[F(T) - K]$. This is the value if exercised, so the value of the call at date T is

$$\max\left(0, P(T, T')[F(T) - K]\right) = \max\left(0, P(T, T')F(T) - P(T, T')K\right). \quad (7.7)$$

We can write this as

$$\max(0, S_1(T) - S_2(T)) \qquad (7.8)$$

if we define

$$S_1(t) = P(t, T')F(t) \quad \text{and} \quad S_2(t) = P(t, T')K \qquad (7.9)$$

for $t = T$ (and more generally for $t \le T$). Thus, the value at maturity of a call option on a forward is the value at maturity of an option to exchange the two assets with prices S_1 and S_2 (we will establish in a moment that S_1 and S_2 are actually asset prices). It follows that the value at date 0 of a call option on a forward is the value at date 0 of an option to exchange the two assets.

Now consider a put option on a forward. Exercising the put and unwinding the short forward position by buying a forward at the market price $F(T)$ will leave one with a net cash flow of $K - F(T)$ to be received at the maturity date T' of the forward. Therefore the value of the put at maturity is

$$\max(0, P(T, T')[K - F(T)]) = \max(S_2(T) - S_1(T)). \qquad (7.10)$$

Therefore, the value at date 0 of the put option on a forward must be the value at date 0 of an exchange option, where asset one in (7.9) is exchanged for asset two.

The key assumption in deriving Margrabe's formula is that the volatility of the ratio of asset prices is a constant. For a call option on a forward, the relevant ratio is $S_1/S_2 = F/K$. Because K is a constant, the volatility of the ratio is the volatility σ of the forward price F, which we have assumed to be constant. For a put option on a forward, the relevant ratio is $S_2/S_1 = K/F$. Itô's formula implies

$$\frac{d(K/F)}{K/F} = -\frac{dF}{F} + \left(\frac{dF}{F}\right)^2,$$
$$= (-\mu + \sigma^2)\, dt - \sigma\, dB$$
$$= (-\mu + \sigma^2)\, dt + \sigma(-dB),$$

The purpose of the last equality displayed here is to emphasize that we should take the volatility of K/F to be the positive number σ. We can do this by using

$-B$ as the Brownian motion instead of B.[3] Thus, we can apply Margrabe's formula to value calls and puts on forwards (once we verify that S_1 and S_2 are indeed asset prices).

To obtain Black's formula (7.6a) for a call on a forward from Margrabe's formula (7.4a), we simply substitute $S_1(0) = P(0, T')F(0)$, $S_2(0) = P(0, T')K$, $q_1 = 0$ and $q_2 = 0$ in Margrabe's formula. A put option is the reverse exchange, so Margrabe's formula gives

$$P(0, T')K \, \mathrm{N}(d_1^m) - P(0, T')F(0) \, \mathrm{N}(d_2^m) \,, \tag{7.11}$$

where

$$d_1^m = \frac{\log\left(\frac{P(0,T')K}{P(0,T')F(0)}\right) + \frac{1}{2}\sigma^2 T}{\sigma\sqrt{T}} \,,$$

$$d_2^m = d_1^m - \sigma\sqrt{T}.$$

We introduce the superscript m here to distinguish these numbers in Margrabe's formula from the d_1 and d_2 defined in (7.6c) and (7.6d). Notice that

$$d_1^m = -\frac{\log\left(\frac{F(0)}{K}\right) - \frac{1}{2}\sigma^2 T}{\sigma\sqrt{T}} = -d_2$$

$$d_2^m = -\frac{\log\left(\frac{F(0)}{K}\right) + \frac{1}{2}\sigma^2 T}{\sigma\sqrt{T}} = -d_1,$$

so Margrabe's formula (7.11) is the same as Black's formula (7.6b) for a put option on a forward.

We still need to explain why S_1 and S_2 defined in (7.9) are asset prices, in fact the prices of non-dividend-paying assets since we have taken $q_1 = q_2 = 0$ in applying Margrabe's formula. The case of S_2 should be clear: it is the price of K units of the discount bond maturing at T'. The case of S_1 is more subtle. It is the price of the following portfolio constructed at date 0 and held until date T: go long one forward contract and buy $F(0)$ units of the discount bond maturing at T'. The value at date t of the bonds in the portfolio is $F(0)P(t, T')$. The value at date t of the long forward contract can be seen by considering unwinding it by selling a forward at date t at the market price $F(t)$. This cancels the delivery/receipt obligations on the underlying and results in a net cash flow of $F(t) - F(0)$ to be received at date T'. The value at date t of this future cash flow is $P(t, T')[F(t) - F(0)]$ and when we add this to the value of the bonds we obtain $P(t, T')F(t) = S_1(t)$.

Put-call parity for options on forwards is

$$\text{Call Price} + P(0, T')K = \text{Put Price} + P(0, T')F(0) \,.$$

---

[3] This is really nothing more than the usual convention of defining the standard deviation of a random variable to be the positive square root of the variance.

The left-hand side is the cost of the call and K units of the discount bond, which have value $\max(F(T), K)P(T, T')$ at time T. The right-hand side is the cost of the put option and $F(0)$ units of the discount bond, which, together with a long forward contract initiated at date 0, also have value $\max(F(T), K)P(T, T')$ at time T.

7.3 Merton's Formula

Now we reconsider the Black-Scholes model but without assuming there is a constant risk-free rate. We assume instead that there is a discount bond maturing at the same date as the option. Letting T denote the maturity date of the option and discount bond, we write the price of the discount bond at dates $t \leq T$ as $P(t, T)$. We continue to assume that the stock has a constant dividend yield q but we make a different assumption about volatility—instead of assuming that the volatility of the stock is constant, we assume that the volatility of its forward price is constant. We relax this to allow time-varying but non-random volatility of the forward price in Sect. 7.9.

The forward contract we consider is a forward contract for the stock maturing at the date T that the option matures. Let $F(t)$ denote the forward price for this contract at dates $0 \leq t \leq T$. Because the forward price must equal the spot price at the maturity of a forward contract, we have $F(T) = S(T)$. Consider a call option on the forward, with the call maturing at T also. In the notation of the previous section, we have $T' = T$ and hence $P(T, T') = 1$ (the discount bond is worth \$1 at maturity). Therefore the value (7.7) of the call on the forward at its maturity T is

$$\max(0, F(T) - K) = \max(0, S(T) - K) ,$$

which is the same as the value of the call on the stock. Therefore, the value at date 0 of the call on the stock must equal the value at date 0 of the call on the forward, and we can use Black's formula (7.6a) for a call option on a forward to price a call option on the stock, assuming the forward price has a constant volatility.

Likewise, the value at the maturity date T of a put option on the same forward contract is, from (7.10),

$$\max(0, K - F(T)) = \max(0, K - S(T)) .$$

Hence, we can use Black's formula (7.6b) to price a put option on the stock, assuming the forward price has a constant volatility.

It is not necessary that the forward contract be traded, because we can create a synthetic forward using the stock. To create a synthetic forward at date t we buy $e^{-q(T-t)}$ shares of the stock at cost $e^{-q(T-t)}S(t)$. With reinvestment of dividends, this will accumulate to one share at date T. We finance the purchase of the stock by shorting $e^{-q(T-t)}S(t)/P(t, T)$ units of the discount bond. This results in a liability of $e^{-q(T-t)}S(t)/P(t,T)$ dollars at date T, so the forward purchase is arranged by promising to pay $e^{-q(T-t)}S(t)/P(t,T)$ dollars at the delivery date; i.e., the forward price is[4]

$$F(t) = \frac{e^{-q(T-t)}S(t)}{P(t,T)}.$$ (7.12)

The assumption we need to apply Black's formulas is that

$$\frac{\mathrm{d}F}{F} = \mu\,\mathrm{d}t + \sigma\,\mathrm{d}B,$$ (7.13)

where B is a Brownian motion, μ can be a quite general random process, and σ is a constant. At the end of this section, we will discuss the meaning of this assumption in terms of the volatilities of the stock and bond and their correlation.

Under this assumption, the following formulas originally due to Merton [51] follow immediately from Black's formulas (7.6) by substituting $F(0) = e^{-qT}S(0)/P(0,T)$.

Assuming the forward price has a constant volatility σ, the values at date 0 of European calls and puts maturing at date T on a stock with a constant dividend yield q are

$$\text{Call Price} = e^{-qT}S(0)\,\mathrm{N}(d_1) - P(0,T)K\,\mathrm{N}(d_2),$$ (7.14a)
$$\text{Put Price} = P(0,T)K\,\mathrm{N}(-d_2) - e^{-qT}S(0)\,\mathrm{N}(-d_1),$$ (7.14b)

where

$$d_1 = \frac{\log\left(\frac{S(0)}{KP(0,T)}\right) - qT + \frac{1}{2}\sigma^2 T}{\sigma\sqrt{T}},$$ (7.14c)

$$d_2 = d_1 - \sigma\sqrt{T}.$$ (7.14d)

These formulas are clearly similar to the Black-Scholes formulas. The similarities are made more apparent by writing the discount bond price in terms of its yield. The yield y of the discount bond is defined as

$$y = \frac{-\log P(0,T)}{T} \qquad \Longleftrightarrow \qquad P(0,T) = e^{-yT}.$$

---

[4] If there is a constant risk-free rate r, then it must be that $P(t,T) = e^{-r(T-t)}$, so (7.12) becomes

$$F(t) = e^{(r-q)(T-t)}S(t),$$

which is the same as the covered interest parity condition (6.2)—recall that we interpret the exchange rate as the price of an asset with dividend yield $q = r_f$.

Substituting this into the expressions above, we have:

Assuming the forward price has a constant volatility σ, the values at date 0 of European calls and puts maturing at date T on a stock with a constant dividend yield q are

$$\text{Call Price} = e^{-qT} S(0) \, N(d_1) - e^{-yT} K \, N(d_2) \,, \tag{7.15a}$$

$$\text{Put Price} = e^{-yT} K \, N(-d_2) - e^{-qT} S(0) \, N(-d_1), \tag{7.15b}$$

$$d_1 = \frac{\log\left(\frac{S(0)}{K}\right) + \left(y - q + \frac{1}{2}\sigma^2\right) T}{\sigma\sqrt{T}} \,, \tag{7.15c}$$

$$d_2 = d_1 - \sigma\sqrt{T} \,. \tag{7.15d}$$

This shows that the Merton call and put formulas can be calculated from the Black-Scholes call and put functions given in Chap. 3 by inputting the yield of the discount bond as the risk-free rate and by inputting the volatility of the forward price as σ.

If one wants to assume that there is a constant risk-free rate, then the discount bond price will have to be e^{-rT} and its yield will be the risk-free rate r. In this case, the forward price is $e^{(r-q)T} S(0)$ and it has the same volatility as S. Making these substitutions, the Merton formulas (7.15) are the same as the Black-Scholes formulas. However, the Merton formulas are an important generalization. It is common practice to use the yield of the discount bond as the risk-free rate that is input into the Black-Scholes formulas. The Merton formulas justify this practice. It is less common to attempt to estimate the volatility of the forward price and use this (as one should since the risk-free rate really is not constant) as the volatility in the Black-Scholes-Merton formulas. However, this does little damage for pricing short-term options, because the volatility of the forward price—see (7.16) below—will be approximately the same as the volatility of the underlying for short-term options, due to the low volatility of short-term bond prices. Moreover, when one computes an implied volatility from the Black-Scholes formula (using the discount bond yield as the risk-free rate), it should be regarded as the market's view of the forward price volatility, and it is perfectly appropriate to input it into the Black-Scholes-Merton formulas to price another option (assuming of course that the forward price volatility can be regarded as constant).

The volatility of the forward price can be computed in terms of the volatilities and correlation of the stock and discount bond as follows. Assume that

$$\frac{dS}{S} = \mu_s \, dt + \sigma_s \, dB_s \,,$$

$$\frac{dP}{P} = \mu_p \, dt + \sigma_p \, dB_p,$$

where B_s and B_p are Brownian motions with correlation ρ. Then (2.38) and (2.40) show that the volatility of $F(t) = e^{-q(T-t)} S(t)/P(t,T)$ is σ defined as

$$\sigma = \sqrt{\sigma_s^2 + \sigma_p^2 - 2\rho\sigma_s\sigma_p} \,. \tag{7.16}$$

As mentioned before, we will consider in Sect. 7.9 that the the volatility (7.16) may vary over time in a non-random way.

7.4 Deferred Exchange Options

A call option on a forward can be viewed as an option to exchange K dollars (or, equivalently, K units of the discount bond maturing at the maturity date of the forward) for the underlying asset, with the exchange taking place at the maturity date of the forward. Therefore, it is an exchange option in which the exchange takes place at a fixed date after the option matures. We can easily extend Margrabe's formula to value options to exchange other assets when the option maturity precedes the date of the exchange.

As in Sect. 7.1, consider two assets with prices S_i and constant dividend yields q_i and assume the prices satisfy

$$\frac{dS_i}{S_i} = \mu_i \, dt + \sigma_i \, dB_i \,,$$

where the drifts μ_i, the volatilities σ_i and the correlation ρ of the two Brownian motions can be general random processes. However, also as in Sect. 7.1, assume that the volatility

$$\sigma = \sqrt{\sigma_1^2 + \sigma_2^2 - 2\rho\sigma_1\sigma_2}$$

of the ratio of asset prices is constant.

Consider an option maturing at date T to exchange the second asset for the first asset at date $T' \geq T$. To understand the value of the option at date T, suppose it is exercised. To unwind the positions in the two assets, one can sell a forward contract on the asset to be received and buy a forward contract on the asset to be delivered, with the forward contracts maturing at the date of the exchange. Then the difference $F_1(T) - F_2(T)$ in the forward prices is a cash flow to be received/paid at the exchange date T' and its value at date T is $P(T,T')[F_1(T)-F_2(T)]$. Therefore, the value of the option at its maturity T is

$$\max(0, P(T,T')F_1(T) - P(T,T')F_2(T)) \,.$$

As in Sect. 7.2, this valuation does not require the existence of traded forward contracts, because synthetic forwards can be created. Also as in Sect. 7.2 we know that

$$S_1^*(t) = P(t,T')F_1(t) \quad \text{and} \quad S_2^*(t) = P(t,T')F_2(t)$$

are the prices of non-dividend-paying assets. Therefore, the option to exchange the assets at date T' must have the same value as an option to exchange at date T the assets with prices S_i^*.

We recall here the arbitrage formula (7.12) for the forward prices (making the change that here the forwards mature at T'):

$$F_i(t) = \frac{e^{-q_i(T'-t)} S_i(t)}{P(t, T')} \, .$$

Thus,

$$S_i^*(t) = e^{-q_i(T'-t)} S_i(t) \, .$$

This implies that the volatility of the ratio S_1^*/S_2^* is the same as the volatility of the ratio S_1/S_2. Therefore, we can price a deferred exchange option from Margrabe's formula, inputting the prices $S_i^*(0) = e^{-q_i T'} S_i(0)$ as the initial asset prices and zero as their dividend yields. This formula is:

The value of a European option maturing at date T to exchange two assets at date T' is

$$e^{-q_1 T'} S_1(0) \, \mathrm{N}(d_1) - e^{-q_2 T'} S_2(0) \, \mathrm{N}(d_2) \, , \qquad (7.17a)$$

where

$$d_1 = \frac{\log\left(\frac{S_1(0)}{S_2(0)}\right) + (q_2 - q_1)T' + \frac{1}{2}\sigma^2 T}{\sigma\sqrt{T}} \, , \qquad (7.17b)$$

$$d_2 = d_1 - \sigma\sqrt{T} \, , \qquad (7.17c)$$

7.5 Calculations in VBA

We could of course write entirely separate programs for the options discussed so far in this chapter but it seems useful to emphasize their common structure. As discussed in the introduction to this chapter, each is the present value of what is received upon exercise multiplied by $\mathrm{N}(x)$ minus the present value of what is delivered upon exercise multiplied by $\mathrm{N}(y)$ and x in each case is the logarithm of the ratio of present values plus one-half $\sigma^2 T$, all divided by $\sigma\sqrt{T}$. In the case of options on forwards, the present values are the present values of what is received or delivered at the maturity of the forward contract. We can do this calculation in the following program.

```
Function Generic_Option(P1, P2, sigma, T)
'
' Inputs are P1 = present value of asset to be received
'            P2 = present value of asset to be delivered
```

```
'              sigma = volatility
'              T = time to maturity
'
Dim x, y, N1, N2
x = (Log(P1 / P2) + 0.5 * sigma * sigma * T) / (sigma * Sqr(T))
y = x - sigma * Sqr(T)
N1 = Application.NormSDist(x)
N2 = Application.NormSDist(y)
Generic_Option = P1 * N1 - P2 * N2
End Function
```

Now we can use the following one-line programs to value exchange and forward options. We will explain in Sect. 7.8 why (and in what circumstance) the Black call and put functions are appropriate for options on futures.

```
Function Margrabe(S1, S2, sigma, q1, q2, T)
'
' Inputs are S1 = price of asset to be received
'            S2 = price of asset to be delivered
'            sigma = volatility of ratio of prices
'            q1 = dividend yield of asset to be received
'            q2 = dividend yield of asset to be delivered
'            T = time to maturity
'
Margrabe = Generic_Option(Exp(-q1*T)*S1,Exp(-q2*T)*S2,sigma,T)
End Function


Function Black_Call(F, K, P, sigma, T)
'
' Inputs are F = forward price
'            K = strike price
'            P = price of bond maturing when forward matures
'            sigma = volatility of forward price
'            T = time to maturity
'
' To value a futures option, input F = futures price and
' P = price of bond maturing when option matures.
'
Black_Call = Generic_Option(P * F, P * K, sigma, T)
End Function


Function Black_Put(F, K, P, sigma, T)
'
' Inputs are F = forward price
'            K = strike price
'            P = price of bond maturing when forward matures
'            sigma = volatility of forward price
'            T = time to maturity
'
```

```
' To value a futures option, input F = futures price and
' P = price of bond maturing when option matures.
'
Black_Put = Generic_Option(P * K, P * F, sigma, T)
End Function


Function Margrabe_Deferred(S1, S2, sigma, q1, q2, Tm, Te)
'
' Inputs are S1 = price of asset to be received
'            S2 = price of asset to be delivered
'            sigma = volatility of ratio of prices
'            q1 = dividend yield of asset to be received
'            q2 = dividend yield of asset to be delivered
'            Tm = time to maturity of option
'            Te = time until exchange >= Tm
'
Margrabe_Deferred = _
        Generic_Option(Exp(-q1*Te)*S1,Exp(-q2*Te)*S2,sigma,Tm)
End Function
```

We could also have calculated the Black-Scholes call formula as

```
Generic_Option(Exp(-q*T)*S, Exp(-r*T)*K, sigma, T)
```

and the Black-Scholes put formula as

```
Generic_Option(Exp(-r*T)*K, Exp(-q*T)*S, sigma, T).
```

7.6 Greeks and Hedging

The Greeks for the Margrabe and Black formulas can be calculated in the same way that we calculated them in Chap. 3 for the Black-Scholes formula. In analogy with (3.8), it can be shown for the Margrabe formula that

$$e^{-q_1 T} S_1(0) \operatorname{n}(d_1) = e^{-q_2 T} S_2(0) \operatorname{n}(d_2) ,$$

and again this simplifies the calculations. This equation applies to the Black call formula by taking $q_1 = q_2 = 0$, $S_1(0) = P(0, T')F(0)$, and $S_2(0) = P(0, T')K$, leading to

$$F(0) \operatorname{n}(d_1) = K \operatorname{n}(d_2) .$$

The Greeks for the Black call formula and the Margrabe formula are:

Black Call	Margrabe
$\dfrac{\partial}{\partial F} = P(0,T')\,N(d_1)$	$\dfrac{\partial}{\partial S_1} = e^{-q_1 T}\,N(d_1)$
$\dfrac{\partial}{\partial P} = F(0)\,N(d_1) - K\,N(d_2)$	$\dfrac{\partial}{\partial S_2} = -e^{-q_2 T}\,N(d_2)$
$\dfrac{\partial^2}{\partial F^2} = \dfrac{P(0,T')\,n(d_1)}{\sigma\sqrt{T}F(0)}$	$\dfrac{\partial^2}{\partial S_1^2} = \dfrac{e^{-q_1 T}\,n(d_1)}{\sigma\sqrt{T}S_1(0)}$
$\dfrac{\partial^2}{\partial P^2} = 0$	$\dfrac{\partial^2}{\partial S_2^2} = \dfrac{e^{-q_2 T}\,n(d_2)}{\sigma\sqrt{T}S_2(0)}$
$\dfrac{\partial^2}{\partial F \partial P} = N(d_1)$	$\dfrac{\partial^2}{\partial S_1 \partial S_2} = -\dfrac{e^{-q_1 T}\,n(d_1)}{\sigma\sqrt{T}S_2(0)}$
$-\dfrac{\partial}{\partial T} = -\dfrac{\sigma P(0,T')F(0)\,n(d_1)}{2\sqrt{T}}$	$-\dfrac{\partial}{\partial T} = q_1 e^{-q_1 T}S_1(0)\,N(d_1)$ $\qquad - q_2 e^{-q_2 T}S_2(0)\,N(d_2)$ $\qquad - \dfrac{\sigma e^{-q_1 T}S_1(0)\,n(d_1)}{2\sqrt{T}}$
$\dfrac{\partial}{\partial \sigma} = \sqrt{T}P(0,T')F(0)\,n(d_1)$	$\dfrac{\partial}{\partial \sigma} = \sqrt{T}e^{-q_1 T}S_1(0)\,n(d_1)$

Hedging for the Margrabe formula is much the same as for the Black-Scholes formula. We would delta-hedge a written exchange option by holding $\delta_1 = e^{-q_1 T}\,N(d_1)$ shares of the first asset and $\delta_2 = -e^{-q_2 T}\,N(d_2)$ shares of the second asset (which means shorting the second asset). Note that selling the option will exactly finance the hedge, so the overall portfolio has zero cost. The same argument that we used in Sect. 3.5 shows that this zero-cost portfolio will have a zero return if continuously rebalanced.

Because the Black formulas are a special case of the Margrabe formula, we can delta-hedge options on forwards in the same way. Putting $q_1 = q_2 = 0$, $S_1(0) = P(0,T')F(0)$ and $S_2(0) = P(0,T')K$, we would delta-hedge a written call option by buying $N(d_1)$ shares of the first asset and shorting $N(d_2)$ shares of the second, where d_1 and d_2 are defined in the Margrabe formulas (7.4b) and (7.4c) and equivalently in the Black formulas (7.6c) and (7.6d). The first asset here consists of a long forward contract plus $F(0)$ units of the discount bond, and the second asset is K units of the discount bond. Therefore, we should buy $N(d_1)F(0) - N(d_2)\,K$ units of the discount bond and go long $N(d_1)$ forward contracts.

A more direct analysis of hedging options on forwards is possible and instructive. We will consider that topic further in Sect. 7.10.

7.7 The Relation of Futures Prices to Forward Prices

The difference between futures and forward contracts is that futures are marked to market, which means that daily gains and losses are posted to the investor's margin account. Thus, there are interim cash flows on a futures contract, whereas the only cash flows on a forward contract are at the maturity of the forward. We will establish three useful facts in this section, the last of which follows from the first two:

(1) Futures prices are martingales under the risk-neutral measure.
(2) Forward prices are martingales when we use the discount bond with the same maturity as the forward as the numeraire.
(3) When interest rates are non-random, futures prices equal forward prices.

We consider the idealized case in which the futures contract is continuously marked to market. Assume there is an instantaneously risk-free asset with rate of return r, which could vary randomly over time, and define, as in Sect. 1.1,

$$R(t) = \exp\left(\int_0^t r(s)\,ds\right),$$

which is the value at date t of a \$1 investment in the asset at date 0, with interest continuously reinvested. Let T' denote the maturity of the futures contract, and let $F^*(t)$ denote the futures price at dates $t \leq T'$ (the $*$ notation is to distinguish the futures price from the forward price F). Consider the portfolio strategy that starts with zero dollars and one long futures contract at price $F^*(0)$ and which continuously invests and withdraws from the risk-free asset the gains and losses on the futures contract. Let $V(t)$ denote the value of this portfolio at date t. The change in the value of the portfolio at any instant is the interest earned (or paid, if $V < 0$) on the risk-free asset plus the gain/loss on the futures. This means that

$$dV = rV\,dt + dF^*.$$

Because all gains and losses on this portfolio are reinvested, V is the price of a non-dividend-paying asset. Therefore, under the risk-neutral measure (i.e., using R as the numeraire), the ratio V/R must be a martingale and hence have zero drift. From Itô's formula,

$$\frac{d(V/R)}{V/R} = \frac{dV}{V} - \frac{dR}{R}$$
$$= \frac{rV\,dt + dF^*}{V} - r\,dt$$
$$= \frac{dF^*}{V}.$$

Thus, the drift of V/R being zero implies the drift of F^* is zero. We need to assume (and can assume) that F^* is an Itô process with finite expected quadratic variation—cf. condition (2.2) in which case the absence of a drift implies that it is a martingale.

Now we turn to fact (2). Consider a forward contract with maturity T' and a discount bond also maturing at T'. Let $F(t)$ denote the forward price and let $P(t, T')$ denote the price of the discount bond at dates $t \leq T'$. We observed in Sect. 7.2 that there is a non-dividend paying asset with price $P(t, T')F(t)$. When we use the discount bond as the numeraire, the ratio $P(t, T')F(t)/P(t, T') = F(t)$ must be a martingale, which is fact (2). Because of this fact, a probability measure corresponding to a discount bond being the numeraire is called a "forward measure."

Suppose now that interest rates are deterministic, that is, even if r varies over time, it does so in a non-random way. Then the discount bond price at date 0 must be the discount factor

$$P(0, T') = \exp\left(-\int_0^{T'} r(t)\,dt\right) .$$

Equation (1.11) gives the probability of any event A when the discount bond is used as the numeraire as

$$\text{prob}^P[A] = E\left[1_A \phi(T') \frac{P(T', T')}{P(0, T')}\right] = \exp\left(-\int_0^{T'} r(t)\,dt\right) E[1_A \phi(T')] ,$$

where ϕ denotes the state prices. On the other hand, the same equation gives the probability of A when R is used as the numeraire as

$$\text{prob}^R[A] = E\left[1_A \phi(T') \frac{R(T')}{R(0)}\right] = \exp\left(-\int_0^{T'} r(t)\,dt\right) E[1_A \phi(T')] .$$

Therefore, the two probability measures are the same, and consequently the expectations E^P and E^R are the same. Now using the fact that both the futures price and the forward price must equal the spot price at maturity, we have $F^*(T') = F(T')$, and, from the martingale properties,

$$F^*(t) = E_t^P[F^*(T')] = E_t^P[F(T')] = E_t^R[F(T')] = F(t) ,$$

which is fact (3).

7.8 Futures Options

Now we consider options on futures contracts, assuming that interest rates are deterministic. We just showed that in this circumstance the futures price will equal the forward price for a contract of the same maturity. However, the values of options on a futures contract do not equal the values of options on the corresponding forward contract.

The difference is due to marking to market. Consider futures and forward contracts with maturity T' and options maturing at $T \leq T'$. Exercise of a call option on a futures contract will roll the investor into a long futures contract with futures price equal to the market futures price at that date. The difference $F^*(T) - K$ between the market futures price and the strike price of the option is immediately credited to the investor's margin account. On the other hand, exercise of an option on a forward and sale of the forward results in a cash flow of $F(T) - K$ that is received at the maturity date T' of the forward. Therefore, the value at maturity of a call option on a futures contract is $\max(0, F^*(T) - K)$, whereas, as noted before, the value of a call option on a forward at the maturity of the option is $P(T, T') \max(0, F(T) - K)$

As in the analysis of options on forwards, an options on a futures contract can be viewed as an exchange option, where one exchanges the asset with price $S_2(t) = P(t, T)K$ at date $t \leq T$ for the asset with price $S_1(t) = P(t, T)F^*(t)$. The asset with price S_2 is of course K units of the discount bond maturing at T. Assuming interest rates are deterministic, we have $F^*(t) = F(t)$, and we noted in Sect. 7.2 that $P(t, T)F(t)$ is the price of a non-dividend-paying asset. Thus, we can apply Margrabe's formula to price call (and put) options on futures when interest rates are deterministic. Compared to options on forwards, the difference is that the discount bonds defining the prices S_1 and S_2 mature at the maturity date of the option rather than at the maturity date of the futures or forward contract. The result is Black's formula:

When interest rates are deterministic and the futures price F^* has a constant volatility σ, the values of European calls and puts on a futures contract are

$$\text{Call Price} = P(0, T)F^*(0) \, \text{N}(d_1) - P(0, T)K \, \text{N}(d_2) \, , \tag{7.18a}$$

$$\text{Put Price} = P(0, T)K \, \text{N}(-d_2) - P(0, T)F^*(0) \, \text{N}(-d_1) \, , \tag{7.18b}$$

where

$$d_1 = \frac{\log\left(\frac{F^*(0)}{K}\right) + \frac{1}{2}\sigma^2 T}{\sigma\sqrt{T}} \, , \tag{7.18c}$$

$$d_2 = d_1 - \sigma\sqrt{T} \, . \tag{7.18d}$$

We can calculate these values from the `Black_Call` and `Black_Put` functions by inputting the price of the discount bond maturing when the option matures rather than the price of the discount bond maturing when the forward/futures matures. We will derive delta hedges for futures options in Sect. 7.10.

7.9 Time-Varying Volatility

All of the option pricing formulas in this chapter were derived from Margrabe's formula, the main assumption of which is that the logarithm of the ratio of asset prices at date T is normally distributed with variance equal to $\sigma^2 T$. As in Sect. 3.8 regarding the Black-Scholes formulas, the formulas in this chapter can easily be adapted to allow a time-varying but non-random volatility. If the volatility is a non-random function $\sigma(t)$ of time, then we define σ_{avg} to be the number such that

$$\sigma_{\text{avg}}^2 = \frac{1}{T} \int_0^T \sigma^2(t)\,\mathrm{d}t\,. \tag{7.19}$$

We should input σ_{avg} as (i) the volatility of the ratio of asset prices in Margrabe's formula and the deferred exchange option formula if $\sigma(t)$ is the volatility of the ratio at date t or as (ii) the volatility of the forward price in Black's and Merton's formulas if $\sigma(t)$ is the volatility of the forward price at date t.

As in Sect. 3.8, equation (7.19) enables one to interpret and apply different implied volatilities computed at different maturities. Another circumstance in which it can be useful is in conjunction with bond price models such as the Vasicek and extended Vasicek models described in Chap. 13 that imply a time-varying non-random volatility for discount bond prices.[5] If we assume a constant volatility for the price of the underlying and a constant correlation between the underlying and the discount bond, then we will have a time-varying non-random forward price volatility via (7.16), and we should input the "average volatility" σ_{avg} defined in (7.19) for the forward price volatility in Black's and Merton's formulas. As mentioned in Sect. 7.3, this will be more important for long-term options than for short-term options.

7.10 Hedging with Forwards and Futures

In Chap. 6, we considered hedging quanto contracts with currency forwards. In Sect. 7.6, we considered hedging options on forwards with forwards. To present a more complete analysis of these topics, we need to discuss the gains and losses that accrue from trading forwards.

Consider dates $t < u$ and a forward contract with maturity T. Suppose we purchase $x(t)$ forwards at date t and then change our position in forwards to $x(u)$ at time u. The purchase/sale of $x(u) - x(t)$ new contracts does not affect the portfolio value, so the change in the value of the portfolio is the change in the value of the $x(t)$ contracts purchased at date t. These contracts were worth zero at date t, because forward contracts have zero value at initiation. Selling them at date u cancels the obligation to deliver/receive the underlying, leaving one with a cash flow of $x(t)[F(u) - F(t)]$ dollars to be received at date T. The

---

[5] The volatility of a discount bond price cannot be constant because it must go to zero as the bond approaches maturity.

value of this cash flow at date u is $x(t)P(u,T)[F(u) - F(t)]$. We can write this as

$$x(t)P(u,T)[F(u) - F(t)]$$
$$= x(t)\big[P(t,T)[F(u) - F(t)] + [P(u,T) - P(t,T)][F(u) - F(t)]\big]$$
$$= x(t)\big[P(t,T)\,\Delta F + (\Delta P)(\Delta F)\big] \,.$$

This motivates the following definition:

The change in the value of a portfolio of forward contracts at date t is

$$x(t)\big[P(t,T)\,\mathrm{d}F(t) + \mathrm{d}P(t,T) \times \mathrm{d}F(t)\big]\,, \tag{7.20}$$

where $x(t)$ denotes the number of forward contracts held, $F(t)$ denotes the forward price, $P(t,T)$ denotes the price of a discount bond maturing at T, and T is the maturity of the forward contract.

Hedging with futures is a bit simpler, because the gains and losses are received instantaneously (daily, at least) rather than being deferred to the contract maturity. Letting $x(t)$ denote the number of futures contracts held at date t and $F^*(t)$ the futures price, the cash flow from the contracts is $x(t)\,\mathrm{d}F^*(t)$. This is also the change in the value of the portfolio, because marking to market means that the contracts always have zero value.

To compare hedging with futures and forwards, assume there is a constant risk-free rate r. Let T denote the maturity of the futures and forward contracts. Because there is a constant risk-free rate, we have $P(t,T) = \mathrm{e}^{-r(T-t)}$, which implies $(\mathrm{d}P)(\mathrm{d}F) = 0$. Moreover, futures prices equal forward prices. Thus,

---

If there is a constant risk-free rate r, the change in the value of a portfolio of forward contracts at date t is

$$x(t)\mathrm{e}^{-r(T-t)}\,\mathrm{d}F(t) \tag{7.21}$$

and the change in the value of a portfolio of futures contracts is

$$x(t)\,\mathrm{d}F(t)\,, \tag{7.22}$$

where $x(t)$ denotes the number of futures/forward contracts held at date t, T is the maturity of the futures and forward contracts and $F(t)$ is the futures ($=$ forward) price at date t.

Comparing (7.21) and (7.22), we see that if $x(t)$ is the number of forward contracts that should be held in a hedge, then

$$y(t) = e^{-r(T-t)}x(t) \tag{7.23}$$

is the number of futures contracts that should be held, because with this number of contracts we have

$$\text{Change in Forward Portfolio} = x(t)e^{-r(T-t)}\,\mathrm{d}F(t)$$
$$= y(t)\,\mathrm{d}F(t)$$
$$= \text{Change in Futures Portfolio}\,.$$

In short, we don't require as many futures contracts as forward contracts, and the scaling factor to convert from forwards to futures is just the present value factor $e^{-r(T-t)}$.

For example, the result of Sect. 6.6 on replicating the payoff $\bar{X}S(T)$ with forward contracts leads to the following:

> To replicate the payoff $\bar{X}S(T)$ at date T, where $\bar{X}$ is a fixed exchange rate and S is the foreign currency price of an asset, one should invest $V(t)$ units of domestic currency in the foreign asset and be short $e^{(r_f-r)(T-t)}V(t)/X(t)$ currency futures at date t, where $V(t)$ is defined in (6.7) and $X(t)$ is the spot exchange rate.

We can use (7.23) to determine how to delta hedge futures options. As explained in Sect. 7.8, assuming non-random interest rates, futures options are more valuable than options on forwards because futures are marked to market upon exercise of an option. Specifically, Black's formulas (7.6) for options on forwards and (7.18) for options on futures show that the values are the same except for the maturity of the discount bond appearing in the equations. With a constant risk-free rate r, options maturing at T and futures/forwards maturing at T', the relation is

$$\text{Value of Futures Option} = e^{r(T'-T)} \times \text{Value of Forward Option}\,.$$

Because the scaling factor $e^{r(T'-T)}$ does not change as time passes, this implies that as time passes we have

Change in Futures Option Value
$$= e^{r(T'-T)} \times \text{Change in Forward Option Value}\,. \tag{7.24}$$

We can combine (7.23) and (7.24) to convert from a hedge of a forward option using forward contracts, which we discussed in Sect. 7.6, to a hedge of a futures option using futures contracts. For example, we concluded in Sect. 7.6 that we should be long $N(d_1)$ forwards to hedge a short call option on a forward contract. Consequently, (7.23) shows that we can hedge a short call option on a forward by being long $e^{-r(T'-t)}N(d_1)$ futures, and then we see from (7.24) that the hedge for a short call on a futures is being long

$$e^{r(T'-T)}e^{-r(T'-t)}\,\mathrm{N}(d_1) = e^{-r(T-t)}\,\mathrm{N}(d_1)$$

futures contracts.

In Sect. 7.6, we derived the hedges for forward options by considering them as exchange options. We can use the definition (7.20) to confirm that our calculations were correct. Consider hedging a short call maturing at T on a forward contract maturing at T'. We can assume interest rates vary randomly and use discount bonds in the hedge. We stated in Sect. 7.6 that we should hold $F(0)\,\mathrm{N}(d_1) - K\,\mathrm{N}(d_2)$ units of the discount bond maturing at T' and we should go long $\mathrm{N}(d_1)$ forwards to hedge the short call. This is a zero-cost portfolio when we include the proceeds from selling the call. Using (7.20), we see that the change in the value of the portfolio will be

$$-\mathrm{d}C + [F(0)\,\mathrm{N}(d_1) - K\,\mathrm{N}(d_2)]\mathrm{d}P + \mathrm{N}(d_1)[P(0,T')\,\mathrm{d}F + (\mathrm{d}P)(\mathrm{d}F)]\,. \quad (7.25)$$

The value of the call at date t will be a function of t, $P(t,T')$ and $F(t)$, which we write as $C(t,P,F)$. From Itô's formula,

$$\begin{aligned}
\mathrm{d}C &= \frac{\partial C}{\partial t}\,\mathrm{d}t + \frac{\partial C}{\partial P}\,\mathrm{d}P + \frac{\partial C}{\partial F}\,\mathrm{d}F \\
&\quad + \frac{1}{2}\frac{\partial^2 C}{\partial P^2}\,(\mathrm{d}P)^2 + \frac{1}{2}\frac{\partial^2 C}{\partial F^2}\,(\mathrm{d}F)^2 + \frac{\partial^2 C}{\partial F \partial P}(\mathrm{d}P)(\mathrm{d}F)\,. \\
&= \Theta\,\mathrm{d}t + \delta_P\,\mathrm{d}P + \delta_F\,\mathrm{d}F + \frac{1}{2}\Gamma_{PP}(\mathrm{d}P)^2 + \frac{1}{2}\Gamma_{FF}(\mathrm{d}F)^2 + \Gamma_{FP}\,(\mathrm{d}P)(\mathrm{d}F)\,,
\end{aligned}$$

where the δ's and Γ's denote the first and second partial derivatives indicated by the subscripts. Inserting this formula into (7.25) and making use of the formulas in the table in Sect. 7.6, we see that the $\mathrm{d}P$ terms cancel because $\delta_P = F(0)\,\mathrm{N}(d_1) - K\,\mathrm{N}(d_2)$. Furthermore, the $\mathrm{d}F$ terms cancel because $\delta_F = \mathrm{N}(d_1)P(0,T')$. Thus, there is no exposure in the portfolio to the two risky asset prices P and F. Furthermore, $\Gamma_{PP} = 0$ and the $(\mathrm{d}P)(\mathrm{d}F)$ terms cancel because $\Gamma_{FP} = \mathrm{N}(d_1)$. These substitutions simplify the change (7.25) in the value of the portfolio to

$$-\Theta\,\mathrm{d}t - \frac{1}{2}\Gamma_{FF}(\mathrm{d}F)^2 = \frac{\sigma P(0,T')F(0)\,\mathrm{n}(d_1)}{2\sqrt{T}}\,\mathrm{d}t - \frac{P(0,T')\,\mathrm{n}(d_1)}{2\sigma\sqrt{T}F(0)}\,(\mathrm{d}F)^2\,,$$

which we can see to be zero because $(\mathrm{d}F)^2 = \sigma^2 F^2\,\mathrm{d}t$. Thus, the hedge is perfect when continuously rebalanced.

7.11 Market Completeness

A formal definition of market completeness must specify which state-contingent claims (random variables depending on the history of prices) can be replicated by trading the marketed assets—for example, one might want all of the claims with finite means to be replicable, or only all of the claims with finite variances, etc. A formal analysis of market completeness is not presented in this book, except for the binomial and trinomial models in Chap. 1. However, we have stated that stochastic volatility models are incomplete. This follows intuitively from the fact that a portfolio containing only one risky asset (the underlying) cannot be perfectly correlated with the two Brownian motions that determine the value of a derivative asset (the Brownian motions driving the price of the underlying and its volatility). In general, a market must include an instantaneously risk-free asset and as many risky assets as there are Brownian motions in order to be complete.

The exchange-option model of Margrabe—with two risky assets, two Brownian motions, and no risk-free asset—is obviously incomplete. For example, it is impossible to have exactly \$100 at date T. With no risk-free asset, there is simply no way to store money. This may seem far-fetched, but we might be interested in payoffs in "real" (i.e., inflation-adjusted) dollars, in which case the absence of a risk-free asset may be a normal situation. In any case, we have not assumed a risk-free asset exists, but we have priced options without appealing to "equilibrium" arguments. This deserves some clarification.

As mentioned above, a formal definition of market completeness must specify which contingent claims are to be replicable. The Margrabe model is complete for a certain set of contingent claims. Contingent claims of the form $S_2(T)X(T)$ where $X(T)$ is a random variable depending on the relative prices $S_1(t)/S_2(t)$ for $0 \leq t \leq T$ can be replicated. Likewise, contingent claims of the form $S_1(T)X(T)$ can be replicated. The payoffs of exchange options are of this form, so they can be priced by arbitrage, even though there are other contingent claims (for example, receiving exactly \$100 at date T) that cannot be replicated and hence cannot be priced by arbitrage. Likewise, the Black and Merton models in which there is a zero-coupon bond but no instantaneously risk-free asset are examples of incomplete markets that are still sufficiently complete to price options by arbitrage (the options can be replicated). The proof that the Margrabe model is complete in the sense stated here follows from the change of numeraire argument used to derive Margrabe's formula from the Black-Scholes formula (recall that the second asset is risk-free when we use it as the numeraire, so there is a risk-free asset under the new numeraire) and a proof that the Black-Scholes model is complete (which we have omitted, except to show that European options can be replicated).

We will conclude with a proof of the Margrabe formula that does not depend on the Black-Scholes formula. Let x denote the random variable taking the value 1 when $S_1(T) > S_2(T)$ and which is 0 otherwise. Then the value of the exchange option at maturity is $xS_1(T) - xS_2(T)$. Let V_i denote the value of the portfolio beginning with $\mathrm{e}^{-q_i T}$ units of asset i at date 0 and reinvesting dividends, to accumulate to one share at date T. Then $V_i(T) = S_i(T)$ and from the fundamental pricing formula (1.17) the value at date 0 of receiving $xS_i(T)$ at date T is

$$V_i(0)E^{V_i}\left[x\frac{S_i(T)}{V_i(T)}\right] = e^{-q_iT}S_i(0)E^{V_i}[x]$$

$$= e^{-q_iT}S_i(0) \times \mathrm{prob}^{V_i}\left(V_1(T) > V_2(T)\right).$$

We can write the value of receiving $xS_1(T)$ as

$$e^{-q_1T}S_1(0) \times \mathrm{prob}^{V_1}\left(\frac{V_2(T)}{V_1(T)} < 1\right)$$

and the value of receiving $xS_2(T)$ as

$$e^{-q_2T}S_2(0) \times \mathrm{prob}^{V_2}\left(\frac{V_1(T)}{V_2(T)} > 1\right).$$

Note that V_2/V_1 is a martingale when we use V_1 as the numeraire and V_1/V_2 is a martingale when we use V_2 as the numeraire. Because they are martingales, they have no drifts. The volatility of the ratios is given in (7.3). Therefore, we have

$$\frac{\mathrm{d}(V_2/V_1)}{V_2/V_1} = \sigma\,\mathrm{d}B_1^*,$$

$$\frac{\mathrm{d}(V_1/V_2)}{V_1/V_2} = \sigma\,\mathrm{d}B_2^*,$$

where B_i^* is a Brownian motion when V_i is used as the numeraire. Margrabe's formula now follows from the tail probability formulas (2.34)–(2.36).

Problems

7.1. Derive the Greeks of a call option on a futures contract.

7.2. Using the results of the previous exercise, show that the delta hedge of a written call on a futures contract that consists of $e^{-r(T-t)}\,\mathrm{N}(d_1)$ long futures contracts and the value of the call invested in the risk-free asset is a riskless hedge.

7.3. Derive a formula (like put-call parity) for the value of an option to exchange asset 1 for asset 2 in terms of the value of an option to do the reverse exchange.

7.4. Create a VBA function `Black_Call_ImpliedVol` that uses bisection to compute an implied forward price volatility from Black's formula and the market price of a call option on a forward.

7.5. Using a "synthetic forward" argument, derive the forward price for a forward contract on a stock, where the forward matures at T' and the stock pays a single known cash dividend D at date $T < T'$.

7.6. Using the result of the previous exercise and Black's formula, derive a formula for the value of a European call option on a stock that pays a single known cash dividend before the option matures.

7.7. Modify the function `Simulated_Delta_Hedge_Profits` to compute the percentiles of gains and losses for an investor who writes a call option on a forward contract and uses a discretely-rebalanced delta hedge. As in Prob. 6.6, you will need to create a variable to keep track of the net asset/liability from trading forwards and include it in the valuation at date T.

7.8. Consider the portfolio that promises to pay $\bar{X}S(T)$ at date T and replicates the payoff using currency forwards described in Sect. 6.6, where $\bar{X}$ is a fixed exchange rate and S is the foreign currency price of an asset. Using the definition (7.20) of gains and losses from trading forwards, verify that the portfolio is riskless.

7.9. Repeat the previous exercise using the futures hedge described in Sect. 7.10.

7.10. It has been observed empirically that implied volatilities of stocks are upward biased estimates of future volatility. Given that there is not really a constant risk-free rate, implied volatilities should be interpreted as implied forward-price volatilities, whereas the empirical literature has measured "future volatility" as the subsequent volatility of the stock. What assumptions about bond volatilities and the correlation of bonds and stocks could explain the empirical finding; i.e., what assumptions imply that the volatility of the forward price exceeds the volatility of the stock?

7.11. In the continuous-time Ho-Lee model described in Chap. 13, the volatility of a discount bond with time τ to maturity is $\sigma_r \tau$ for a constant σ_r. Under this assumption, calculate the average volatility of the forward price of a stock from date 0 to date T, where T is the maturity of the forward contract. Assume the stock has a constant volatility σ_s and the correlation between the stock and bond is a constant ρ.

7.12. Making the same assumptions as in the previous exercise, and using the result of that exercise and Merton's formula, write a VBA function to calculate the value of a call option on a stock. The inputs should be S, K, P, σ_s, σ_r, ρ, q, and T.

8

Exotic Options

We will only discuss a few exotic options. The reason for studying the derivation of an option pricing formula is that it may better equip one to analyze new products. Hopefully, this chapter will be of some assistance in that regard. Of course, one cannot expect to derive a closed-form solution for the value of every product, and often numerical methods will be necessary. For a much more comprehensive presentation of exotic option pricing formulas, the book by Haug [32] and the Excel software that accompanies it are highly recommended. Zhang [65] is also a comprehensive reference for exotics.

The valuation of an American call option on an asset paying a discrete cash dividend (rather than a continuous dividend) is considered in Sect. 8.3. Under a particular assumption on the volatility, valuing this option is very similar to valuing a compound option. Except for the assumption of a discrete dividend in Sect. 8.3, we will make the Black-Scholes assumptions throughout this chapter: there is a constant risk-free rate r and the underlying asset has a constant volatility σ and a constant dividend yield q.

The order in which exotics are presented in this chapter is based on the simplicity of the analysis—the chapter begins with the easiest to analyze and works towards the more difficult. This order is roughly the inverse of importance, the most important exotics in practice being barriers, baskets, spreads and Asians. In the cases of Asian and basket options, we will explain why there are no simple closed-form formulas (sums of lognormally distributed random variables are not lognormal). We will use these cases and discretely-sampled barriers and lookbacks as examples in the next chapter.

8.1 Forward-Start Options

A forward-start option is an option for which the strike price is set equal to the stock price at some later date. In essence, it is issued at the later date, with the strike price set at the money. For example, an executive may know that he is to be given an option grant at some later date with the strike price set equal to the stock price at that date.

Forward-Start Call Payoff

A forward-start call is defined by its maturity date T' and the date $T < T'$ at which the strike price is set. The value of a forward-start call at maturity is

$$\max(0, S(T') - S(T)) .$$

Let

$$x = \begin{cases} 1 & \text{if } S(T') > S(T) , \\ 0 & \text{otherwise} . \end{cases}$$

Then, the value of the call at maturity can be written as

$$xS(T') - xS(T) .$$

Numeraires

1. Use $V(t) = e^{qt}S(t)$ as numeraire to price the payoff $xS(T')$. From the fundamental pricing formula (1.17), the value at date 0 is

$$e^{-qT'} S(0) E^V [x] = e^{-qT'} S(0) \times \text{prob}^V(S(T') > S(T)) .$$

2. To price the payoff $xS(T)$, use the following portfolio as numeraire:[1] purchase e^{-qT} shares of the stock at date 0 and reinvest dividends until date T. This will result in the ownership of one share at date T, worth $S(T)$ dollars. At date T, sell the share and invest the proceeds in the risk-free asset and hold this position until date T'. At date T', the portfolio will be worth $e^{r(T'-T)}S(T)$. Let $Z(t)$ denote the value of this portfolio for each $0 \leq t \leq T'$. The fundamental pricing formula (1.17) implies that the value of receiving $xS(T)$ at date T' is

$$Z(0)E^Z \left[\frac{xS(T)}{Z(T')} \right] = e^{-qT} S(0) E^Z \left[\frac{xS(T)}{e^{r(T'-T)}S(T)} \right]$$

$$= e^{-qT-r(T'-T)} S(0) E^Z [x]$$

$$= e^{-qT-r(T'-T)} S(0) \times \text{prob}^Z (S(T') > S(T)) .$$

---

[1] We are going to use equation (1.11) at date T' to define the probabilities, because it will not be known until date T' whether the event $S(T') > S(T)$ is true. Thus, we need the price of a numeraire asset at date T'. We would like this price to be a constant times $S(T)$, which is what we will obtain. An equivalent numeraire is to make a smaller investment in the same portfolio: start with $e^{-r(T'-T)-qT}$ shares. This results in a final value of $S(T)$ at date T'. As will be seen, this is useful for deriving the put-call parity relation for forward-start options.

Calculating Probabilities

1. As in the case of a share digital, we know that

$$\log S(t) = \log S(0) + \left(r - q + \frac{1}{2}\sigma^2 \right) t + \sigma B^*(t)$$

for all $t > 0$, where B^* is a Brownian motion when V is used as the numeraire. Taking $t = T'$ and $t = T$ and subtracting yields

$$\log S(T') - \log S(T) = \left(r - q + \frac{1}{2}\sigma^2 \right) (T' - T) + \sigma \left[B^*(T') - B^*(T) \right].$$

Hence, $S(T') > S(T)$ if and only if

$$-\frac{B^*(T') - B^*(T)}{\sqrt{T' - T}} < \frac{\left(r - q + \frac{1}{2}\sigma^2 \right) (T' - T)}{\sigma\sqrt{T' - T}}.$$

The random variable on the left hand side is a standard normal, so

$$\mathrm{prob}^V(S(T') > S(T)) = \mathrm{N}(d_1),$$

where

$$d_1 = \frac{\left(r - q + \frac{1}{2}\sigma^2 \right) (T' - T)}{\sigma\sqrt{T' - T}} = \frac{\left(r - q + \frac{1}{2}\sigma^2 \right) \sqrt{T' - T}}{\sigma}. \tag{8.1a}$$

2. To calculate the probability $\mathrm{prob}^Z(S(T') > S(T))$, note that between T and T', the portfolio with price Z earns the risk-free rate r. The same argument presented in Sect. 2.9 shows that between T and T' we have

$$\frac{\mathrm{d}S}{S} = (r - q)\,\mathrm{d}t + \sigma\,\mathrm{d}B^*,$$

where now B^* denotes a Brownian motion when Z is used as the numeraire. This implies as usual that

$$\mathrm{d}\log S = \left(r - q - \frac{1}{2}\sigma^2 \right)\,\mathrm{d}t + \sigma\,\mathrm{d}B^*,$$

which means that

$$\log S(T') - \log S(T) = \left(r - q - \frac{1}{2}\sigma^2 \right) (T' - T) + \sigma(B^*(T') - B^*(T)).$$

Hence, $S(T') > S(T)$ if and only if

$$-\frac{B^*(T') - B^*(T)}{\sqrt{T' - T}} < \frac{\left(r - q - \frac{1}{2}\sigma^2 \right) (T' - T)}{\sigma\sqrt{T' - T}}.$$

As before, the random variable on the left hand side is a standard normal, so

$$\mathrm{prob}^Z(S(T') > S(T)) = \mathrm{N}(d_2),$$

where

$$d_2 = \frac{\left(r - q - \frac{1}{2}\sigma^2 \right) \sqrt{T' - T}}{\sigma} = d_1 - \sigma\sqrt{T' - T}. \tag{8.1b}$$

Forward-Start Call Pricing Formula

Combining these results, we have:

The value of a forward-start call at date 0 is

$$e^{-qT'}S(0)\,\mathrm{N}(d_1) - e^{-qT-r(T'-T)}S(0)\,\mathrm{N}(d_2)\,, \qquad (8.2)$$

where d_1 and d_2 are defined in (8.1).

Put-Call Parity

Forward-strike calls and puts satisfy a somewhat unusual form of put-call parity. The usual put-call parity is of the form:

$$\text{Call} + \text{Cash} \quad = \quad \text{Put} + \text{Underlying}\,.$$

The amount of cash is the amount that will accumulate to the exercise price at maturity; i.e., it is $e^{-rT'}K$. For forward-start calls and puts, the effective exercise price is $S(T)$, which is not known at date 0. However, the portfolio used as numeraire to value the second part of the payoff will be worth $e^{r(T'-T)}S(T)$ at date T', and by following the same strategy but starting with $e^{-r(T'-T)-qT}$ instead of e^{-qT} shares, we will have $S(T)$ dollars at date T'. The date–0 value of this portfolio should replace "Cash" in the above. Thus:

Put-call parity for forward-start calls and puts is

$$\text{Call Price} + e^{-r(T'-T)-qT}S(0) = \text{Put Price} + e^{-qT'}S(0)\,. \qquad (8.3)$$

8.2 Compound Options

A compound option is an option on an option, for example a call option on a call option or a call on a put. These options are useful for hedging when there is some uncertainty about the need for hedging which may be resolved by the exercise date of the compound option. As speculative trades, they have the benefit of higher leverage than ordinary options. These options were first discussed by Geske [28].

Call-on-a-Call Payoff

Let the underlying call option have exercise price K' and maturity T'. Consider an option maturing at $T < T'$ to purchase the underlying call at price K.

Let $C(t, S)$ denote the value at date t of the underlying call when the stock price is S (i.e., C is the Black-Scholes formula). It is of course rational to exercise the compound call at date T if the value of the underlying call exceeds K; i.e., if $C(T, S(T)) > K$. Let S^* denote the critical price such that $C(T, S^*) = K$. To calculate S^*, we need to solve

Black_Scholes_Call(Sstar,Kprime,r,sigma,q,Tprime-T) = K.

for S^*. We can do this by bisection or one of the other methods mentioned in Sect. 3.7. It is rational to exercise the compound option when $S(T) > S^*$.

When $S(T) > S^*$, exercise of the compound option generates a cash flow of $-K$ at date T. There is a cash flow (of $S(T') - K'$) at date T' only if the compound call is exercised and the underlying call finishes in the money. This is equivalent to:

$$S(T) > S^* \quad \text{and} \quad S(T') > K' . \tag{8.4}$$

Let

$$x = \begin{cases} 1 & \text{if } S(T) > S^* , \\ 0 & \text{otherwise} , \end{cases}$$

and let

$$y = \begin{cases} 1 & \text{if } S(T) > S^* \text{ and } S(T') > K' , \\ 0 & \text{otherwise} . \end{cases}$$

The cash flows of the compound option are $-xK$ at date T and $yS(T') - yK'$ at date T'. We can value the compound option at date 0 by valuing these separate cash flows.

The cash flow $-xK$ is the cash flow from being short K digital options on the underlying asset with strike price S^* and maturity T. Therefore the value at date 0 of this cash flow is $-\mathrm{e}^{-rT} K \, \mathrm{N}(d_2)$, where

$$d_1 = \frac{\log\left(\frac{S(0)}{S^*}\right) + \left(r - q + \frac{1}{2}\sigma^2\right) T}{\sigma\sqrt{T}} , \qquad d_2 = d_1 - \sigma\sqrt{T} . \tag{8.5}$$

Numeraires

The payoffs $yS(T)$ and yK' are similar to share digitals and digitals, respectively, except that the event $y = 1$ is more complex than we have previously encountered. However, we know from the analysis of share digitals and digitals that the values at date 0 of these payoffs are

$$\mathrm{e}^{-qT'} S(0) \times \mathrm{prob}^V(y = 1) \quad \text{and} \quad \mathrm{e}^{-rT'} K' \times \mathrm{prob}^R(y = 1) ,$$

where $V(t) = \mathrm{e}^{qt} S(t)$ and $R(t) = \mathrm{e}^{rt}$.

Calculating Probabilities

We will calculate the two probabilities in terms of the bivariate normal distribution function.

1. The event $y = 1$ is equivalent to

$$\log S(0) + \left(r - q + \frac{1}{2}\sigma^2 \right) T + \sigma B^*(T) > \log S^*$$

and

$$\log S(0) + \left(r - q + \frac{1}{2}\sigma^2 \right) T' + \sigma B^*(T') > \log K' \, ,$$

where B^* is a Brownian motion when the underlying asset (V) is used as the numeraire. These conditions can be rearranged as

$$-\frac{B^*(T)}{\sqrt{T}} < d_1 \quad \text{and} \quad -\frac{B^*(T')}{\sqrt{T'}} < d_1' \, , \tag{8.6}$$

where d_1 is defined in (8.5), and

$$d_1' = \frac{\log\left(\frac{S(0)}{K'} \right) + \left(r - q + \frac{1}{2}\sigma^2 \right) T'}{\sigma\sqrt{T'}} \, , \qquad d_2' = d_1' - \sigma\sqrt{T'} \, . \tag{8.7}$$

The two standard normal variables on the left-hand sides in (8.6) have a covariance equal to

$$\frac{1}{\sqrt{TT'}} \operatorname{cov}(B(T), B(T')) = \frac{1}{\sqrt{TT'}} \operatorname{cov}(B(T), B(T)) = \sqrt{\frac{T}{T'}} \, ,$$

the first equality following from the fact that $B(T)$ is independent of $B(T') - B(T)$ and the second from the fact that the covariance of a random variable with itself is its variance. Hence, $\operatorname{prob}^V(y = 1)$ is the probability that $a \leq d_1$ and $b \leq d_1'$, where a and b are standard normal random variables with covariance (= correlation coefficient) of $\sqrt{T/T'}$. We will write this probability as $\mathrm{M}\left(d_1, d_1', \sqrt{T/T'} \right)$. A program to approximate the bivariate normal distribution function M is provided in Sect. 8.10.

2. The calculation for $\operatorname{prob}^R(y = 1)$ is similar. The event $y = 1$ is equivalent to

$$\log S(0) + \left(r - q + \frac{1}{2}\sigma^2 \right) T + \sigma B^*(T) > \log S^* \, ,$$

and

$$\log S(0) + \left(r - q + \frac{1}{2}\sigma^2 \right) T' + \sigma B^*(T') > \log K' \, ,$$

where B^* now denotes a Brownian motion under the risk-neutral measure. These are equivalent to

$$-\frac{B^*(T)}{\sqrt{T}} < d_2 \quad \text{and} \quad -\frac{B^*(T')}{\sqrt{T'}} < d_2' \,. \tag{8.8}$$

Hence, $\mathrm{prob}^R(y = 1) = \mathrm{M}\left(d_2, d_2', \sqrt{T/T'}\right)$.

Call-on-a-Call Pricing Formula

We conclude:

> The value of a call on a call is
>
> $$-\,\mathrm{e}^{-rT} K\,\mathrm{N}(d_2) + \mathrm{e}^{-qT'} S(0)\,\mathrm{M}\left(d_1, d_1', \sqrt{T/T'}\right)$$
> $$-\,\mathrm{e}^{-rT'} K'\,\mathrm{M}\left(d_2, d_2', \sqrt{T/T'}\right)\,, \tag{8.9}$$
>
> where d_1 and d_2 are defined in (8.5) and d_1' and d_2' are defined in (8.7).

Put-Call Parity

European compound options with the same underlyings and strikes satisfy put-call parity in the usual way:

$$\mathrm{Cash} + \mathrm{Call} = \mathrm{Underlying} + \mathrm{Put}\,.$$

The portfolio on each side of this equation gives the owner the maximum of the strike and the value of the underlying at the option maturity. In the case of options on calls, put-call parity is specifically

$$\mathrm{e}^{-rT} K + \text{Value of call on call}$$
$$= \text{Value of underlying call} + \text{Value of put on call}\,,$$

where K is the strike price of the compound options and T is their maturity date. Likewise, for options on puts, we have

$$\mathrm{e}^{-rT} K + \text{Value of call on put}$$
$$= \text{Value of underlying put} + \text{Value of put on put}\,.$$

Thus, the value of a put on a call can be derived from the value of a call on a call. The value of a put on a put can be derived from the value of a call on a put, which we will now consider.

Call-on-a-Put Pricing Formula

Consider a call option maturing at T with strike K with the underlying being a put option with strike K' and maturity $T' > T$. The underlying of the put is the asset with price S and constant volatility σ. The call on the put will never be in the money at T and hence is worthless if $K > e^{-r(T'-T)}K'$, because the maximum possible value of the put option at date T is $e^{-r(T'-T)}K'$. So assume $K < e^{-r(T'-T)}K'$.

Let S^* again denote the critical value of the stock price such that the call is at the money at date T when $S(T) = S^*$. This means that S^* solves

```
Black_Scholes_Put(Sstar,Kprime,r,sigma,q,Tprime-T) = K.
```

We leave it as an exercise to confirm the following.

The value of a call on a put is

$$
-e^{-rT} K \, N(-d_2) + e^{-rT'} K' \, M\left(-d_2, -d_2', \sqrt{T/T'}\right)
$$
$$
-e^{-qT'} S(0) \, M\left(-d_1, -d_1', \sqrt{T/T'}\right) , \quad (8.10)
$$

where d_1 and d_2 are defined in (8.5) and d_1' and d_2' are defined in (8.7).

8.3 American Calls with Discrete Dividends

It can be optimal to exercise an American call option early if the underlying asset pays a dividend. The optimal exercise date will be immediately prior to the asset going "ex-dividend." Consider a call option maturing at T' on an asset that will pay a known cash dividend D at a known date $T < T'$. We assume there is no continuous dividend payment, so $q = 0$. For simplicity, we assume that the date of the dividend payment is also the date that the asset goes ex-dividend; i.e., ownership of the asset at any date $t < T$ entitles the owner to receive the dividend at date T. Under this assumption, it is reasonable also to assume that the stock price drops by D when the dividend is paid.

There is some ambiguity about how to define the asset price at the instant the dividend is paid—whether to include or exclude the dividend. We will let $S(T)$ denote the price including the dividend and denote the price excluding the dividend by $Z(T)$, so $Z(T) = S(T) - D$. In fact, it is convenient to let $Z(t)$ denote the price stripped of the dividend value at all dates $t \leq T$, so we will define

$$
Z(t) = \begin{cases} S(t) - e^{-r(T-t)}D & \text{if } t \leq T , \\ S(t) & \text{if } t > T . \end{cases}
$$

Note that Z is the price of the following non-dividend-paying portfolio: buy one unit of the asset at date 0, borrow $e^{-rT}D$ at date 0 to help finance the purchase, and use the dividend D at date T to retire the debt.

If we assume as usual that the asset price S has a constant volatility, then, using formula (2.21) for a geometric Brownian motion and letting B^* denote a Brownian motion under the risk-neutral measure, we have

$$
\begin{aligned}
S(T') &= [S(T) - D]\exp\left\{(r - \sigma^2/2)(T' - T) + \sigma B^*(T') - \sigma B^*(T)\right\} \\
&= \left[S(0)\exp\left\{(r - \sigma^2/2)T + \sigma B^*(T)\right\} - D\right] \\
&\quad \times \exp\left\{(r - \sigma^2/2)(T' - T) + \sigma B^*(T') - \sigma B^*(T)\right\} \\
&= S(0)\exp\left\{(r - \sigma^2/2)T' + \sigma B^*(T')\right\} \\
&\quad - D\exp\left\{(r - \sigma^2/2)(T' - T) + \sigma B^*(T') - \sigma B^*(T)\right\}.
\end{aligned}
$$

Thus, S will be a sum of lognormal random variables. A sum of lognormals is not itself lognormal, so S will not be lognormal, and we are unable to calculate the option value in a simple way.

We will assume instead that Z has a constant volatility σ. Thus, Z is the price of a non-dividend-paying portfolio, it satisfies the Black-Scholes assumptions, and we have $S(T') = Z(T')$. To value a European option, we would simply use $Z(0) = S(0) - e^{-rT}D$ as the initial asset price and σ as the volatility.

American Call Payoff

If the call is not exercised before the dividend is paid at date T, then its value at date T will be

$$\texttt{Black\_Scholes\_Call(Z,K,r,sigma,0,Tprime-T)}$$

where $Z = Z(T)$. Hence, exercise is optimal when

$$\texttt{Z + D - K > Black\_Scholes\_Call(Z,K,r,sigma,0,Tprime-T)}.$$

A lower bound for the Black-Scholes call value on the right-hand side is $Z(T) - e^{-r(T'-T)}K$. If $Z(T) + D - K$ is less than or equal to this lower bound, then exercise cannot be optimal. Thus, if $D - K$ is less than or equal to $-e^{-r(T'-T)}K$, then exercise will never be optimal. In this circumstance, the dividend is simply too small to offset the time value of money on the exercise price K, and the value of the American call written on the asset with price S is the same as the value of the European call written on the non-dividend-paying portfolio with price Z.

On the other hand, if $D - K > -e^{-r(T'-T)}K$, then exercise will be optimal for sufficiently large $Z(T)$. In this case, there is some Z^* such that the owner of the call will be indifferent about exercise, and exercise will be optimal for all $Z(T) > Z^*$. This Z^* is defined by

$$\texttt{Zstar + D - K = Black\_Scholes\_Call(Zstar,K,r,sigma,0,Tprime-T)}.$$

As in the previous section, we can compute Z^* by bisection.

Define

$$x = \begin{cases} 1 & \text{if } Z(T) > Z^*, \\ 0 & \text{otherwise}, \end{cases}$$

$$y = \begin{cases} 1 & \text{if } Z(T) \leq Z^* \text{ and } Z(T') > K, \\ 0 & \text{otherwise}. \end{cases}$$

Then the American call option will pay $[Z(T) + D - K]x$ at date T (due to early exercise) and $[Z(T') - K]y$ at date T' (due to exercise at maturity), if $D - K > -e^{-r(T'-T)}K$.

Numeraires

Assume for now that $D - K > -e^{-r(T'-T)}K$. The payoff $(D - K)x$ is the payoff of $D - K$ digital options maturing at T, and the payoff $Z(T)x$ is the payoff of one share digital on the portfolio with price Z. Therefore, the value of receiving $[Z(T) + D - K]x$ at date T is

$$Z(0)\,\mathrm{N}(d_1) + e^{-rT}(D - K)\,\mathrm{N}(d_2),$$

where

$$d_1 = \frac{\log\left(\frac{Z(0)}{Z^*}\right) + \left(r + \frac{1}{2}\sigma^2\right)T}{\sigma\sqrt{T}}$$

$$= \frac{\log\left(\frac{S(0) - e^{-rT}D}{Z^*}\right) + \left(r + \frac{1}{2}\sigma^2\right)T}{\sigma\sqrt{T}}, \tag{8.11a}$$

$$d_2 = d_1 - \sigma\sqrt{T}. \tag{8.11b}$$

As in the previous section,[2] the value of receiving $[Z(T) - K]y$ at date T' is

$$Z(0) \times \mathrm{prob}^Z(y = 1) - e^{-rT'}K \times \mathrm{prob}^R(y = 1).$$

Calculating Probabilities

The calculations are very similar to the calculations we did for a call option on a call. In fact, they are exactly the same as we would do for a put option on a call.

---

[2] The only difference is that here Z is the price of a non-dividend-paying portfolio, so, in the notation of the previous section, we have $V(t) = Z(t)$.

1. The event $y = 1$ is equivalent to

$$\log Z(0) + \left(r + \frac{1}{2}\sigma^2 \right) T + \sigma B^*(T) \leq \log Z^*$$

and

$$\log Z(0) + \left(r + \frac{1}{2}\sigma^2 \right) T' + \sigma B^*(T') > \log K \,,$$

where B^* is a Brownian motion when the underlying asset (with price Z) is used as the numeraire. We can write this as

$$\frac{B^*(T)}{\sqrt{T}} \leq -d_1 \quad \text{and} \quad -\frac{B^*(T')}{\sqrt{T'}} < d_1' \,, \tag{8.12}$$

where d_1 is defined in (8.11a),

$$d_1' = \frac{\log\left(\frac{Z(0)}{K} \right) + \left(r + \frac{1}{2}\sigma^2 \right) T'}{\sigma\sqrt{T'}}$$

$$= \frac{\log\left(\frac{S(0) - e^{-rT}D}{K} \right) + \left(r + \frac{1}{2}\sigma^2 \right) T'}{\sigma\sqrt{T'}} \tag{8.13a}$$

$$d_2' = d_1' - \sigma\sqrt{T'} \,. \tag{8.13b}$$

The two standard normal variables on the left-hand sides in (8.12) have a covariance equal to

$$-\frac{1}{\sqrt{TT'}} \text{cov}(B(T), B(T')) = -\frac{1}{\sqrt{TT'}} \text{cov}(B(T), B(T)) = -\sqrt{\frac{T}{T'}} \,.$$

Hence, $\text{prob}^Z(y = 1)$ is the probability that $a \leq -d_1$ and $b \leq d_1'$, where a and b are standard normal random variables with covariance (= correlation coefficient) of $-\sqrt{T/T'}$. We are writing this probability as $M\left(-d_1, d_1', -\sqrt{T/T'} \right)$.

2. The calculation for $\text{prob}^R(y = 1)$ is similar. The event $y = 1$ is equivalent to

$$\log Z(0) + \left(r - \frac{1}{2}\sigma^2 \right) T + \sigma B^*(T) \leq \log Z^*$$

and

$$\log Z(0) + \left(r - \frac{1}{2}\sigma^2 \right) T' + \sigma B^*(T') > \log K \,,$$

where B^* now denotes a Brownian motion under the risk-neutral measure. These are equivalent to

$$\frac{B^*(T)}{\sqrt{T}} \leq -d_2 \quad \text{and} \quad -\frac{B^*(T')}{\sqrt{T'}} < d_2' \,. \tag{8.14}$$

Hence, $\text{prob}^R(y = 1) = M\left(-d_2, d_2', -\sqrt{T/T'} \right)$.

American Call Pricing Formula

Under our assumptions, the value of an American call option maturing at T' with a dividend payment of D at date $T < T'$ is as follows. If

$$D - K \leq -\mathrm{e}^{-r(T'-T)}K \,,$$

then the value of the call is given by the Black-Scholes formula

$$[S(0) - \mathrm{e}^{-rT}D]\,\mathrm{N}(d_1') - \mathrm{e}^{-rT}K\,\mathrm{N}(d_2') \,,$$

where d_1' and d_2' are defined in (8.13). On the other hand, if

$$D - K > -\mathrm{e}^{-r(T'-T)}K \,,$$

then the value of the call is

$$\begin{aligned}
[S(0) - \mathrm{e}^{-rT}D]\,\mathrm{N}(d_1) &+ \mathrm{e}^{-rT}(D - K)\,\mathrm{N}(d_2) \\
&+ [S(0) - \mathrm{e}^{-rT}D]\,\mathrm{M}\!\left(-d_1, d_1', -\sqrt{T/T'}\right) \\
&\qquad - \mathrm{e}^{-rT'}K\,\mathrm{M}\!\left(-d_2, d_2', -\sqrt{T/T'}\right) \,, \quad (8.15)
\end{aligned}$$

where d_1 and d_2 are defined in (8.11) and d_1' and d_2' are defined in (8.13).

8.4 Choosers

A "chooser option" allows the holder to choose whether the option will be a put or call at some fixed date before the option maturity. Let T denote the date at which the choice is made, T_c the date at which the call expires, T_p the date at which the put expires, K_c the exercise price of the call, and K_p the exercise price of the put, where $0 < T < T_c$ and $0 < T < T_p$. A "simple chooser" has $T_c = T_p$ and $K_c = K_p$. A chooser is similar in spirit to a straddle: it is a bet on volatility without making a bet on direction. A simple chooser must be cheaper than a straddle with the same exercise price and maturity $T' = T_c = T_p$, because a straddle is always in the money at maturity, whereas a simple chooser has the same value as the straddle if it is in the money but is only in the money at T' when the choice made at T turns out to have been the best one.

Chooser Payoff

The value of the chooser at date T will be the larger of the call and put prices. Let S^* denote the stock price at which the call and put have the same value. We can find S^* by solving

```
Black_Scholes_Call(Sstar,Kc,r,sigma,q,Tc-T)
       = Black_Scholes_Put(Sstar,Kp,r,sigma,q,Tp-T).
```

For a simple chooser with $K_c = \dot{K}_p = K$ and $T_c = T_p = T'$, we can find S^* from the put-call parity relation at T, leading to $S^* = \mathrm{e}^{(q-r)(T'-T)}K$.

The call will be chosen when $S(T) > S^*$ and it finishes in the money if $S(T_c) > K_c$ at date T_c, so the payoff of the chooser is $S(T_c) - K_c$ when

$$S(T) > S^* \quad \text{and} \quad S(T_c) > K_c .$$

The payoff is $K_p - S(T_p)$ at date T_p when

$$S(T) < S^* \quad \text{and} \quad S(T_p) < K_p .$$

Let

$$x = \begin{cases} 1 & \text{if } S(T) > S^* \text{ and } S(T_c) > K_c , \\ 0 & \text{otherwise} . \end{cases}$$

Likewise, let

$$y = \begin{cases} 1 & \text{if } S(T) < S^* \text{ and } S(T_p) < K_p , \\ 0 & \text{otherwise} . \end{cases}$$

Then the payoff of the chooser is $xS(T_c) - xK_c$ at date T_c and $yK_p - yS(T_p)$ at date T_p.

Numeraires

As in the analysis of compound options, the value of the chooser at date 0 must be

$$\mathrm{e}^{-qT_c}S(0) \times \mathrm{prob}^V(x=1) \; - \; \mathrm{e}^{-rT_c}K_c \times \mathrm{prob}^R(x=1)$$
$$+ \; \mathrm{e}^{-rT_p}K_p \times \mathrm{prob}^R(y=1) \; - \; \mathrm{e}^{-qT_p}S(0) \times \mathrm{prob}^V(y=1) , \quad (8.16)$$

where we use $V(t) = \mathrm{e}^{qt}S(t)$ and $R(t) = \mathrm{e}^{rt}$ as numeraires.

Chooser Pricing Formula

Equation (8.16) and calculations similar to those of the previous two sections lead us to:

> The value of a chooser option is
>
> $$e^{-qT_c}S(0)\,M\!\left(d_1, d_{1c}, \sqrt{T/T_c}\right) - e^{-rT_c}K_c\,M\!\left(d_2, d_{2c}, \sqrt{T/T_c}\right)$$
> $$+ e^{-rT_p}K_p\,M\!\left(-d_2, -d_{2p}, \sqrt{T/T_p}\right)$$
> $$- e^{-qT_p}S(0)\,M\!\left(-d_1, -d_{1p}, \sqrt{T/T_p}\right)\;,\quad (8.17)$$
>
> where
>
> $$d_1 = \frac{\log\!\left(\frac{S(0)}{S^*}\right) + \left(r - q + \frac{1}{2}\sigma^2\right)T}{\sigma\sqrt{T}}\;,\qquad d_2 = d_1 - \sigma\sqrt{T}\;,$$
>
> $$d_{1c} = \frac{\log\!\left(\frac{S(0)}{K_c}\right) + \left(r - q + \frac{1}{2}\sigma^2\right)T_c}{\sigma\sqrt{T_c}}\;,\qquad d_{2c} = d_{1c} - \sigma\sqrt{T_c}\;,$$
>
> $$d_{1p} = \frac{\log\!\left(\frac{S(0)}{K_p}\right) + \left(r - q + \frac{1}{2}\sigma^2\right)T_p}{\sigma\sqrt{T_p}}\;,\qquad d_{2p} = d_{1p} - \sigma\sqrt{T_p}\;.$$

8.5 Options on the Max or Min

We will consider here an option written on the maximum or minimum of two asset prices; for example, a call on the maximum pays

$$\max(0, \max(S_1(T), S_2(T)) - K) = \max(0, S_1(T) - K, S_2(T) - K)$$

at maturity T. There are also call options on $\min(S_1(T), S_2(T))$ and put options on the maximum and minimum of two (or more) asset prices. Pricing formulas for these options are due to Stulz [58], who also discusses applications. We will assume the two assets have constant dividend yields q_i, constant volatilities σ_i, and a constant correlation ρ.

Call-on-the-Max Payoff

To value a call on the maximum, define the random variables:

$$x = \begin{cases} 1 & \text{if } S_1(T) > S_2(T) \text{ and } S_1(T) > K \text{ ,} \\ 0 & \text{otherwise ,} \end{cases}$$

$$y = \begin{cases} 1 & \text{if } S_2(T) > S_1(T) \text{ and } S_2(T) > K \text{ ,} \\ 0 & \text{otherwise ,} \end{cases}$$

$$z = \begin{cases} 1 & \text{if } S_1(T) > K \text{ or } S_2(T) > K \text{ ,} \\ 0 & \text{otherwise .} \end{cases}$$

Then the value of the option at maturity is

$$x S_1(T) + y S_2(T) - z K \text{ .}$$

Numeraires

Consider numeraires $V_1(t) = e^{q_1 t} S_1(t)$, $V_2(t) = e^{q_2 t} S_2(t)$, and $R(t) = e^{rt}$. By familiar arguments, the value of the option at date 0 is

$$e^{-q_1 T} S_1(0) \times \text{prob}^{V_1}(x = 1) + e^{-q_2 T} S_2(0) \times \text{prob}^{V_2}(y = 1)$$
$$- e^{-rT} K \times \text{prob}^R(z = 1) \text{ .}$$

Calculating Probabilities

1. We will begin by calculating $\text{prob}^{V_1}(x = 1)$. From the second and third examples in Sect. 2.9, the asset prices satisfy

$$\frac{dS_1}{S_1} = (r - q_1 + \sigma_1^2) \, dt + \sigma_1 \, dB_1^* \text{ ,}$$

$$\frac{dS_2}{S_2} = (r - q_2 + \rho \sigma_1 \sigma_2) \, dt + \sigma_2 \, dB_2^* \text{ ,}$$

where B_1^* and B_2^* are Brownian motions when we use V_1 as the numeraire. Thus,

$$\log S_1(T) = \log S_1(0) + \left(r - q_1 + \frac{1}{2} \sigma_1^2 \right) T + \sigma_1 B_1^*(T) \text{ ,}$$

$$\log S_2(T) = \log S_2(0) + \left(r - q_2 + \rho \sigma_1 \sigma_2 - \frac{1}{2} \sigma_2^2 \right) T + \sigma_2 B_2^*(T) \text{ .}$$

The condition $\log S_1(T) > \log K$ is therefore equivalent to

$$-\frac{1}{\sqrt{T}} B_1^*(T) < d_{11} \text{ ,} \tag{8.18a}$$

and the condition $\log S_1(T) > \log S_2(T)$ is equivalent to

$$\frac{\sigma_2 B_2^*(T) - \sigma_1 B_1^*(T)}{\sigma\sqrt{T}} < d_1 , \qquad (8.18b)$$

where

$$\sigma = \sqrt{\sigma_1^2 - 2\rho\sigma_1\sigma_2 + \sigma_2^2} , \qquad (8.19)$$

and

$$d_1 = \frac{\log\left(\frac{S_1(0)}{S_2(0)}\right) + \left(q_2 - q_1 + \frac{1}{2}\sigma^2\right) T}{\sigma\sqrt{T}} , \qquad d_2 = d_1 - \sigma\sqrt{T} , \quad (8.20a)$$

$$d_{11} = \frac{\log\left(\frac{S_1(0)}{K}\right) + \left(r - q_1 + \frac{1}{2}\sigma_1^2\right) T}{\sigma_1\sqrt{T}} , \qquad d_{12} = d_{11} - \sigma_1\sqrt{T} . \quad (8.20b)$$

The random variables on the left-hand sides of (8.18) have standard normal distributions and their correlation is

$$\rho_1 = \frac{\sigma_1 - \rho\sigma_2}{\sigma} .$$

Therefore,
$$\mathrm{prob}^{V_1}(x = 1) = \mathrm{M}(d_{11}, d_1, \rho_1) ,$$

where M again denotes the bivariate normal distribution function.

2. The probability $\mathrm{prob}^{V_2}(y = 1)$ is exactly symmetric to $\mathrm{prob}^{V_1}(x = 1)$, with the roles of S_1 and S_2 interchanged. Note that the mirror image of d_1 defined in (8.20a) is

$$\frac{\log\left(\frac{S_2(0)}{S_1(0)}\right) + \left(q_1 - q_2 + \frac{1}{2}\sigma^2\right) T}{\sigma\sqrt{T}} ,$$

which equals $-d_2$. Therefore,

$$\mathrm{prob}^{V_2}(y = 1) = \mathrm{M}(d_{21}, -d_2, \rho_2) ,$$

where

$$d_{21} = \frac{\log\left(\frac{S_2(0)}{K}\right) + \left(r - q_2 + \frac{1}{2}\sigma_2^2\right) T}{\sigma_2\sqrt{T}} , \qquad d_{22} = d_{21} - \sigma_2\sqrt{T} , \quad (8.20c)$$

and

$$\rho_2 = \frac{\sigma_2 - \rho\sigma_1}{\sigma} .$$

3. As usual, we have

$$\log S_1(T) = \log S_1(0) + \left(r - q_1 - \frac{1}{2}\sigma_1^2\right) T + \sigma_1 B_1^*(T) \,,$$

$$\log S_2(T) = \log S_2(0) + \left(r - q_2 - \frac{1}{2}\sigma_2^2\right) T + \sigma_2 B_2^*(T) \,,$$

where B_1^* and B_2^* now denote Brownian motions under the risk-neutral measure. The event $z = 1$ is the complement of the event

$$S_1(T) \le K \quad \text{and} \quad S_2(T) \le K \,,$$

which is equivalent to

$$\frac{1}{\sqrt{T}} B_1^*(T) < -d_{12} \,, \tag{8.21a}$$

and

$$\frac{1}{\sqrt{T}} B_2^*(T) < -d_{22} \,. \tag{8.21b}$$

The random variables on the left-hand sides of (8.21a) and (8.21b) are standard normals and have correlation ρ. Therefore,

$$\text{prob}^R(z = 1) = 1 - M(-d_{12}, -d_{22}, \rho) \,.$$

Call-on-the-Max Pricing Formula

The value of a call option on the maximum of two risky asset prices with volatilities σ_1 and σ_2 and correlation ρ is

$$e^{-q_1 T} S_1(0) \, M\left(d_{11}, d_1, \frac{\sigma_1 - \rho\sigma_2}{\sigma}\right) + e^{-q_2 T} S_2(0) \, M\left(d_{21}, -d_2, \frac{\sigma_2 - \rho\sigma_1}{\sigma}\right)$$

$$+ e^{-rT} K \, M(-d_{12}, -d_{22}, \rho) - e^{-rT} K \,, \tag{8.22}$$

where σ is defined in (8.19) and d_1, d_2, d_{11}, d_{12}, d_{21} and d_{22} are defined in (8.20).

8.6 Barrier Options

A "down-and-out" call pays the usual call value at maturity if and only if the stock price does not hit a specified lower bound during the lifetime of the option. If it does breach the lower barrier, then it is "out." Conversely, a "down-and-in" call pays off only if the stock price *does* hit the lower bound. Up-and-out and up-and-in calls are defined similarly, and there are also put options of this sort. The "out" versions are called "knock-outs" and the "in" versions are called "knock-ins."

Knock-ins can be priced from knock-outs and vice-versa. For example, the combination of a down-and-out call and a down-and-in call creates a standard European call, so the value of a down-and-in can be obtained by subtracting the value of a down-and-out from the value of a standard European call. Likewise, up-and-in calls can be valued by subtracting the value of an up-and-out from the value of a standard European call. Both knock-outs and knock-ins are of course less expensive than comparable standard options.

We will describe the pricing of a down-and-out call. The pricing of up-and-out calls and knock-out puts is similar. Often there are rebates associated with the knocking-out of a barrier option, but we will not include that feature here (see Sect. 10.7 however).

A down-and-out call provides a hedge against an increase in an asset price, just as does a standard call, for someone who is short the asset. The difference is that the down-and-out is knocked out when the asset price falls sufficiently. Presumably this is acceptable to the buyer because the need to hedge against high prices diminishes when the price falls. In fact, in this circumstance the buyer may want to establish a new hedge at a lower strike. However, absent re-hedging at a lower strike, the buyer of a knock-out call obviously faces the risk that the price may reverse course after falling to the knock-out boundary, leading to regret that the option was knocked out. The rationale for accepting this risk is that the knock-out is cheaper than a standard call. Thus, compared to a standard call, a down-and-out call provides cheaper but incomplete insurance.

The combination of a knock-out call and a knock-in call (or knock-out puts) with the same barrier and different strikes creates an option with a strike that is reset when the barrier is hit. This is a hedge that adjusts automatically to the market. An example is given in Probs. 8.9 and 8.10.

Down-and-Out Call Payoff

Let L denote the lower barrier for the down-and-out call and assume it has not yet been breached at the valuation date, which we are calling date 0. Denote the minimum stock price realized during the remaining life of the contract by $z = \min_{0 \le t \le T} S(t)$. In practice, this minimum is calculated at discrete dates (for example, based on daily closing prices), but we will assume here that the stock price is monitored continuously for the purpose of calculating the minimum.

The down-and-out call will pay $\max(0, S(T) - K)$ if $z > L$ and 0 otherwise, at its maturity T. Let

$$x = \begin{cases} 1 & \text{if } S(T) > K \text{ and } z > L \,, \\ 0 & \text{otherwise} \,. \end{cases}$$

Then the value of the down-and-out call at maturity is

$$xS(T) - xK \,.$$

Numeraires

As in other cases, the value at date 0 can be written as

$$e^{-qT} S(0) \times \text{prob}^V (x = 1) - e^{-rT} K \times \text{prob}^R (x = 1) \,,$$

where $V(t) = e^{qt} S(t)$ and $R(t) = e^{rt}$.

Calculating Probabilities

To calculate $\text{prob}^V (x = 1)$ and $\text{prob}^R (x = 1)$, we consider two cases.

1. Suppose $K > L$. Define

$$y = \begin{cases} 1 & \text{if } S(T) > K \text{ and } z \le L \\ 0 & \text{otherwise} \,. \end{cases}$$

The event $S(T) > K$ is equal to the union of the disjoint events $x = 1$ and $y = 1$. Therefore,

$$\text{prob}^V (x = 1) = \text{prob}^V (S(T) > K) - \text{prob}^V (y = 1) \,,$$
$$\text{prob}^R (x = 1) = \text{prob}^R (S(T) > K) - \text{prob}^R (y = 1) \,.$$

As in the derivation of the Black-Scholes formula, we have

$$\text{prob}^V (S(T) > K) = N(d_1) \quad \text{and} \quad \text{prob}^R (S(T) > K) = N(d_2) \,, \quad (8.23)$$

where

$$d_1 = \frac{\log\left(\frac{S(0)}{K}\right) + \left(r - q + \frac{1}{2}\sigma^2\right) T}{\sigma\sqrt{T}} \,, \qquad d_2 = d_1 - \sigma\sqrt{T} \,. \quad (8.24a)$$

Furthermore, defining

$$d_1' = \frac{\log\left(\frac{L^2}{KS(0)}\right) + \left(r - q + \frac{1}{2}\sigma^2\right) T}{\sigma\sqrt{T}} \,, \qquad d_2' = d_1' - \sigma\sqrt{T} \,, \quad (8.24b)$$

it can be shown (see Appendix B.2) that

$$\text{prob}^V (y = 1) = \left(\frac{L}{S(0)}\right)^{2(r - q + \frac{1}{2}\sigma^2)/\sigma^2} N(d_1') \,, \quad (8.25a)$$

$$\text{prob}^R (y = 1) = \left(\frac{L}{S(0)}\right)^{2(r - q - \frac{1}{2}\sigma^2)/\sigma^2} N(d_2') \,. \quad (8.25b)$$

2. Suppose $K \leq L$. Then the condition $S(T) > K$ in the definition of the event $x = 1$ is redundant: if $z > L \geq K$, then it is necessarily true that $S(T) > K$. Therefore, the probability (under either numeraire) of the event $x = 1$ is the probability that $z > L$. Define

$$y = \begin{cases} 1 & \text{if } S(T) > L \text{ and } z \leq L, \\ 0 & \text{otherwise}. \end{cases}$$

The event $S(T) > L$ is the union of the disjoint events $x = 1$ and $y = 1$. Therefore, as in the previous case (but now with K replaced by L),

$$\text{prob}^V(x = 1) = \text{prob}^V(S(T) > L) - \text{prob}^V(y = 1),$$
$$\text{prob}^R(x = 1) = \text{prob}^R(S(T) > L) - \text{prob}^R(y = 1).$$

Also as before, we know that

$$\text{prob}^V(S(T) > L) = N(d_1) \quad \text{and} \quad \text{prob}^R(S(T) > L) = N(d_2), \quad (8.26)$$

where now

$$d_1 = \frac{\log\left(\frac{S(0)}{L}\right) + \left(r - q + \frac{1}{2}\sigma^2\right)T}{\sigma\sqrt{T}}, \qquad d_2 = d_1 - \sigma\sqrt{T}. \quad (8.27a)$$

Moreover, $\text{prob}^V(y = 1)$ and $\text{prob}^R(y = 1)$ are given by (8.25) but with K replaced by L, which means that

$$d_1' = \frac{\log\left(\frac{L}{S(0)}\right) + \left(r - q + \frac{1}{2}\sigma^2\right)T}{\sigma\sqrt{T}}, \qquad d_2' = d_1' - \sigma\sqrt{T}. \quad (8.27b)$$

Down-and-Out Call Pricing Formula

The value of a continuously-sampled down-and-out call option with barrier L is

$$e^{-qT}S(0)\left[N(d_1) - \left(\frac{L}{S(0)}\right)^{2(r-q+\frac{1}{2}\sigma^2)/\sigma^2} N(d_1')\right]$$

$$- e^{-rT}K\left[N(d_2) - \left(\frac{L}{S(0)}\right)^{2(r-q-\frac{1}{2}\sigma^2)/\sigma^2} N(d_2')\right], \quad (8.28)$$

where

(1) if $K > L$, d_1, d_2, d_1' and d_2' are defined in (8.24),
(2) if $K \leq L$, d_1, d_2, d_1' and d_2' are defined in (8.27).

8.7 Lookbacks

A "floating-strike lookback call" pays the difference between the asset price at maturity and the minimum price realized during the life of the contract. A "floating-strike lookback put" pays the difference between the maximum price over the life of the contract and the price at maturity. Thus, the floating-strike lookback call allows one to buy the asset at its minimum price, and the floating-strike lookback put allows one to sell the asset at its maximum price. Of course, one pays upfront for this opportunity to time the market. These options were first discussed by Goldman, Sosin and Gatto [30].

A "fixed-strike lookback put" pays the difference between a fixed strike price and the minimum price during the lifetime of the contract. Thus, a fixed-strike lookback put and a floating-strike lookback call are similar in one respect: both enable one to buy the asset at its minimum price. However, the put allows one to sell the asset at a fixed price whereas the call allows one to sell it at the terminal asset price. A "fixed-strike lookback call" pays the difference between the maximum price and a fixed strike price and is similar to a floating-strike lookback put in the sense that both enable one to sell the asset at its maximum price. Fixed-strike lookback options were first discussed by Conze and Viswanathan [18]. We will discuss the valuation of floating-strike lookback calls. As in the discussion of barrier options, we will assume that the price is continuously sampled for the purpose of computing the minimum.

Floating-Strike Lookback Call Payoff

As in the previous section, let z denote the minimum stock price realized over the *remaining lifetime of the contract*. This is not necessarily the minimum stock price realized during the entire lifetime of the contract. Let $S_{\min}$ denote the minimum stock price realized during the lifetime of the contract up to and including date 0, which is the date at which we are valuing the contract. The minimum stock price during the *entire lifetime of the contract* will be the smaller of z and $S_{\min}$. The payoff of the floating strike lookback call is $S(T) - \min(z, S_{\min})$.

Calculations

The value at date 0 of the piece $S(T)$ is simply $e^{-qT}S(0)$. Using the result in Appendix B.2 on the distribution of z, it can be shown (see, e.g., Musiela and Rutkowski [54] for the details) that the value at date 0 of receiving

$$\min(z, S_{\min})$$

at date T is

$$\mathrm{e}^{-rT} S_{\min} \mathrm{N}(d_2) - \frac{\sigma^2}{2(r-q)} \left(\frac{S_{\min}}{S(0)} \right)^{2(r-q)/\sigma^2} \mathrm{e}^{-rT} S(0) \mathrm{N}(d_2') \cdot$$

$$+ \left(1 + \frac{\sigma^2}{2(r-q)} \right) \mathrm{e}^{-qT} S(0) \mathrm{N}(-d_1) \,.$$

where

$$d_1 = \frac{\log\left(\frac{S(0)}{S_{\min}} \right) + \left(r - q + \frac{1}{2}\sigma^2 \right) T}{\sigma\sqrt{T}} \,, \qquad d_2 = d_1 - \sigma\sqrt{T} \,, \qquad (8.29\mathrm{a})$$

$$d_1' = \frac{\log\left(\frac{S_{\min}}{S(0)} \right) + \left(r - q + \frac{1}{2}\sigma^2 \right) T}{\sigma\sqrt{T}} \,, \qquad d_2' = d_1' - \sigma\sqrt{T} \,. \qquad (8.29\mathrm{b})$$

Using the fact that $[1 - \mathrm{N}(-d_1)]\mathrm{e}^{-qT} S(0) = \mathrm{e}^{-qT} S(0) \mathrm{N}(d_1)$, this implies:

Floating-Strike Lookback Call Pricing Formula

The value at date 0 of a continuously-sampled floating-strike lookback call, given that the minimum price during the lifetime of the contract through date 0 is $S_{\min}$ and the remaining time to maturity is T, is

$$\mathrm{e}^{-qT} S(0) \mathrm{N}(d_1) - \mathrm{e}^{-rT} S_{\min} \mathrm{N}(d_2)$$

$$+ \frac{\sigma^2}{2(r-q)} \left(\frac{S_{\min}}{S(0)} \right)^{2(r-q)/\sigma^2} \mathrm{e}^{-rT} S(0) \mathrm{N}(d_2')$$

$$- \frac{\sigma^2}{2(r-q)} \mathrm{e}^{-qT} S(0) \mathrm{N}(-d_1) \,, \qquad (8.30)$$

where d_1, d_2, and d_2' are defined in (8.29).

8.8 Basket and Spread Options

A "spread option" is a call or a put written on the difference of two asset prices. For example, a spread call will pay at maturity T the larger of zero and $S_1(T) - S_2(T) - K$, where the S_i are the asset prices and K is the strike price of the call. Spread options can be used by producers to hedge the difference between an input price and an output price. They are also useful for hedging basis risk. For example, someone may want to hedge an asset by selling a futures contract on a closely related but not identical asset. This exposes the hedger to basis risk: the difference in value between the asset and the underlying asset on the futures contract. A spread call can hedge the basis risk: take S_1 to be the value of the asset underlying the futures contract and S_2 the value of the asset being hedged.

A spread option is actually an exchange option. Assuming constant dividend yields q_1 and q_2, we can take the assets underlying the exchange option to be as follows

1. At date 0, purchase $e^{-q_1 T}$ units of the asset with price S_1 and reinvest dividends, leading to a value of $S_1(T)$ at date T,
2. At date 0, purchase $e^{-q_2 T}$ units of the asset with price S_2 and invest $e^{-rT}K$ in the risk-free asset. Reinvesting dividends and accumulating interest means that we will have $S_2(T) + K$ dollars at date T.

However, we cannot apply Margrabe's formula to price spread options, because the second portfolio described above will have a stochastic volatility. To see this, note that if the price $S_2(t)$ falls to a low level, then the portfolio will consist primarily of the risk-free asset, so the portfolio volatility will be near the volatility of the risk-free asset, which is zero. On the other hand, if $S_2(t)$ becomes very high, then the portfolio will be weighted very heavily on the stock investment, and its volatility will approach the volatility of S_2.

A "basket option" is an option written on a portfolio of assets. For example, someone may want to hedge the change in the value of the dollar relative to a basket of currencies. A basket option is an alternative to purchasing separate options on each currency. Generally, the basket option would have a lower premium than the separate options, because an option on a portfolio is cheaper (and pays less at maturity) than a portfolio of options.

Letting $S_1, \ldots, S_n$ denote the asset prices and $w_1, \ldots, w_n$ the weights specified by the contract, a basket call would pay

$$\max\left(0, \sum_{i=1}^{n} w_i S_i(T) - K\right)$$

at maturity T. A spread option is actually a special case of a basket option, with $n = 2$, $w_1 = 1$, and $w_2 = -1$. The difficulty in valuing basket options is the same as that encountered in valuing spread options. The volatility of the basket price $\sum_{i=1}^{n} w_i S_i(t)$ will vary over time, depending on the relative volatilities of the assets and the price changes in the assets. For example, consider the case $n = 2$ and write $S(t)$ for the basket price $w_1 S_1(t) + w_2 S_2(t)$. Then

$$\frac{dS}{S} = \frac{w_1 \, dS_1}{S} + \frac{w_2 \, dS_2}{S}$$

$$= \frac{w_1 S_1}{S} \times \frac{dS_1}{S_1} + \frac{w_2 S_2}{S} \times \frac{dS_2}{S_2} \, .$$

Let $x_i(t) = w_i S_i(t)/S(t)$. This is the fraction of the portfolio value that the i-th asset contributes. It will vary randomly over time as the prices change.

Letting σ_i denote the volatilities of the individual assets and ρ their correlation, the formula just given for $\mathrm{d}S/S$ shows that the instantaneous volatility of the basket price at any date t is

$$\sqrt{x_1^2(t)\sigma_1^2 + 2x_1(t)x_2(t)\rho\sigma_1\sigma_2 + x_2^2(t)\sigma_2^2}\ .$$

Hence, the volatility will vary randomly over time as the x_i change. As in the case of spread options, there is no simple closed-form solution for the value of a basket option.

8.9 Asian Options

An "Asian option" is an option the value of which depends on the average underlying asset price during the lifetime of the option. "Average-price" calls and puts are defined like standard calls and puts but with the final asset price replaced by the average price. "Average-strike" calls and puts are defined like standard calls and puts but with the exercise price replaced by the average asset price. A firm that must purchase an input at frequent intervals or will sell a product in a foreign currency at frequent intervals can use an average price option as an alternative to buying multiple options with different maturity dates. The average-price option will generally be both less expensive and a better hedge than purchasing multiple options.

In practice, the average price is computed by averaging over the prices sampled at a finite number of discrete dates. First, we consider the case of continuous sampling. With continuous sampling, the average price at date T for an option written at date 0 will be denoted by $A(T)$ and is defined as

$$A(T) = \frac{1}{T}\int_0^T S(t)\,\mathrm{d}t\ .$$

To obtain a closed-form solution for the value of an option on the average price, we face essentially the same problem as for basket and spread options: a sum of lognormally distributed variables is not itself lognormally distributed. In this case, the integral, which is essentially a continuous sum of the prices at different dates, is not lognormally distributed.

An alternative contract would replace the average price with the "geometric average." This is defined as the exponential of the average logarithm. We will denote this by $A_{\mathrm{g}}(T)$. The average logarithm is

$$\frac{1}{T}\int_0^T \log S(t)\,\mathrm{d}t\ ,$$

and the geometric average is

$$A_{\mathrm{g}}(T) = \exp\left(\frac{1}{T}\int_0^T \log S(t)\,\mathrm{d}t\right)\ .$$

The concavity of the logarithm function guarantees that

$$\log \frac{1}{T} \int_0^T S(t) > \frac{1}{T} \int_0^T \log S(t) \, dt \, .$$

Therefore,

$$A(T) = \exp\left(\log \frac{1}{T} \int_0^T S(t)\right)$$

$$> \exp\left(\frac{1}{T} \int_0^T \log S(t) \, dt\right)$$

$$= A_g(T) \, .$$

Consequently, approximating the value of an average-price or average-strike option by substituting $A_g(T)$ for $A(T)$ will produce a biased estimate of the value. Nevertheless, the geometric average$A_g(T)$ and the arithmetic average $A(T)$ will be highly correlated, so $A_g(T)$ forms a very useful control variate for Monte-Carlo valuation of average-price and average-strike options, as will be discussed in Chap. 9. To implement the idea, we need a valuation formula for options written on $A_g(T)$. We will derive this for an average-price call, in which $A_g(T)$ substitutes for $A(T)$.

Specifically, consider a contract that pays

$$\max(0, A_g(T) - K)$$

at its maturity T. This is a "geometric-average-price call,", and we will analyze it in the same way that we analyzed quanto options in Chap. 6. Let $V(t)$ denote the value at date t of receiving $A_g(T)$ at date T. This can be calculated, and the result will be given below. $V(t)$ is the value of a non-dividend-paying portfolio, and, by definition, $V(T) = A_g(T)$, so the geometric-average-price call is equivalent to a standard call with V being the price of the underlying. We will show that V has a time-varying but non-random volatility. Therefore, we can apply the Black-Scholes formula, inputting the average volatility as described in Sect. 3.8, to value the geometric-average-price call. We could attempt the same route to price average-price options, but we would find that the volatility of the corresponding value V would vary randomly, just as we found the basket portfolio to have a random volatility in the previous section.

The value $V(t)$ can be calculated as

$$V(t) = e^{-r(T-t)} E_t^R [A_g(T)] \, .$$

Define

$$A_g(t) = \exp\left(\frac{1}{t} \int_0^t \log S(u) \, du\right) \, .$$

We will verify at the end of this section that

$$V(t) = e^{-r(T-t)} A_g(t)^{\frac{t}{T}} S(t)^{\frac{T-t}{T}} \exp\left(\frac{(r-q-\sigma^2/2)(T-t)^2}{2T} + \frac{\sigma^2(T-t)^3}{6T^2}\right).$$

(8.31)

Two points are noteworthy. First, the value at date 0 is

$$V(0) = e^{-rT} S(0) \exp\left(\frac{(r-q-\sigma^2/2)T}{2} + \frac{\sigma^2 T}{6}\right)$$

$$= \exp\left(-\frac{6r+6q+\sigma^2}{12}T\right) S(0).$$

(8.32)

Second, the volatility comes from the factor

$$S(t)^{\frac{T-t}{T}},$$

and, by Itô's formula,

$$\frac{dS^{\frac{T-t}{T}}}{S^{\frac{T-t}{T}}} = \text{something } dt + \left(\frac{T-t}{T}\right)\sigma\, dB.$$

This implies that the average volatility, in the sense of Sect. 3.8, is

$$\sigma_{\text{avg}} = \sqrt{\frac{1}{T}\int_0^T \left(\frac{T-t}{T}\right)^2 \sigma^2\, dt} = \frac{\sigma}{\sqrt{3}}.$$

Applying the Black-Scholes formula yields:

The value at date 0 of a continuously-sampled geometric-average-price call written at date 0 and having T years to maturity is

$$V(0)\, N(d_1) - e^{-rT} K\, N(d_2),$$

where

$$d_1 = \frac{\log\left(\frac{V(0)}{K}\right) + \left(r + \frac{1}{2}\sigma_{\text{avg}}^2\right)T}{\sigma_{\text{avg}}\sqrt{T}}, \qquad d_2 = d_1 - \sigma_{\text{avg}}\sqrt{T},$$

$V(0)$ is defined in (8.32), and $\sigma_{\text{avg}} = \sigma/\sqrt{3}$.

We can also value a discretely-sampled geometric-average-price call by the same arguments. Consider dates $0 < t_0 < t_1 < \cdots t_N = T$, where $t_i - t_{i-1} = \Delta t$ for each i and suppose the price is to be sampled at the dates $t_1, \ldots, t_N$. Now let $V(t)$ denote the value at date t of the contract that pays

$$\exp\left(\frac{1}{N}\sum_{i=1}^N \log S(t_i)\right) = \left(\prod_{i=1}^N S(t_i)\right)^{1/N}$$

(8.33)

at date T. The call option will pay $\max(0, V(T) - K)$ at date T. Let k denote the integer such that $t_{N-k-1} \leq t < t_{N-k}$. This means that we have already observed the prices $S(t_1), \ldots, S(t_{N-k-1})$ and we have yet to observe the $k+1$ prices $S(t_{N-k}), \ldots, S(t_N)$. Define $\varepsilon = (t_{N-k} - t)/\Delta t$, which is fraction of the interval Δt that must pass before we reach the next sampling date t_{N-k}. We will show at the end of this section that

$$
V(t) = \mathrm{e}^{-r(T-t)} S(t)^{\frac{k+1}{N}} \prod_{i=1}^{N-k-1} S(t_i)^{\frac{1}{N}}
$$

$$
\times \exp\left(\left[\frac{(k+1)\varepsilon\nu}{N} + \frac{k(k+1)\nu}{2N} + \frac{(k+1)^2\sigma^2\varepsilon}{2N^2} + \frac{k(k+1)(2k+1)\sigma^2}{12N^2}\right]\Delta t\right),
$$

$$\tag{8.34}$$

where $\nu = r - q - \sigma^2/2$

Again, two points are noteworthy. Assume the call was written at date 0 and the first observation date t_1 is Δt years away. Then, we have $k + 1 = N$ and $\varepsilon = 1$ so

$$
V(0) = \mathrm{e}^{-rT} S(0) \exp\left(\frac{(N+1)\nu\Delta t}{2} + \frac{(N+1)(2N+1)\sigma^2\Delta t}{12N}\right). \tag{8.35}
$$

Second, the volatility of $V(t)$ comes from the factor $S(t)^{(k+1)/N}$, and

$$
\frac{\mathrm{d}S^{\frac{k+1}{N}}}{S^{\frac{k+1}{N}}} = \text{something } \mathrm{d}t + \left(\frac{k+1}{N}\right)\sigma\,\mathrm{d}B.
$$

This implies that the average volatility, in the sense of Sect. 3.8, is

$$
\sigma_{\mathrm{avg}} = \sqrt{\frac{1}{N}\sum_{k=0}^{N-1}\left(\frac{k+1}{N}\right)^2\sigma^2\,\mathrm{d}t}
$$

$$
= \frac{\sigma}{N^{3/2}}\sqrt{\frac{N(N+1)(2N+1)}{6}}, \tag{8.36}
$$

where we have used the fact that $\sum_{i=1}^{N} i^2 = N(N+1)(2N+1)/6$ to obtain the second equality. Thus, the Black-Scholes formula implies:

The value at date 0 of a discretely-sampled geometric-average-price call written at date 0 and having T years to maturity is

$$V(0)\,\mathrm{N}(d_1) - \mathrm{e}^{-rT}K\,\mathrm{N}(d_2)\,, \tag{8.37}$$

where

$$d_1 = \frac{\log\left(\frac{V(0)}{K}\right) + \left(r + \frac{1}{2}\sigma_{\mathrm{avg}}^2\right)T}{\sigma_{\mathrm{avg}}\sqrt{T}}\,, \qquad d_2 = d_1 - \sigma_{\mathrm{avg}}\sqrt{T}\,,$$

$V(0)$ is defined in (8.35), and σ_{avg} is defined in (8.36).

This formula will be used in Sect. 9.5 as a control variate for pricing discretely-sampled average-price calls (even average-price calls that were written before the date of valuation).

We will now derive equations (8.31) and (8.34). We will begin with (8.31). The random variable $A_g(T)$ is normally distributed under the risk-neutral measure given information at time t. To establish this, and to calculate the mean and variance of $A_g(T)$, the key is to change the order of integration in the integral in the second line below to obtain the third line:

$$\int_t^T \log S(u)\,du = \int_t^T \left\{ \log S(t) + \left(r - q - \frac{1}{2}\sigma^2\right)(u - t) + \sigma[B(u) - B(t)] \right\} du$$

$$= (T - t)\log S(t) + \left(r - q - \frac{1}{2}\sigma^2\right)\frac{(T - t)^2}{2} + \sigma\int_t^T \int_t^u dB(s)\,du$$

$$= (T - t)\log S(t) + \left(r - q - \frac{1}{2}\sigma^2\right)\frac{(T - t)^2}{2} + \sigma\int_t^T \int_s^T du\,dB(s)$$

$$= (T - t)\log S(t) + \left(r - q - \frac{1}{2}\sigma^2\right)\frac{(T - t)^2}{2} + \sigma\int_t^T (T - s)\,dB(s)$$

and then to note that $\int_t^T (T - s)\,dB(s)$ is normally distributed with mean zero and variance equal to

$$\int_t^T (T - s)^2\,ds = \frac{(T - t)^3}{3}\,.$$

Therefore $E_t^R[A_g(T)]$ is the expectation of the exponential of a normally distributed random variable. Equation (8.31) now follows from the fact that if x is normally distributed with mean μ and variance σ^2 then $E[\mathrm{e}^x] = \mathrm{e}^{\mu + \sigma^2/2}$.

To establish (8.34), note that the discounted risk-neutral expectation of (8.33), conditional on having observed $S(t_1), \ldots, S(t_{N-k-1})$, is

$$V(t) = e^{-r(T-t)} E_t^R \left[\exp\left(\frac{1}{N} \sum_{i=1}^{N} \log S(t_i) \right) \right]$$

$$= e^{-r(T-t)} \exp\left(\frac{1}{N} \sum_{i=1}^{N-k-1} \log S(t_i) \right) \times E_t^R \left[\exp\left(\frac{1}{N} \sum_{i=N-k}^{N} \log S(t_i) \right) \right]$$

$$= \left(\prod_{i=1}^{N-k-1} S(t_i)^{\frac{1}{N}} \right) \times e^{-r(T-t)} E_t^R \left[\exp\left(\frac{1}{N} \sum_{i=N-k}^{N} \log S(t_i) \right) \right]. \qquad (8.38)$$

Let $\Delta_0 B = B(t_{N-k}) - B(t)$ and $\Delta_i B = B(t_{N-k+i}) - B(t_{N-k+i-1})$ for $i \geq 1$. We can write the sum of logarithms inside the expectation above as

$$\sum_{i=0}^{k} \left\{ [\log S(t) + (t_{N-k+i} - t)\nu + \sigma[B(t_{N-k+i}) - B(t)]] \right\}$$

$$= (k+1)\log S(t) + \sum_{i=0}^{k}(\varepsilon + i)\nu \Delta t + \sigma \sum_{i=0}^{k}[\Delta_0 B + \Delta_1 B + \cdots + \Delta_i B]$$

$$= (k+1)\log S(t) + (k+1)\varepsilon\nu \Delta t + \frac{k(k+1)}{2}\nu \Delta t + \sigma \sum_{i=0}^{k}(k+1-i)\Delta_i B \,,$$

where to obtain the last equality we used the fact that $\sum_{i=0}^{k} i = k(k+1)/2$. The random variables $\Delta_i B$ are normally distributed with mean zero and variance Δt (the variance is $\varepsilon \Delta t$ for $i = 0$). Thus, the sum of logarithms is a normally distributed random variable with mean

$$(k+1)\log S(t) + (k+1)\varepsilon\nu \Delta t + \frac{k(k+1)}{2}\nu \Delta t$$

and variance

$$(k+1)^2 \sigma^2 \varepsilon \Delta t + \sigma^2 \sum_{i=1}^{k}(k+1-i)^2 \Delta t = (k+1)^2 \sigma^2 \varepsilon \Delta t + \frac{k(k+1)(2k+1)\sigma^2}{6} \,,$$

using the fact that $\sum_{i=1}^{k} i^2 = k(k+1)(2k+1)/6$. The expectation of the exponential of a normally distributed random variable equals the exponential of its mean plus one-half of its variance, and the exponential of $(k+1)\log S(t)/N$ is $S(t)^{(k+1)/N}$. Therefore the conditional expectation in (8.38) is

$$S(t)^{\frac{k+1}{N}} \exp\left(\left[\frac{(k+1)\varepsilon\nu}{N} + \frac{k(k+1)\nu}{2N} + \frac{(k+1)^2 \sigma^2 \varepsilon}{2N^2} + \frac{k(k+1)(2k+1)\sigma^2}{12N^2} \right] \Delta t \right) \,,$$

which implies (8.34).

8.10 Calculations in VBA

The new features in the option pricing formulas in this chapter are the use of the bivariate normal distribution function and sometimes the need to compute a critical (at-the-money) value of the underlying asset price. We will compute the critical values by bisection, in the same way that we computed implied volatilities for the Black-Scholes formula in Chap. 3.

Bivariate Normal Distribution Function

The following is a fast approximation of the bivariate cumulative normal distribution function, accurate to six decimal places, due to Drezner [23]). For given numbers a and b, this function gives the probability that $\xi_1 < a$ and $\xi_2 < b$ where ξ_1 and ξ_2 are standard normal random variables with a given correlation ρ, which we must input.

```
Function BiNormalProb(a, b, rho)
Dim a1, b1, sum, z1, Z2, z3, rho1, rho2, Delta, x, y, i, j
x = Array(0.24840615, 0.39233107, 0.21141819, _
          0.03324666, 0.00082485334)
y = Array(0.10024215, 0.48281397, 1.0609498, _
          1.7797294, 2.6697604)
a1 = a / Sqr(2 * (1 - rho ^ 2))
b1 = b / Sqr(2 * (1 - rho ^ 2))
If a <= 0 & b <= 0 & rho <= 0 Then
    sum = 0
    For i = 0 To 4
        For j = 0 To 4
            z1 = a1 * (2 * y(i) - a1)
            Z2 = b1 * (2 * y(j) - b1)
            z3 = 2 * rho * (y(i) - a1) * (y(j) - b1)
            sum = sum + x(i) * x(j) * Exp(z1 + Z2 + z3)
        Next j
    Next i
    BiNormalProb = sum * Sqr(1 - rho ^ 2) / Application.Pi
ElseIf a <= 0 & b >= 0 & rho >= 0 Then
    BiNormalProb = Application.NormSDist(a)-BiNormalProb(a,-b,-rho)
ElseIf a >= 0 & b <= 0 & rho >= 0 Then
    BiNormalProb = Application.NormSDist(b)-BiNormalProb(-a,b,-rho)
ElseIf a >= 0 & b >= 0 & rho <= 0 Then
    sum = Application.NormSDist(a) + Application.NormSDist(b)
    BiNormalProb = sum - 1 + BiNormalProb(-a, -b, rho)
ElseIf a * b * rho > 0 Then
    rho1 = (rho*a-b) * Sgn(a) / Sqr(a^ 2 - 2*rho*a*b + b^ 2)
    rho2 = (rho*b-a) * Sgn(b) / Sqr(a^2 - 2*rho*a*b + b^ 2)
    Delta = (1 - Sgn(a) * Sgn(b)) / 4
    BiNormalProb = BiNormalProb(a,0,rho1) _
                    +BiNormalProb(b,0,rho2) - Delta
End If
End Function
```

Notice that this function calls itself. This is an example of "recursion."

Forward-Start Call

The forward-start call pricing formula is of the same form as the Black-Scholes, Margrabe, Black, and Merton formulas, as discussed in Sect. 7.5. We can compute it with our `Generic_Option` pricing function.

```
Function Forward_Start_Call(S, r, sigma, q, Tset, TCall)
'
' Inputs are S = initial stock price
'            r = risk-free rate
'            sigma = volatility
'            q = dividend yield
'            Tset = time until the strike is set
'            TCall = time until call matures >= Tset
'
Dim P1, P2
P1 = Exp(-q * TCall) * S
P2 = Exp(-q * Tset - r * (TCall - Tset)) * S
Forward_Start_Call = Generic_Option(P1, P2, sigma, TCall - Tset)
End Function
```

Call on a Call

We will use bisection to find the critical price S^*. We can use $e^{q(T'-T)}(K+K')$ as an upper bound for S^* and 0 as a lower bound.[3] The following uses 10^{-6} as the error tolerance in the bisection.

```
Function Call_On_Call(S, Kc, Ku, r, sigma, q, Tc, Tu)
Dim tol, lower, upper, guess, flower, fupper, fguess, Sstar
Dim d1, d2, d1prime, d2prime, rho, N2, M1, M2
'
' Inputs are S = initial stock price
'            Kc = strike price of compound call
'            Ku = strike price of underlying call option
'            r = risk-free rate
'            sigma = volatility
'            q = dividend yield
'            Tc = time to maturity of compound call
'            Tu = time to maturity of underlying call >= Tc
'
' The first step is to find Sstar.
'
tol = 10 ^ -6
lower = 0
upper = exp(q * Tu) * (Kc + Ku)
guess = 0.5 * lower + 0.5 * upper
flower = -Kc
fupper = Black_Scholes_Call(upper, Ku, r, sigma, q, Tu - Tc) - Kc
fguess = Black_Scholes_Call(guess, Ku, r, sigma, q, Tu - Tc) - Kc
```

---

[3] We set the value of the call to be zero when the stock price is zero. The upper bound works because (by put-call parity and the fact that the put value is nonnegative) $C(T,S) \geq e^{-q(T'-T)}S - e^{-r(T'-T)}K'$. Therefore, when $S = e^{q(T'-T)}(K + K')$, we have $C(T,S) \geq K + K' - e^{-r(T'-T)}K' > K$.

```
    Do While upper - lower > tol
        If fupper * fguess < 0 Then
            lower = guess
            flower = fguess
            guess = 0.5 * lower + 0.5 * upper
            guess = Black_Scholes_Call(guess,Ku,r,sigma,q,Tu-Tc)-Kc
        Else
            upper = guess
            fupper = fguess
            guess = 0.5 * lower + 0.5 * upper
            fguess = Black_Scholes_Call(guess,Ku,r,sigma,q,Tu-Tc)-Kc
        End If
    Loop
    Sstar = guess
    '
    ' Now we calculate the probabilities.
    '
    d1 = (Log(S/Sstar) + (r-q+sigma^2/2)*Tc) / (sigma*Sqr(Tc))
    d2 = d1 - sigma * Sqr(Tc)
    d1prime = (Log(S/Ku) + (r-q+sigma^2/2) * Tu) / (sigma*Sqr(Tu))
    d2prime = d1prime - sigma * Sqr(Tu)
    rho = Sqr(Tc / Tu)
    N2 = Application.NormSDist(d2)
    M1 = BiNormalProb(d1, d1prime, rho)
    M2 = BiNormalProb(d2, d2prime, rho)
    '
    ' Now we calculate the option price.
    '
    Call_On_Call = -Exp(-r * Tc) * Kc * N2 + Exp(-q * Tu) * S * M1 _
                   -Exp(-r * Tu) * Ku * M2
End Function
```

Call on a Put

The implementation of the call-on-a-put formula is of course very similar to
that of a call-on-a-call. One difference is that there is no obvious upper bound
for S^*, so we start with $2K'$ (= 2*K2) and double this until the value of the
put is below K. We can take 0 again to be the lower bound. Recall that we
assume $K < e^{-r(T'-T)}K'$ and the right-hand side of this is the value of the
put at date T when $S(T) = 0$.

```
Function Call_On_Put(S, Kc, Ku, r, sigma, q, Tc, Tu)
Dim tol, lower, flower, upper, fupper, guess, fguess, Sstar
Dim d1, d2, d1prime, d2prime, rho, N2, M1, M2
'
' Inputs are S = initial stock price
'            Kc = strike price of compound call
'            Ku = strike price of underlying put option
```

```
'            r = risk-free rate
'            sigma = volatility
'            q = dividend yield
'            Tc = time to maturity of compound call
'            Tu = time to maturity of underlying put >= Tc
'
tol = 10 ^ -6
lower = 0
flower = Exp(-r * (Tu - Tc)) * Ku - Kc
upper = 2 * Ku
'
'  We double upper until the put value is below Kc
'
fupper = Black_Scholes_Put(upper,Ku,r,sigma,q,Tu-Tc)-Kc
Do While fupper > 0
    upper = 2 * upper
    fupper = Black_Scholes_Put(upper,Ku,r,sigma,q,Tu-Tc)-Kc
Loop
'
' Now we do the bisection to find Sstar
'
guess = 0.5 * lower + 0.5 * upper
fguess = Black_Scholes_Put(guess,Ku,r,sigma,q,Tu-Tc)-Kc
Do While upper - lower > tol
    If fupper * fguess < 0 Then
        lower = guess
        flower = fguess
        guess = 0.5 * lower + 0.5 * upper
        fguess = Black_Scholes_Put(guess,Ku,r,sigma,q,Tu-Tc)-Kc
    Else
        upper = guess
        fupper = fguess
        guess = 0.5 * lower + 0.5 * upper
        fguess = Black_Scholes_Put(guess,Ku,r,sigma,q,Tu-Tc)-Kc
    End If
Loop
Sstar = guess
'
' Now we calculate the probabilities.
'
d1 = (Log(S/Sstar) + (r-q+sigma^2/2)*Tc) / (sigma*Sqr(Tc))
d2 = d1 - sigma * Sqr(Tc)
d1prime = (Log(S/Ku) + (r-q+sigma^2/2)*Tu) / (sigma*Sqr(Tu))
d2prime = d1prime - sigma * Sqr(Tu)
rho = Sqr(Tc / Tu)
N2 = Application.NormSDist(-d2)
M1 = BiNormalProb(-d1, -d1prime, rho)
M2 = BiNormalProb(-d2, -d2prime, rho)
'
```

```
' Now we calculate the option price.
'
Call_On_Put = -Exp(-r * Tc) * Kc * N2 + Exp(-r * Tu) * Ku * M2 _
              - Exp(-q * Tu) * S * M1
End Function
```

American Calls with Discrete Dividends

To value an American call when there is one dividend payment before the option matures, we input the initial asset price $S(0)$ and then compute $Z(0) = X(0) - e^{-rT}D$. If $D - K \le -e^{-r(T'-T)}K$, we return the Black-Scholes value of a European call written on Z. Otherwise, we need to compute Z^* and our bisection algorithm requires an upper bound for Z^*, which would be any value of $Z(T)$ such that exercise at T is optimal. It is not obvious what this should be, so we start with K and keep doubling this until we obtain a value of $Z(T)$ at which exercise would be optimal. Then, we use the bisection algorithm to compute Z^* and finally compute the option value (8.15).

```
Function American_Call_Dividend(S, K, r, sigma, Div, TDiv, TCall)
'
' Inputs are S = initial stock price
'            K = strike price
'            r = risk-free rate
'            sigma = volatility
'            Div = cash dividend
'            TDiv = time until dividend payment
'            TCall = time until option matures >= TDiv
'
Dim LessDiv, upper, tol, lower, flower, fupper, guess, fguess
Dim LessDivStar, d1, d2, d1prime, d2prime, rho, N1, N2, M1, M2
LessDiv = S - Exp(-r * TDiv) * Div          ' called Z in text
If Div / K <= 1 - Exp(-r * (TCall - TDiv)) Then
    American_Call_Dividend = _
        Black_Scholes_Call(LessDiv, K, r, sigma, 0, TCall)
    Exit Function
End If
'
' Now we find an upper bound for the bisection.
'
upper = K
Do While upper + Div - K < _
              Black_Scholes_Call(upper,K,r,sigma,0,TCall-TDiv)
   upper = 2 * upper
Loop
'
' Now we use bisection to compute Zstar = LessDivStar.
'
tol = 10 ^ -6
lower = 0
```

```
flower = Div - K
fupper = upper + Div - K _
         - Black_Scholes_Call(upper,K,r,sigma,0,TCall-TDiv)
guess = 0.5 * lower + 0.5 * upper
fguess = guess + Div - K _
         - Black_Scholes_Call(guess,K,r,sigma,0,TCall-TDiv)
Do While upper - lower > tol
    If fupper * fguess < 0 Then
        lower = guess
        flower = fguess
        guess = 0.5 * lower + 0.5 * upper
        fguess = guess + Div - K _
             - Black_Scholes_Call(guess,K,r,sigma,0,TCall- Div)
    Else
        upper = guess
        fupper = fguess
        guess = 0.5 * lower + 0.5 * upper
        fguess = guess + Div - K _
             - Black_Scholes_Call(guess,K,r,sigma,0,TCall-TDiv)
    End If
Loop
LessDivStar = guess
'
' Now we calculate the probabilities and the option value.
'
d1 = (Log(LessDiv / LessDivStar) _
    + (r + sigma ^ 2 / 2) * TDiv) / (sigma * Sqr(TDiv))
d2 = d1 - sigma * Sqr(TDiv)
d1prime = (Log(LessDiv / K) _
        + (r + sigma ^ 2 / 2) * TCall) / (sigma * Sqr(TCall))
d2prime = d1prime - sigma * Sqr(TCall)
rho = -Sqr(TDiv / TCall)
N1 = Application.NormSDist(d1)
N2 = Application.NormSDist(d2)
M1 = BiNormalProb(-d1, d1prime, rho)
M2 = BiNormalProb(-d2, d2prime, rho)
American_Call_Dividend = LessDiv*N1 + Exp(-r*TDiv)*(Div-K)*N2 _
                       + LessDiv*M1 - Exp(-r*TCall)*K*M2
End Function
```

Choosers

To implement the bisection to compute S^*, we can take zero as a lower bound and $e^{qT_c}(K_c + K_p)$ as an upper bound.[4]

---

[4] We take the call value to be zero and the put value to be $e^{-r(T_p-T)}K_p$ at date T when the stock price is zero. To see why the upper bound works, note that when the stock price is S at date T, the call is worth at least $e^{-q(T_c-T)}S - K_c$

```
Function Chooser(S, Kc, Kp, r, sigma, q, T, Tc, Tp)
'
' Inputs are S = initial stock price
'            Kc = strike price of call option
'            Kp = strike price of put option
'            r = risk-free rate
'            sigma = volatility
'            Div = cash dividend
'            T = time until choice must be made
'            Tc = time until call matures >= T
'            Tp = time until put matures >= T
'
Dim tol, lower, upper, guess, flower, CallUpper, PutUpper
Dim fupper, CallGuess, Putguess, fguess, Sstar, d1, d2
Dim d1c, d2c, d1p, d2p, rhoc, rhop, M1c, M2c, M1p, M2p
'
' First we find Sstar by bisection.
'
tol = 10 ^ -6
lower = 0
upper = exp(q * Tc)*(Kc + Kp)
guess = 0.5 * Kc + 0.5 * Kp
flower = -Exp(-r * (Tp - T)) * Kp
fupper = Black_Scholes_Call(upper,Kc,r,sigma,q,Tc-T) _
        - Black_Scholes_Put(upper,Kp,r,sigma,q,Tp-T)
fguess = Black_Scholes_Call(guess,Kc,r,sigma,q,Tc-T) _
        - Black_Scholes_Put(guess,Kp,r,sigma,q,Tp-T)
Do While upper - lower > tol
    If fupper * fguess < 0 Then
        lower = guess
        flower = fguess
        guess = 0.5 * lower + 0.5 * upper
        fguess = Black_Scholes_Call(guess,Kc,r,sigma,q,Tc-T) _
                - Black_Scholes_Put(guess,Kp,r,sigma,q,Tp-T)
    Else
        upper = guess
        fupper = fguess
        guess = 0.5 * lower + 0.5 * upper
        fguess = Black_Scholes_Call(guess,Kc,r,sigma,q,Tc-T) _
                - Black_Scholes_Put(guess,Kp,r,sigma,q,Tp-T)
    End If
Loop
Sstar = guess
'
' Now we compute the probabilities and option value.
'
```

---

and the put is worth no more than K_p. Hence when $S = e^{qT_c}(K_c + K_p)$, we have $C - P \geq 0$.

```
d1 = (Log(S/Sstar) + (r-q+sigma^2/2)*T) / (sigma*Sqr(T))
d2 = d1 - sigma * Sqr(T)
d1c = (Log(S/Kc) + (r-q+sigma^2/2)*Tc) / (sigma*Sqr(Tc))
d2c = d1c - sigma * Sqr(Tc)
d1p = (Log(S/Kp) + (r-q+sigma^2/2)*Tp) / (sigma*Sqr(Tp))
d2p = d1p - sigma * Sqr(Tp)
rhoc = Sqr(T / Tc)
rhop = Sqr(T / Tp)
M1c = BiNormalProb(d1, d1c, rhoc)
M2c = BiNormalProb(d2, d2c, rhoc)
M1p = BiNormalProb(-d1, -d1p, rhop)
M2p = BiNormalProb(-d2, -d2p, rhop)
Chooser = Exp(-q*Tc)*S*M1c - Exp(-r*Tc)*Kc*M2c _
          + Exp(-r*Tp)*Kp*M2p - Exp(-q*Tp)*S*M1p
End Function
```

Call on the Max

```
Function Call_On_Max(S1, S2, K, r, sig1, sig2, rho, q1, q2, T)
'
' Inputs are S1 = price of stock 1
'            S2 = price of stock 2
'            K = strike price
'            r = risk-free rate
'            sig1 = volatility of stock 1
'            sig2 = volatility of stock 2
'            rho = correlation
'            q1 = dividend yield of stock 1
'            q2 = dividend yield of stock 2
'            T = time to maturity
'
Dim sigma, d1, d2, d11, d12, d21, d22, rho1, rho2, M1, M2, M3
sigma = Sqr(sig2 ^ 2 - 2 * rho * sig1 * sig2 + sig1 ^ 2)
d1 = (Log(S1/S2) + (q2-q1+sigma^2/2)*T) / (sigma*Sqr(T))
d2 = d1 - sigma * Sqr(T)
d11 = (Log(S1/K) + (r-q1+sig1^2/2)*T) / (sig1*Sqr(T))
d12 = d11 - sig1 * Sqr(T)
d21 = (Log(S2/K) + (r-q2+sig2^2/2)*T) / (sig2*Sqr(T))
d22 = d21 - sig2 * Sqr(T)
rho1 = (sig1 - rho * sig2) / sigma
rho2 = (sig2 - rho * sig1) / sigma
M1 = BiNormalProb(d11, d1, rho1)
M2 = BiNormalProb(d21, -d2, rho2)
M3 = BiNormalProb(-d12, -d22, rho)
Call_On_Max = Exp(-q1 * T) * S1 * M1 + Exp(-q2 * T) * S2 * M2 _
              + Exp(-r * T) * K * M3 - Exp(-r * T) * K
End Function
```

Down-and-Out Calls

```
Function Down_And_Out_Call(S, K, r, sigma, q, T, Barrier)
'
' Inputs are S = initial stock price
'            K = strike price
'            r = risk-free rate
'            sigma = volatility
'            q = dividend yield
'            T = time to maturity
'            Barrier = knock-out barrier < S
'
Dim a, b, d1, d2, d1prime, d2prime, N1, N2
Dim N1prime, N2prime, x, y, q1, q2
If K > Barrier Then
    a = S / K
    b = Barrier * Barrier / (K * S)
Else
    a = S / Barrier
    b = Barrier / S
End If
d1 = (Log(a) + (r-q+0.5*sigma^2)*T) / (sigma*Sqr(T))
d2 = d1 - sigma * Sqr(T)
d1prime = (Log(b) + (r-q+0.5*sigma^2)*T) / (sigma*Sqr(T))
d2prime = d1prime - sigma * Sqr(T)
N1 = Application.NormSDist(d1)
N2 = Application.NormSDist(d2)
N1prime = Application.NormSDist(d1prime)
N2prime = Application.NormSDist(d2prime)
x = 1 + 2 * (r - q) / (sigma ^ 2)
y = x - 2
q1 = N1 - (Barrier / S) ^ x * N1prime
q2 = N2 - (Barrier / S) ^ y * N2prime
Down_And_Out_Call = Exp(-q * T) * S * q1 - Exp(-r * T) * K * q2
End Function
```

Floating-Strike Lookbacks

```
Function Floating_Strike_Call(S, r, sigma, q, T, SMin)
'
' Inputs are S = initial stock price
'            r = risk-free rate
'            sigma = volatility
'            q = dividend yield
'            T = time to maturity
'            Smin = minimum during past life of contract
'
Dim d1, d2, d1prime, d2prime, N1, N2, N2prime, x, y
d1 = (Log(S/SMin) + (r-q+0.5*sigma^2)*T) / (sigma*Sqr(T))
```

```
d2 = d1 - sigma * Sqr(T)
d2prime = (Log(SMin/S) + (r-q-0.5*sigma^2)*T) / (sigma*Sqr(T))
N1 = Application.NormSDist(d1)
N2 = Application.NormSDist(d2)
N2prime = Application.NormSDist(d2prime)
x = 2 * (r - q) / (sigma ^ 2)
Floating_Strike_Call = Exp(-q*T)*S*N1 - Exp(-r*T)*SMin*N2 _
                     + (1/x)*(SMin/S)^x * Exp(-r*T)*SMin*N2prime _
                     - (1/x)*Exp(-q*T)*S*(1-N1)
End Function
```

Discretely-Sampled Geometric-Average-Price Calls

```
Function Discrete_Geom_Average_Price_Call(S,K,r,sigma,q,T,N)
'
' Inputs are S = initial stock price
'            K = stock price
'            r = risk-free rate
'            sigma = volatility
'            q = dividend yield
'            T = time to maturity
'            N = number of time periods
'
Dim dt, nu, a, V, sigavg
dt = T / N
nu = r - q - 0.5 * sigma ^ 2
a = N * (N + 1) * (2 * N + 1) / 6
V = Exp(-r*T)*S*Exp(((N+1)*nu/2 + sigma^2*a/(2*N^2))*dt)
sigavg = sigma * Sqr(a) / (N ^ 1.5)
Discrete_Geom_Average_Price_Call = _
              Black_Scholes_Call(V, K, r, sigavg, 0, T)
End Function
```

Problems

8.1. Intuitively, the value of a forward-start call option should be lower the closer is the date T at which the strike is set to the date T' at which the option matures, because then the option has less time to maturity after being "created" at T. Create an Excel worksheet to confirm this. Allow the user to input S, r, σ, q, and T'. Compute and plot the value of the option for $T = 0.1T'$, $T = 0.2T'$, ..., $T = 0.9T'$.

8.2. Create an Excel worksheet to demonstrate the additional leverage of a call-on-a-call relative to a standard call. Allow the user to input S, r, σ, q, and T'. Use the Black-Scholes_Call function to compute and output the value C of a European call with strike $K' = S$ (i.e., the call is at the money) and maturity T'. Use the `Call_on_Call` function to compute and output the value of a call option on the call with strike $K = C$ (i.e., the call-on-a-call is at the money) and maturity $T = 0.5T'$. Compute the percentage returns the standard European call and the call-on-a-call would experience if the stock price S instantaneously increased by 10%.

8.3. Create an Excel worksheet to illustrate the early exercise premium for an American call on a stock paying a discrete dividend. Allow the user to input S, r, σ, and T'. Take the date of the dividend payment to be $T = 0.5T'$ and take the strike price to be $K = S$. As discussed in Sect. 8.3, the value of a European call is given by the Black-Scholes formula with $S - e^{-rT}D$ being the initial asset price and $q = 0$ being the constant dividend yield. Use the function `American_Call_Dividend` to compute the value of an American call for dividends $D = .1S, \ldots D = .9S$. Subtract the value of the European call with the same dividend to obtain the early exercise premium. Plot the early exercise premium against the dividend D.

8.4. Create a VBA function to value a simple chooser (a chooser option in which $K_c = K_p$ and $T_c = T_p$) using put-call parity to compute S^* as mentioned in Sect. 8.4. Verify that the function gives the same result as the function `Chooser`.

8.5. Create an Excel worksheet to compare the cost of a simple chooser to that of a straddle (straddle = call + put with same strike and maturity). Allow the user to input S, r, σ, q, and T'. Take the time to maturity of the underlying call and put to be T' for both the chooser and the straddle. Take the strike prices to be $K = S$. Take the time the choice must be made for the chooser to be $T = 0.5T'$. Compute the cost of the chooser and the cost of the straddle.

8.6. A stock has fallen in price and you are attempting to persuade a client that it is now a good buy. The client believes it may fall further before bouncing back and hence is inclined to postpone a decision. To convince the client to buy now, you offer to deliver the stock to him at the end of two months at which time he will pay you the lowest price it trades during the two months plus a fee for your costs. The stock is not expected to pay a dividend during the next two months. Assuming the stock actually satisfies the Black-Scholes assumptions, find a formula for the minimum fee that you would require. (Hint: It is almost in Sect. 8.7.) Create an Excel worksheet allowing the user to input S, r, and σ and computing the minimum fee.

8.7. Suppose you must purchase 100 units of an asset at the end of a year. Create an Excel worksheet simulating the asset price and comparing the quality of the following hedges (assuming 100 contracts of each):

(a) a standard European call,

(b) a down-and-out call in which the knock-out barrier is 10% below the current price of the asset.

Take both options to be at the money at the beginning of the year. Allow the user to input S, r, σ and q. Generate 500 simulated end-of-year costs (net of the option values at maturity) for each hedging strategy and create histogram charts to visually compare the hedges. Note: to create histograms, you will need the Data Analysis add-in, which may be need to be loaded (click Tools/Add Ins).

8.8. Compute the prices of the options in the previous exercise. Modify the simulations to compare the end-of-year costs including the costs of the options, adding interest on the option prices to put everything on an end-of-year basis.

8.9. Modify Prob. 8.7 by including a third hedge: a combination of a down-and-out call as in part (b) of Prob. 8.7 and a down-and-in call with knockout barrier and strike 10% below the current price of the asset. Note that this combination forms a call option with strike that is reset when the underlying asset price hits a barrier.

8.10. Modify Prob. 8.8 by including the hedge in Prob. 8.9. Value the down-and-in call using the function `Down_And_Out_Call` and the fact that a down-and-out and down-and-in with the same strikes and barriers form a standard option.

8.11. Each week you purchase 100 units of an asset, and you want to hedge your total quarterly (13-week) cost. Create an Excel worksheet simulating the asset price and comparing the quality of the following hedges:

(a) a standard European call maturing at the end of the quarter ($T = 0.25$) on 1300 units of the asset,
(b) 13 call options maturing at the end of each week of the quarter, each written on 100 units of the asset, and
(c) a discretely sampled average-price call with maturity $T = 0.25$ written on 1300 units of the asset, where the sampling is at the end of each week.
(d) a discretely sampled geometric-average-price call with maturity $T = 0.25$ written on 1300 units of the asset, where the sampling is at the end of each week.

Allow the user to input S, r, σ and q. Assume all of the options are at the money at the beginning of the quarter ($K = S$). Compare the hedges as in Prob. 8.7.

8.12. In the setting of the previous problem, compute the prices of the options in parts (a), (b) and (d). Modify the simulations in the previous problem to compare the end-of-quarter costs including the costs of the options (adding interest on the option prices to put everything on an end-of-quarter basis).

8.13. Using the put-call parity relation, derive a formula for the value of a forward-start put.

8.14. Derive formula (8.10) for the value of a call on a put.

8.15. Complete the derivation of formula (8.17) for the value of a chooser option.

8.16. Derive a formula for the value of a put option on the maximum of two risky asset prices.

8.17. Using the result of the preceding exercise and Margrabe's formula, verify that calls and puts (having the same strike K and maturity T) on the maximum of two risky asset prices satisfy the following put-call parity relation:

$$e^{-rT}K + \text{Value of call on max}$$
$$= e^{-q_2 T}S_2(0) + \text{Value of option to exchange asset 2 for asset 1}$$
$$+ \text{Value of put on max} .$$

9

More on Monte Carlo and Binomial Valuation

This chapter is a continuation of Chap. 5, introducing somewhat more advanced issues and applications of Monte Carlo and binomial models. We will consider some of the exotic options introduced in the previous chapter for which closed-form solutions do not exist: basket options, spread options, discretely-sampled lookback options, and Asian options. The next chapter introduces finite difference methods, which are similar to binomial models in their applications (useful for American options, not so useful for path-dependent options) but generally faster.

9.1 Monte Carlo Models for Path-Dependent Options

A derivative is said to be "path dependent" if its value depends on the path of the underlying asset price rather than just on the price at the time of exercise. Examples of path-dependent options are lookbacks, barrier options, and Asians. To value a path-dependent option by Monte Carlo, we need to simulate an approximate path of the stock price. We do this by considering time periods of length $\Delta t = T/N$ for some integer N. Under the risk-neutral measure, the logarithm of the stock price changes over such a time period by

$$\Delta \log S = \nu \, \Delta t + \sigma \sqrt{\Delta t} \, z \, , \tag{9.1}$$

where $\nu = r - q - \sigma^2/2$ and z is a standard normal. Given that there are N time periods of length Δt, we need to generate N standard normals to generate a stock price path. If we generate M paths to obtain a sample of M option values, then we will need to generate MN standard normals.

Consider for example a floating-strike lookback call. The formula for this option given in Sect. 8.7 assumes the minimum stock price is computed over the entire path of the stock price, i.e., with continuous sampling of the stock price. In practice, the minimum will be computed by recording the price at a discrete number of dates. We can value the discretely sampled lookback using

Monte-Carlo by choosing Δt to be the interval of time (e.g., a day or week) at which the price is recorded. For example, if the contract calls for weekly observation, we will attain maximum precision by setting N to be the number of weeks before the option matures.

Asian and barrier options are also subject to discrete rather than continuous sampling and can be valued by Monte-Carlo in the same way as lookbacks. We will discuss Asian options in Sect. 9.5.

9.2 Binomial Valuation of Basket and Spread Options

By combining binomial models, we can value options or other derivatives on multiple assets. We will illustrate for an option on two assets. This is the most important case, and the extension to more than two assets is straightforward.

Consider two stocks with constant dividend yields q_i and constant volatilities σ_i. Suppose the two Brownian motions driving the two stocks have a constant correlation coefficient ρ. We will denote the price of stock i $(i = 1, 2)$ in the up state in each period by $u_i S_i$ and the price in the down state by $d_i S_i$, where S_i is the price at the beginning of the period, and u_i and d_i are parameters to be specified. In each period, there are four possible combinations of returns on the two stocks: up for both stocks, up for stock 1 and down for stock 2, down for stock 1 and up for stock 2, and down for both stocks. Denote the probabilities of these four combinations by p_{uu}, p_{ud}, p_{du}, and p_{dd} respectively. Thus, there are eight parameters in the binomial model: the number N of periods (which defines the length of each period as $\Delta t = T/N$ where T is the option maturity), the up and down parameters u_i and d_i for each stock, and three probabilities (the fourth probability being determined by the condition that the probabilities sum to one).

Given the period length Δt, we want to choose the up and down parameters and the probabilities to match (or approximately match in an appropriate sense) the means, variances and covariances of the returns $\Delta S_i / S_i$ or the continuously-compounded returns $\Delta \log S_i$. There are two means, two variances and one covariance, so there are five restrictions to be satisfied and seven parameters. As in Chap. 5, it is convenient to take $d_i = 1/u_i$, leaving five restrictions and five free parameters.

As discussed in Sect. 5.4, there are multiple ways to define the binomial model so that it converges to the continuous-time model as the number of periods is increased. As an example, we will describe here the suggestion of Trigeorgis [61], which matches the means, variances and covariance of the continuously-compounded returns. Letting p_i denote the probability of the up state for stock i, matching the means and variances implies, as in Sect. 5.4,

$$\log u_i = \sqrt{\sigma_i^2 \Delta t + \nu_i^2 (\Delta t)^2} \,,$$

$$p_i = \frac{1}{2} + \frac{\nu_i \Delta t}{2 \log u_i} \,.$$

where $\nu_i = r - q_i - \sigma_i^2/2$. In terms of the notation p_{uu}, p_{ud}, p_{du}, and p_{dd}, the probability of the up state for stock 1 is $p_1 = p_{uu} + p_{ud}$ and the probability of the up state for stock 2 is $p_2 = p_{uu} + p_{du}$. Therefore,

$$p_{uu} + p_{ud} = \frac{1}{2} + \frac{\nu_1 \Delta t}{2 \log u_1} , \tag{9.2a}$$

$$p_{uu} + p_{du} = \frac{1}{2} + \frac{\nu_2 \Delta t}{2 \log u_2} . \tag{9.2b}$$

In the continuous time model, over a discrete time period Δt, the covariance of $\Delta \log S_1$ and $\Delta \log S_2$ is $\rho \sigma_1 \sigma_2 \Delta t$. In the binomial model, with $d_i = 1/u_i$, we have

$$E\left[\Delta \log S_1 \times \Delta \log S_2\right] = (p_{uu} - p_{ud} - p_{du} + p_{dd}) \log u_1 \log u_2 .$$

Given that $E[\Delta \log S_i] = \nu_i \Delta t$, this implies a covariance of

$$(p_{uu} - p_{ud} - p_{du} + p_{dd}) \log u_1 \log u_2 - \nu_1 \nu_2 (\Delta t)^2 .$$

Matching the covariance in the binomial model to the covariance in the continuous-time model therefore implies

$$p_{uu} - p_{ud} - p_{du} + p_{dd} = \frac{\rho \sigma_1 \sigma_2 \Delta t + \nu_1 \nu_2 (\Delta t)^2}{\log u_1 \log u_2} . \tag{9.2c}$$

We can solve the system (9.2), together with the condition that the probabilities sum to one, to obtain the probabilities p_{uu}, p_{ud}, p_{du}, and p_{dd}. This solution and a VBA function for valuing an American spread call option are given in Sect. 9.7. This function operates much like the binomial valuation of American options described in Chap. 5. The primary difference is that the value of the option at maturity depends on both stock prices, so we have to consider each possible combination of stock prices. In an N–period model, there are $N+1$ nodes at the final date for each of the two stocks, and hence $(N+1)^2$ possible combinations of nodes. In fact, at each date n $(n = 0, \ldots, N)$ there are $(n+1)^2$ combinations of nodes to be considered. The computation time required for a spread call option is therefore roughly the square of the time required for a standard call.

Likewise, in an N–period model for a basket option written on three assets, there are $(n+1)^3$ combinations of nodes to be considered at date n; if there are five assets, there are $(n+1)^5$ combinations, etc. Thus, the computation time required increases exponentially with the number of assets. This can be a serious problem. For example, with five assets and $N = 99$, we would have 100^5 (10 billion) combinations. As this suggests, problems with multiple assets quickly become intractable in a binomial framework. This is called the "curse of dimensionality."

9.3 Monte Carlo Valuation of Basket and Spread Options

In this section, we will consider the valuation of European spread and basket options by the Monte Carlo method. As noted in Sect. 8.8, there are no simple formulas for these options. In each simulation, we will generate a terminal price for each of the underlying assets and compute the value of the option at its maturity. Discounting the average terminal value gives the estimate of the option value as usual.

The difference between binomial and Monte Carlo methods for options written on multiple assets can be understood as follows. Both methods attempt to estimate the discounted expected value of the option (under the risk-neutral measure). In an N–period model, the binomial model produces $N + 1$ values for the terminal price of each underlying asset. Letting k denote the number of underlying assets, this produces $(N + 1)^k$ combinations of asset prices. Of course, each combination has an associated probability. In contrast, the Monte Carlo method produces M combinations of terminal prices, where M is the number of simulations. Each combination is given the same weight $(1/M)$ when estimating the expected value.

With a single underlying asset, the binomial model is more efficient, as discussed in Sect. 5.2, because the specifically chosen terminal prices in the binomial model sample the set of possible terminal prices more efficiently than randomly generated terminal prices. However, this advantage disappears, and the ranking of the methods can be reversed, when there are several underlying assets. The reason is that many of the $(N + 1)^k$ combinations of prices in the binomial model will have very low probabilities. For example, with two assets that are positively correlated, it is very unlikely that one asset will be at its highest value in the binomial model and the other asset simultaneously at its lowest. It is computationally wasteful to evaluate the option for such a combination, because the probability-weighted value will be very small and hence contribute little to the estimate of the expected value. On the other hand, each set of terminal prices generated by the Monte Carlo method will be generated from a distribution having the assumed correlation. Thus, only relatively likely combinations will typically be generated, and time is not wasted on evaluating unlikely combinations. However, it should not be concluded that Monte Carlo valuation of a derivative on multiple assets will be quick and easy—even though the computation time required for more underlying assets does not increase as much with Monte Carlo as for binomial models, it can nevertheless be substantial.

To implement Monte Carlo valuation of options on multiple assets, we must first explain how to simulate correlated asset prices. As observed in Sect. 4.5, we can simulate the changes in two Brownian motions B_1 and B_2 that have correlation ρ by generating two independent standard normals Z_1 and Z_2 and defining

$$\Delta B_1 - \sqrt{\Delta t}\, Z_1\, , \qquad \text{and} \qquad \Delta B_2 = \sqrt{\Delta t}\, Z\, ,$$

where Z is defined as

$$Z = \rho Z_1 + \sqrt{1 - \rho^2}\, Z_2\, .$$

The random variable Z is also a standard normal, and the correlation between Z_1 and Z is ρ. Thus, we can simulate the changes in the logarithms of two correlated asset prices as

$$\Delta \log S_1 = \nu_1 \Delta t + \sigma_1 \sqrt{\Delta t} Z_1\, ,$$

$$\Delta \log S_2 = \nu_2 \Delta t + \sigma_2 \rho \sqrt{\Delta t} Z_1 + \sigma_2 \sqrt{1 - \rho^2} \sqrt{\Delta t} Z_2\, ,$$

where $\nu_i = r - q_1 - \sigma_i^2/2$ and the Z_i are independent standard normals.

To generalize this idea to more than two assets, we introduce some additional notation. The simulation for the case of two assets can be written as

$$\Delta \log S_1 = \nu_1 \Delta t + a_{11} \sqrt{\Delta t} Z_1 + a_{12} \sqrt{\Delta t} Z_2\, , \qquad (9.3\text{a})$$

$$\Delta \log S_2 = \nu_2 \Delta t + a_{21} \sqrt{\Delta t} Z_1 + a_{22} \sqrt{\Delta t} Z_2\, , \qquad (9.3\text{b})$$

where

$$a_{11} = \sigma_1\, , \quad a_{12} = 0\, ,$$
$$a_{21} = \sigma_2 \rho\, , \quad a_{22} = \sigma_2 \sqrt{1 - \rho^2}\, .$$

These are not the only possible choices for the constants a_{ij}. Given that Z_1 and Z_2 are independent standard normals, the conditions the a_{ij} must satisfy in order to match the variances $\sigma_i^2 \Delta t$ and correlation ρ of the changes in the logarithms are

$$a_{11}^2 + a_{12}^2 = \sigma_1^2\, , \qquad (9.4\text{a})$$

$$a_{21}^2 + a_{22}^2 = \sigma_2^2\, , \qquad (9.4\text{b})$$

$$a_{11} a_{21} + a_{12} a_{22} = \sigma_1 \sigma_2 \rho\, . \qquad (9.4\text{c})$$

These three equations in the four coefficients a_{ij} leave one degree of freedom. We choose to take $a_{12} = 0$ and then solve for the other three.

In matrix notation, the system (9.4) plus the condition $a_{12} = 0$ can be written as the equation

$$\begin{pmatrix} a_{11} & 0 \\ a_{21} & a_{22} \end{pmatrix} \begin{pmatrix} a_{11} & 0 \\ a_{21} & a_{22} \end{pmatrix}^{\top} = \begin{pmatrix} \sigma_1^2 & \rho \sigma_1 \sigma_2 \\ \rho \sigma_1 \sigma_2 & \sigma_2^2 \end{pmatrix}\, ,$$

where $^{\top}$ denotes the matrix transpose. The matrix on the right hand side is the covariance matrix of the continuously-compounded annual returns (changes in log asset prices). Choosing the a_{ij} so that the "lower triangular" matrix

$$A \equiv \begin{pmatrix} a_{11} & 0 \\ a_{21} & a_{22} \end{pmatrix}$$

satisfies

$$AA^\top = \text{covariance matrix}$$

is called the the "Cholesky decomposition" of the covariance matrix. Given any number L of assets, provided none of the assets is redundant (perfectly correlated with a portfolio of the others), the Cholesky decomposition of the $L \times L$ covariance matrix always exists. An algorithm for computing the Cholesky decomposition is given in Sect. 9.7.

We can use the Cholesky decomposition to perform Monte-Carlo valuation of a basket or spread option.[1] If there were some path dependency in the option value, we would simulate the paths of the asset prices as in (9.3). However a standard basket option is not path dependent, so we only need to simulate the asset prices at the option maturity date T, as in Sect. 5.1. The value of a basket call option at its maturity T is

$$\max\left(0, \sum_{i=1}^{L} w_i S_i(T) - K\right),$$

where L is the number of assets in the basket (portfolio) and w_i is the weight of the i–th asset in the basket. The logarithm of the i–th asset price at maturity is simulated as

$$\log S_i(T) = \log S_i(0) + \nu_i T + \sqrt{T} \sum_{j=1}^{L} a_{ij} Z_j,$$

where the Z_j are independent standard normals. Given the simulated values of the $\log S_i(T)$, the value at maturity of the basket option is readily computed. The estimate of the date–0 value is then computed as the discounted average of the simulated values at maturity.

9.4 Antithetic Variates in Monte Carlo

In this and the following section, we will discuss two methods to increase the efficiency of the Monte Carlo method. These are two of the simplest methods. They are used extensively, but there are other important methods that are also widely used. Jäckel [39] and Glasserman [29] provide a wealth of information on this topic.

The Monte Carlo method estimates the mean μ of a random variable x as the sample average of randomly generated values of x. An antithetic variate is a random variable y with the same mean as x and a negative correlation with x. It follows that the random variable $z = (x + y)/2$ will have the same mean as x and a lower variance. Therefore the sample mean of M simulations of z will be an unbiased estimate of μ and will have a lower standard error

---

[1] For a spread option, take $L = 2$, $w_1 = 1$ and $w_2 = -1$.

than the sample mean of M simulations of x. Thus, we should obtain a more efficient estimator of μ by simulating z instead of x.[2]

In the context of derivative valuation, the standard application of this idea is to generate two negatively correlated underlying asset prices (or price paths, if the derivative is path dependent). The terminal value of the derivative written on the first asset serves as x and the terminal value of the derivative written on the second serves as y. Because both asset prices have the same distribution, the means of x and y will be the same, and the discounted mean is the date–0 value of the derivative.

Consider for example a non-path-dependent option in a world with constant volatility. In each simulation i $(i = 1, \ldots, M)$, we would generate a standard normal Z_i and compute

$$\log S_i(T) = \log S(0) + \left(r - q - \frac{1}{2}\sigma^2\right)T + \sigma\sqrt{T}Z_i\,,$$

$$\log S_i'(T) = \log S(0) + \left(r - q - \frac{1}{2}\sigma^2\right)T - \sigma\sqrt{T}Z_i\,.$$

Given the first terminal price, the value of the derivative will be some number x_i and given the second it will be some number y_i. The date–0 value of the derivative is estimated as

$$\mathrm{e}^{-rT}\frac{1}{M}\sum_{i=1}^{M}\frac{x_i + y_i}{2}\,.$$

We will illustrate this method for the floating-strike lookback call in Sect. 9.7.

9.5 Control Variates in Monte Carlo

Another approach to increasing the efficiency of the Monte Carlo method is to adjust the estimated mean (option value) based on the known mean of another related variable. We can explain this in terms of linear regression in statistics. Suppose we have a random sample $\{x_1, \ldots, x_M\}$ of a variable x with unknown mean μ, and suppose we have a corresponding sample $\{y_1, \ldots, y_M\}$ of another variable y with known mean ϕ. Then an efficient estimate of μ is $\hat{\mu} = \bar{x} + \hat{\beta}(\phi - \bar{y})$, where $\bar{x}$ and $\bar{y}$ denote the sample means of x and y, and where $\hat{\beta}$ is the coefficient of y in the linear regression of x on y (i.e., the

---

[2] The negative correlation between x and y is essential for this method to generate a real gain in efficiency. To generate M simulations of z, one must generate M simulations of x and M of y, which will generally require about as much computation time as generating $2M$ simulations of x. If x and y were independent, the standard error from M simulations of z would be the same as the standard error from $2M$ simulations of x, so using the antithetic variate would be no better than just doubling the sample size for x.

estimate of β in the linear model $x = \alpha + \beta y + \varepsilon$). The standard Monte Carlo method, which we have described thus far, simply estimates the mean of x as $\bar{x}$. The "control variate" method adjusts the estimate by adding $\hat{\beta}(\phi - \bar{y})$. To understand this correction, assume for example that the true β is positive. If the random sample is such that $\bar{y} < \phi$, then it must be that small values of y were over-represented in the sample. Since x and y tend to move up and down together (this is the meaning of a positive β) it is likely that small values of x were also over-represented in the sample. Therefore, one should adjust the sample mean of x upwards in order to estimate μ. The best adjustment will take into account the extent to which small values of y were over-represented (i.e., the difference between $\bar{y}$ and ϕ) and the strength of the relation between x and y (which the estimate $\hat{\beta}$ represents). The efficient correction of this sort is also the simplest: just add $\hat{\beta}(\phi - \bar{y})$ to $\bar{x}$. In practice, the estimation of $\hat{\beta}$ may be omitted and one may simply take $\hat{\beta} = 1$, if the relationship between x and y can be assumed to be "one-for-one." If β is to be estimated, the estimate (by ordinary least squares) is

$$\hat{\beta} = \frac{\sum_{i=1}^{M} x_i y_i - M\bar{x}\bar{y}}{\sum_{i=1}^{M} y_i^2 - M\bar{y}^2} .$$

In general, the correction term $\hat{\beta}(\phi - \bar{y})$ will have a nonzero mean, which introduces a bias in the estimate of μ. To eliminate the bias, one can compute $\hat{\beta}$ from a "pre-sample" of $\{x, y\}$ values.

As an example, consider the classic case of estimating the value of a discretely-sampled average-price call, using a discretely-sampled geometric-average-price call as a control variate. Let τ denote the amount of time that has elapsed since the call was issued and T the amount of time remaining before maturity, so the total maturity of the call is $T + \tau$. To simplify somewhat, assume date 0 is the beginning of a period between observations. Let $t_1, \ldots, t_N$ denote the remaining sampling dates, with $t_1 = \Delta t$, $t_i - t_{i-1} = \Delta t = T/N$ for each i, and $t_N = T$. We will input the average price $A(0)$ computed up to date 0, assuming this average includes the price $S(0)$ at date 0. The average price at date T will be

$$A(T) = \frac{\tau}{T+\tau} A(0) + \frac{T}{T+\tau} \left(\frac{\sum_{i=1}^{N} S(t_i)}{N} \right) .$$

The average-price call pays $\max(0, A(T) - K)$ at its maturity T, and we can write this as

$$\max(A(T) - K, 0) = \max \left(\frac{T}{T+\tau} \left(\frac{\sum_{i=1}^{N} S(t_i)}{N} \right) - \left(K - \frac{\tau}{T+\tau} A(0) \right), 0 \right)$$

$$= \frac{T}{T+\tau} \max \left(\frac{\sum_{i=1}^{N} S(t_i)}{N} - K^*, 0 \right) ,$$

where

$$K^* = \frac{T+\tau}{T} K - \frac{\tau}{T} A(0) .$$

Therefore, the value at date 0 of the discretely-sampled average-price call is

$$\frac{T}{T+\tau} e^{-rT} E^R \left[\max \left(\frac{\sum_{i=1}^N S(t_i)}{N} - K^*, 0 \right) \right] .$$

In terms of the discussion above, the random variable the mean of which we want to estimate is

$$x = e^{-rT} \max \left(\frac{\sum_{i=1}^N S(t_i)}{N} - K^*, 0 \right) .$$

A random variable y that will be closely correlated to x is

$$y = e^{-rT} \max \left(e^{\sum_{i=1}^N \log S(t_i)/N} - K^*, 0 \right) .$$

The mean ϕ of y under the risk-neutral measure is given in the pricing formula (8.37). We can use the sample mean of y and its known mean ϕ to adjust the sample mean of x as an estimator of the value of the average-price call. Generally, the estimated adjustment coefficient $\hat{\beta}$ will be quite close to 1.

9.6 Accelerating Binomial Convergence

Broadie and Detemple [13] show that a modified binomial model is a quite efficient way to value American put options. They modify the binomial model as follows: (i) the Black-Scholes formula is used to value the option at the penultimate date, and (ii) Richardson extrapolation is used to estimate what the option value would be with an infinite number of periods.

  If an option is not exercised at date $N-1$ in an N–period binomial model (i.e., one date from the end), then, because in the binomial model there are no further opportunities for early exercise, the American option at date $N-1$ is equivalent to a European option at that date. The value of a European option is given by the Black-Scholes formula. Therefore, the estimate of the option value can be improved by replacing

```
PutV(j) = max(S-K, dpd*PutV(j)+dpu*PutV(j+1))
```
with

```
PutV(j) = max(S-K,Black_Scholes_Put(S,K,r,sigma,q,dt))
```

at date $N-1$ (of course this also means that we do not need to compute the intrinsic value at date N). This idea can be effectively used in binomial valuation of any option for which there is a closed-form solution (like the

Black-Scholes formula) for the value of the corresponding European option in a continuous-time model.

Broadie and Detemple combine the use of the Black-Scholes formula at date $N - 1$ with Richardson extrapolation. Richardson extrapolation is a method that may improve the efficiency of any algorithm by "extrapolating to the limit." In the case of a binomial model, the idea is to extrapolate the values calculated for different numbers of periods (different N's) to try to estimate the value for $N = \infty$.

It is easier to work with convergence to zero than convergence to infinity, so define $x = 1/N$. For any value of N, the binomial model will return a value, which is an estimate of the option value and which we denote as $y = f(x)$. We would like to know the value at $N = \infty$, which in this notation is $f(0)$. Of course, we cannot calculate $f(0)$, because we do not know the function f, but we can approximate f by a known function g and then estimate $f(0)$ by $g(0)$.

A linear approximation is the simplest and is shown by Broadie and Detemple to be quite effective. For a linear approximation, we would take

$$g(x) = a + bx$$

for parameters a and b to be determined. We can input values N_1 and $N_2 = 2N_1$ for the number of periods, run the binomial model for each, set $x_i = 1/N_i$, and define $y_i = f(x_i)$ to be the value of the option returned by the binomial model when the number of periods is N_i. Then we force $g(x_i) = f(x_i)$ for $i = 1, 2$ by solving the equations

$$y_i = a + bx_i$$

for a and b. Of course, $g(0) = a$, so we will return the constant a as our estimate of $f(0)$. This is simpler than it may appear—we put

$$y_1 = a + bx_1 = a + 2bx_2\,,$$
$$y_2 = \qquad\quad a + bx_2\,,$$

and subtracting gives us $y_1 - y_2 = bx_2$, which implies from the bottom equation that $a = 2y_2 - y_1$. We can think of N_2 as being the number of periods we want to use in the binomial model, in which case y_2 would be our estimate of the option value. Richardson extrapolation here means also running the binomial model for half as many periods ($N_1 = N_2/2$) and adding the difference of the estimates $y_2 - y_1$ to the estimate y_2.

Richardson extrapolation can be viewed as cancelling the first-order term in the Taylor series expansion of f. We have

$$y_1 = f(x_1) = f(0) + f'(0)x_1 + \text{higher order terms}$$
$$= f(0) + 2f'(0)x_2 + \text{higher order terms}\,,$$
$$y_2 = f(x_2) = f(0) + f'(0)x_2 + \text{higher order terms}\,.$$

This implies

$$2y_2 - y_1 = f(0) + \text{difference of higher order terms} .$$

Having eliminated the first-order term, one can hope to obtain a closer approximation to $f(0)$.

9.7 Calculations in VBA

Monte Carlo Valuation of Path-Dependent Options

We will illustrate the valuation of path-dependent options by Monte Carlo by valuing a discretely-sampled floating-strike lookback call. The inputs of the following are the same as for the lookback-call pricing function in Sect. 8.7, plus the number of time periods N and the number of simulations M. The logarithm of the initial stock price S and the logarithm of the historical minimum Smin are calculated and stored at the beginning of the program. In each simulation $i = 1, \ldots, M$, the variables logS and logSmin are initialized to equal the stored values. They are then updated at each time period $j = 1, \ldots N$ in accord with (9.1), with LogSmin being changed only when a new minimum is reached.

```
Function Floating_Strike_Call_MC_SE(S, r, sigma, q, SMin, T, N, M)
'
' Inputs are S = initial stock price
'            r = risk-free rate
'            sigma = volatility
'            q = dividend yield
'            T = remaining time to maturity
'            Smin = minimum during previous life of contract
'            N = number of time periods
'            M = number of simulations
'
' This returns the row vector (call value, standard error)
'
Dim dt, LogS0, LogSmin0, LogS, LogSMin, sigsdt, nudt
Dim CallV, SumCall, SumCallSq, i, j, StdError
dt = T / N
nudt = (r - q - 0.5 * sigma * sigma) * dt
sigsdt = sigma * Sqr(dt)
LogS0 = Log(S)              ' store log of initial stock price
LogSmin0 = Log(SMin)       ' store log of historical minimum
SumCall = 0                ' initialize sum of call values
SumCallSq = 0              ' initialize sum of squared values
For i = 1 To M
    LogS = LogS0           ' initialize log of stock price
    LogSMin = LogSmin0     ' initialize log of minimum
```

```
    For j = 1 To N
        LogS = LogS + nudt + sigsdt * RandN()    ' update log price
        LogSMin = Application.Min(LogS, LogSMin) ' update log min
    Next j
    CallV = Exp(LogS) - Exp(LogSMin)             ' call at maturity
    SumCall = SumCall + CallV                    ' sum of call values
    SumCallSq = SumCallSq + CallV * CallV        ' sum of squares
Next i
CallV = Exp(-r * T) * SumCall / M
StdError = Exp(-r*T) * Sqr((SumCallSq-SumCall*SumCall/M)/(M*(M-1)))
Floating_Strike_Call_MC_SE = Array(CallV, StdError)
End Function
```

Binomial Valuation of American Spread Options

To illustrate binomial valuation of basket and spread options, we will consider an American put written on a spread $S_1 - S_2$. The value of the option at maturity is

$$\max\left(0, K - [S_1(T) - S_2(T)]\right) = \max\left(0, K - S_1(T) + S_2(T)\right).$$

We input the stock parameters as vectors $S = \{S(1), S(2)\}$, $q = \{q(1), q(2)\}$, $sigma = \{\sigma(1), \sigma(2)\}$. This can be done by inputting a cell range (e.g., A2:B2) for each variable or by inputting the values enclosed in curly braces (e.g., $\{50, 40\}$). The following declares certain arrays and defines the binomial parameters.

```
Function American_Spread_Put_Binomial(S, K, r, sigma, rho, q, T, N)
'
' Inputs are S = 2-vector of initial stock prices
'            K = strike price
'            r = risk-free rate
'            sigma = 2-vector of volatilities
'            rho = correlation
'            q = 2-vector of dividend yields
'            T = time to maturity
'            N = number of periods in binomial model
'
Dim dt, num, constant, pud, pdu, puu, pdd, disc
Dim dpud, dpdu, dpuu, dpdd, IntrinsicV, DiscV, x, h, i, j
Dim p(1 To 2), u(1 To 2), d(1 To 2), logu(1 To 2)
Dim nu(1 To 2), u2(1 To 2)
Dim Stock() As Double
Dim PutV() As Double
ReDim Stock(2, N)
ReDim PutV(N, N)
dt = T / N                              ' length of time period
For x = 1 To 2                          ' parameters for each stock
    nu(x) = r - q(x) - sigma(x) * sigma(x) / 2
```

```
    logu(x) = Sqr(sigma(x)*sigma(x)*dt + nu(x)*nu(x)*dt*dt)
    u(x) = Exp(logu(x))
    d(x) = 1 / u(x)
    p(x) = 0.5 * (1 + nu(x) * dt / logu(x))
    u2(x) = u(x) * u(x)
Next x
num = rho * sigma(1) * sigma(2) * dt + nu(1) * nu(2) * dt * dt
constant = num / (logu(1) * logu(2))
pud = (p(1) - p(2)) / 2 + (1 - constant) / 4  ' prob 1 up, 2 down
pdu = pud - p(1) + p(2)                       ' prob 1 down, 2 up
puu = p(1) - pud                              ' prob both up
pdd = 1 - puu - pud - pdu                     ' prob both down
disc = Exp(-r * dt)                           ' one-period discount
dpuu = disc * puu                             ' prob x discount factor
dpud = disc * pud                             ' prob x discount factor
dpdu = disc * pdu                             ' prob x discount factor
dpdd = disc * pdd                             ' prob x discount factor
```

In each of the two binomial models, there are $N+1$ nodes at the last date. This implies $(N+1)^2$ possible combinations. We define the put values at the last date by looping over the $N+1$ possibilities for the second stock price, for each possible value for the first stock price. Rather than recalculating the $N+1$ values for the second stock price each time we loop over them, it is more efficient to store them in a vector. For the sake of simplicity, we will do the same for the first stock. We store all the prices in a $2 \times (N+1)$ array denoted Stock. We store the put values in an $(N+1) \times (N+1)$ array. PutV(0,0) is the put value when both stocks are at their bottom nodes, PutV(0,1) is the value when the first stock is at its bottom node and the second stock is one node up from the bottom, etc.

```
For x = 1 To 2                          ' for matrix at last date
    Stock(x, 0) = S(x) * d(x) ^ N       ' stock x at bottom node
    For j = 1 To N                      ' step up in tree x
        Stock(x, j) = Stock(x, j-1)*u2(x) ' stock x at node j
    Next j
Next x
For j = 0 To N                          ' loop over nodes in tree 1
    For h = 0 To N                      ' loop over nodes in tree 2
        PutV(j, h) = Application.Max(K-Stock(1,j)+Stock(2,h),0)
    Next h
Next j
```

Now we back up N times to get to the beginning of the tree, checking the early exercise condition at each date, and return the put value as PutV(0,0).

```
For i = N - 1 To 0 Step -1              ' back up in time to date 0
    For x = 1 To 2
        Stock(x, 0) = S(x) * d(x) ^ i   ' stock x at bottom node
        For j = 1 To i                  ' step up in tree x
            Stock(x, j) = Stock(x, j-1)*u2(x) ' stock x at node j
        Next j
```

```
    Next x
    For j = 0 To i                  ' loop over nodes in tree 1
        For h = 0 To i              ' loop over nodes in tree 2
            IntrinsicV = K - Stock(1, j) + Stock(2, h)
            DiscV = dpdd * PutV(j, h) + dpdu * PutV(j,h+1) _
                    + dpud * PutV(j+1, h) + dpuu * PutV(j+1, h+1)
            PutV(j, h) = Application.Max(IntrinsicV, DiscV)
        Next h
    Next j
Next i
American_Spread_Put_Binomial = PutV(0, 0)
End Function
```

Cholesky Decomposition

An algorithm for computing the decomposition is as follows. We assume the $L \times L$ matrix has been input as cov, with the interpretation that cov(i,j) is $\sigma_i \sigma_j \rho_{ij}$ and cov(i,i) is σ_i^2 for $i, j = 1, \ldots, L$. We start the indices i, j at 1 because this is the convention when inputting arrays from an Excel worksheet. The output matrix is also indexed $i, j = 1, \ldots, L$.

```
Function Cholesky(L, cov)
'
' Inputs are L = number of assets
'           Cov = L x L matrix of covariances
'
Dim SumSq, SumPr, h, i, j
Dim a()
ReDim a(1 To L, 1 To L)
For i = 1 To L
    SumSq = 0
    For h = 1 To i - 1
        SumSq = SumSq + a(i, h) * a(i, h)
    Next h
    a(i, i) = Sqr(cov(i, i) - SumSq)
    For j = i To L
        SumPr = 0
        For h = 1 To i - 1
            SumPr = SumPr + a(i, h) * a(j, h)
        Next h
        a(j, i) = (cov(i, j) - SumPr) / a(i, i)
    Next j
Next i
Cholesky = a
End Function
```

Monte Carlo Valuation of European Basket Options

We input S as the vector of initial asset prices, q as the vector of dividend yields, cov as the covariance matrix, w as the vector of weights defining the basket option, and L as the number of assets in the basket. The following function can also be used to value a spread option, by inputting $L = 2$ and w $= [1, -1]$.

```
Function European_Basket_Call_MC(S, K, r, cov, q, w, T, L, M)
'
' Inputs are S = L-vector of initial stock prices
'            K = strike price
'            r = risk-free rate
'            Cov = L x L matrix of covariances
'            q = L-vector of dividend yields
'            w = L-vector of basket weights
'            T = time to maturity
'            L = number of assets in the basket
'            M = number of simulations
'
Dim BasketValue, CallV, SumCall, h, i, j
Dim a(), Mean(), z(), Multiplier(), LogS()
ReDim Mean(1 To L)
ReDim z(1 To L)
ReDim Multiplier(1 To L, 1 To L)
ReDim LogS(L)
a = Cholesky(L, cov)
For i = 1 To L
    Mean(i) = Log(S(i))+(r-q(i)-0.5*cov(i,i))*T ' expected log S(i)
    For j = 1 To L
        Multiplier(i, j) = Sqr(T) * a(i, j)
    Next j
Next i
SumCall = 0                    ' initialize sum of option values
For h = 1 To M
    BasketValue = 0            ' initialize portfolio value
    For j = 1 To L
        z(j) = RandN()
    Next j
    For i = 1 To L             ' calculate each stock separately
        LogS(i) = Mean(i)      ' start at expected log S(i)
        For j = 1 To L         ' add L random terms
            LogS(i) = LogS(i) + Multiplier(i, j) * z(j)
        Next j
        BasketValue = BasketValue + w(i) * Exp(LogS(i))
    Next i
    CallV = Application.Max(BasketValue - K, 0)    ' call value
    SumCall = SumCall + CallV                      ' update sum
Next h
```

```
European_Basket_Call_MC = Exp(-r * T) * SumCall / M
End Function
```

Monte Carlo Valuation with an Antithetic Variate

We will illustrate antithetic variates by modifying our previous valuation of a floating strike call. We denote $\log S$ by LogS(1) and $\log S^*$ by LogS(2). We simulate

$$\Delta \, \text{LogS}(1) = \nu \, \Delta t + \sigma \Delta B$$

and

$$\Delta \, \text{LogS}(2) = \nu \, \Delta t - \sigma \Delta B \,.$$

We compute the minimums LogSMin(1) and LogSMin(2) of the logarithms of the two stock prices for each simulated path, and compute at the end of each simulated path the average call value

$$0.5 \times \max \left(0, e^{\text{LogS}(1)} - e^{\text{LogSMin}(1)} \right) + 0.5 \times \max \left(0, e^{\text{LogS}(2)} - e^{\text{LogSMin}(2)} \right) \,.$$

```
Function Floating_Strike_Call_MC_AV_SE(S,r,sigma,q,SMin,T,N,M)
'
' Inputs are S = initial stock price
'            r = risk-free rate
'            sigma = volatility
'            q = dividend yield
'            T = remaining time to maturity
'            Smin = minimum during previous life of contract
'            N = number of time periods
'            M = number of simulations
'
' This returns the row vector (call value, standard error)
'
Dim LogS0, LogSMin0, sigsdt, nudt, CallV, SumCall, SumCallSq
Dim dt, z, StdError, LogS(1 To 2), LogSMin(1 To 2), i, j
dt = T / N
nudt = (r - q - 0.5 * sigma * sigma) * dt
sigsdt = sigma * Sqr(dt)
LogS0 = Log(S)                  ' store log of initial stock price
LogSMin0 = Log(SMin)            ' store log of historical minimum
SumCall = 0                     ' initialize sum of values
SumCallSq = 0                   ' initialize sum of squared values
For i = 1 To M
    LogS(1) = LogS0             ' initialize log stock price
    LogS(2) = LogS0             ' initialize log stock price
    LogSMin(1) = LogSMin0       ' initialize historical log min
    LogSMin(2) = LogSMin0       ' initialize historical log min
    For j = 1 To N
        z = RandN()
```

```
        LogS(1) = LogS(1) + nudt + sigsdt * z
        LogS(2) = LogS(2) + nudt - sigsdt * z
        LogSMin(1) = Application.Min(LogS(1), LogSMin(1))
        LogSMin(2) = Application.Min(LogS(2), LogSMin(2))
      Next j
      CallV = 0.5 * (Exp(LogS(1)) - Exp(LogSMin(1))) _
            + 0.5 * (Exp(LogS(2)) - Exp(LogSMin(2)))
      SumCall = SumCall + CallV                ' sum of call values
      SumCallSq = SumCallSq + CallV * CallV    ' sum of squares
  Next i
  CallV = Exp(-r * T) * SumCall / M
  StdError = Exp(-r*T)*Sqr((SumCallSq-SumCall*SumCall/M)/(M*(M-1)))
  Floating_Strike_Call_MC_AV_SE = Array(CallV, StdError)
  End Function
```

Monte Carlo Valuation with a Control Variate

We use the geometric average as a control variate for the arithmetic average in an average-price call, using a pre-sample to estimate the beta. For pedagogic purposes, we return the beta in addition to the call value.

```
Function Average_Price_Call_MC(S,K,r,sigma,q,Avg,TPast,TFut,N,M1,M2)
'
' Inputs are S = initial stock price
'            K = strike price
'            r = risk-free rate
'            sigma = volatility
'            q = dividend yield
'            Avg = average price during past life of contract
'            TPast = time since creation of contract
'            TFut = remaining time to maturity
'            N = number of time periods
'            M1 = number of simulations in the pre-sample
'            M2 = number of simulations in the sample
'                 and geometric averages
'
' This returns the row vector (call value, beta)
'
Dim Kstar, dt, nudt, sigsdt, disc, x, y, LogS0, LogS, SumS, SumLogS
Dim Sumx, Sumx2, Sumy, Sumy2, Sumxy, beta, phi, CallV, i, j
Kstar = (TFut + TPast) * K / TFut - TPast * Avg / TFut
dt = TFut / N
nudt = (r - q - 0.5 * sigma * sigma) * dt
sigsdt = sigma * Sqr(dt)
disc = Exp(-r * TFut)
LogS0 = Log(S)          ' store log stock price
'
' First we compute the known mean for the geometric average
'
```

```
phi = Discrete_Geom_Average_Price_Call(S,Kstar,r,sigma,q,TFut,N)
'
' Now we run the pre-sample to estimate the regression beta
'
Sumx = 0                        ' sum of arithmetic option values
Sumx2 = 0                       ' sum of squared arithmetic values
Sumy = 0                        ' sum of geometric option values
Sumy2 = 0                       ' sum of squared geometric values
Sumxy = 0                       ' sum of products
For i = 1 To M1
    LogS = LogS0                ' initialize log stock price
    SumS = 0                    ' initialize sum of stock prices
    SumLogS = 0                 ' initialize sum of log stock prices
    For j = 1 To N
        LogS = LogS + nudt + sigsdt * RandN() ' update log price
        SumS = SumS + Exp(LogS)              ' update sum of prices
        SumLogS = SumLogS + LogS             ' update sum of logs
    Next j
    x = disc*Application.Max(SumS/N-Kstar,0)        ' arithmetic
    y = disc*Application.Max(Exp(SumLogS/N)-Kstar,0) ' geometric
    Sumx = Sumx + x            ' sum of arithmetic values
    Sumx2 = Sumx2 + x * x      ' sum of squared arithmetic values
    Sumy = Sumy + y            ' sum of geometric values
    Sumy2 = Sumy2 + y * y      ' sum of squared geometric values
    Sumxy = Sumxy + x * y      ' sum of products
Next i
beta = (M1*Sumxy-Sumx*Sumy)/(M1*Sumy2-Sumy*Sumy) ' regression beta
'
' Now we compute sample arithmetic and geometric averages
'
Sumx = 0                        ' sum of arithmetic option values
Sumy = 0                        ' sum of geometric option values
For i = 1 To M2
    LogS = LogS0                ' initialize log stock price
    SumS = 0                    ' initialize sum of stock prices
    SumLogS = 0                 ' initialize sum of log stock prices
    For j = 1 To N
        LogS = LogS + nudt + sigsdt * RandN() ' update log price
        SumS = SumS + Exp(LogS)              ' update sum ofprices
        SumLogS = SumLogS + LogS             ' update sum of logs
    Next j
    x = disc*Application.Max(SumS/N-Kstar,0)        ' arithmetic
    y = disc*Application.Max(Exp(SumLogS/N)-Kstar,0) ' geometric
    Sumx = Sumx + x            ' total of arithmetic values
    Sumy = Sumy + y            ' total of geometric values
Next i
'
' Now we adjust the sample arithmetic average
'
```

```
CallV = (TFut/(TFut+TPast))*(Sumx/M2+beta*(phi-Sumy/M2))
Average_Price_Call_MC = Array(CallV, beta)
End Function
```

Accelerated Binomial Valuation of American Puts

First we create a binomial valuation program that replaces (i) calculation
of the intrinsic value at maturity and (ii) calculation of the value at the
penultimate date as the larger of intrinsic value and the discounted value
at maturity with (iii) calculation of the value at the penultimate date as the
larger of intrinsic value and the Black-Scholes value of a European option with
one period to maturity.

```
Function American_Put_Binomial_BS(S0, K, r, sigma, q, T, N)
'
' Inputs are S0 = initial stock price
'            K = strike price
'            r = risk-free rate
'            sigma = volatility
'            q = dividend yield
'            T = time to maturity
'            N = number of time periods
'
Dim dt, u, d, pu, dpu, dpd, u2, S, i, j
Dim PutV() As Double
ReDim PutV(N)
dt = T / N                              ' length of time period
u = Exp(sigma * Sqr(dt))                ' size of up step
d = 1 / u                               ' size of down step
pu = (Exp((r - q) * dt) - d) / (u - d)  ' prob of up step
dpu = Exp(-r * dt) * pu                 ' discount x up prob
dpd = Exp(-r * dt) * (1 - pu)           ' discount x down prob
u2 = u * u
'
' First we value at the penultimate date
'
S = S0 * d ^ (N - 1)                    ' bottom stock price
PutV(0) = Application.Max(K-S,Black_Scholes_Put(S,K,r,sigma,q,dt))
For j = 1 To N - 1                      ' step up over nodes
    S = S * u2
    PutV(j) = _
    Application.Max(K-S,Black_Scholes_Put(S,K,r,sigma,q,dt))
Next j
'
' Now we back up to date 0 as before
'
For i = N - 2 To 0 Step -1
    S = S0 * d ^ i                      ' bottom stock price
```

```
        PutV(0) = Application.Max(K-S,dpd*PutV(0)+dpu*PutV(1))
        For j = 1 To i                       ' step up over nodes
            S = S * u2
            PutV(j) = Application.Max(K-S,dpd*PutV(j)+dpu*PutV(j+1))
        Next j
    Next i
    American_Put_Binomial_BS = PutV(0)
    End Function
```

Now we create a program that uses Richardson extrapolation from a binomial model with N periods and a binomial model with $N/2$ periods to estimate the value from a binomial model with an infinite number of periods. We use the previous program as our binomial model.

```
Function American_Put_Binomial_BS_RE(S, K, r, sigma, q, T, N)
'
' Inputs are S = initial stock price
'            K = strike price
'            r = risk-free rate
'            sigma = volatility
'            q = dividend yield
'            T = time to maturity
'            N = number of time periods
'
Dim y2, y1
If Not (N / 2 = Round(N / 2, 0)) Then
    MsgBox ("Number of periods N should be a multiple of two.")
    Exit Function
End If
y2 = American_Put_Binomial_BS(S, K, r, sigma, q, T, N)
y1 = American_Put_Binomial_BS(S, K, r, sigma, q, T, N / 2)
American_Put_Binomial_BS_RE = 2 * y2 - y1
End Function
```

Problems

9.1. Create an Excel worksheet in which the user inputs S, r, σ, q, $S_{\min}$, T, N and M. Compute the value of a floating strike call by Monte Carlo with and without using an antithetic variate. Compare the standard errors.

9.2. Create an Excel worksheet to compare the estimates of the value of an American put given by the functions

$$\texttt{American\_Put\_Binomial}$$

and

$$\texttt{American\_Put\_Binomial\_BS\_RE}$$

for various values of N. Allow the user to input S, K, r, σ, q, and T. To assess the quality of the estimates, provide also the estimate given by the function `American_Put_Binomial` with a large value of N (say $N = 1000$).

9.3. Estimate the value of option (c) in Prob. 8.11 using the function

`Average_Price_Call_MC`.

Complete the simulations in Prob. 8.12 by including the cost of the options for hedge (c).

9.4. Create a VBA function `European_Basket_Call_Binomial` to value a basket option on two assets.

9.5. Create a VBA function `European_Basket_Call_Binomial_RE` that calls the function from Prob. 9.4 and uses Richardson extrapolation to estimate the value of a basket call in a binomial model with $N = \infty$.

9.6. Create a VBA function `European_Basket_Call_MC_AV` that uses Monte Carlo with an antithetic variate to value a basket call option. Create an Excel worksheet to compare the standard errors of the estimates from the functions `European_Basket_Call_MC` and `European_Basket_Call_MC_AV` for a call on a basket of three assets. Allow the user to input S, K, r, cov, q, w, T, and M. Recall that S, cov, q and w are arrays.

9.7. Create an Excel worksheet to compare the estimates of the value of a basket call on two assets given by the binomial model (with and without Richardson extrapolation) and Monte Carlo (with and without an antithetic variate) for various values of N and M. To assess the quality of the estimates, provide also the estimate given by the binomial model without Richardson extrapolation with $N = 100$.

9.8. Suppose you must purchase 100 units of each of two assets at the end of the quarter ($T = 0.25$). You want to hedge the cost at the beginning of the quarter. Use simulation in an Excel worksheet to compare the quality of the following hedges (assuming 100 contracts for each option in (a) and 200 contracts in (b) and (c)):

(a) standard European calls for both assets,
(b) a basket call written on both assets (with $w_1 = w_2 = 0.5$),
(c) a call on the maximum of the two asset prices

Assume the two assets have the same initial price S and the options are all at the money at the beginning of the quarter. Allow the user to input S, r, σ_1, σ_2, ρ, q_1, and q_2. Compare the quality of the hedges as in Prob. 8.7.

9.9. Compute the prices of the options in the previous exercise (using a binomial model or Monte Carlo for the basket option). Compare the hedges including the costs of the options as in Prob. 8.8.

9.10. Create a VBA function `American_Spread_Put_Binomial_RE` that calls `American_Spread_Put_Binomial` and uses Richardson extrapolation to estimate the value of an American spread put for $N = \infty$. Create an Excel spreadsheet to compare the estimates of the functions as in Prob. 9.7.

9.11. Create a VBA function `Down_And_Out_Call_MC` using Monte Carlo to value a discretely sampled down-and-out call option. The inputs should be S, K, r, σ, q, N, M and `Barrier`.

9.12. Create a VBA function `Down_And_Out_Call_MC_CV` that uses Monte Carlo to value a discretely sampled down-and-out call option and that uses a standard European call as a control variate.

10

Finite Difference Methods

In this chapter we will see how to estimate derivative values by numerically solving the partial differential equation (pde) that the derivative value satisfies, using finite difference methods. More advanced discussions of this topic can be found in Wilmott, DeWynne and Howison [64], Wilmott [63], and Tavella [60], among other places. We will only consider derivatives written on a single underlying asset, but the ideas generalize to derivatives written on multiple underlying assets (e.g., basket and spread options) in much the same way that binomial models can be applied to derivatives on multiple underlying assets. The curse of dimensionality is the same for finite difference methods as for binomial models—the computation time increases exponentially with the number of underlying assets.

10.1 Fundamental PDE

Consider an asset with price S and constant dividend yield q. Set $X = \log S$. Then we have

$$\mathrm{d}X = \nu \, \mathrm{d}t + \sigma \, \mathrm{d}B \, ,$$

where $\nu = r - q - \sigma^2/2$ and B is a Brownian motion under the risk-neutral measure.

Let T denote the maturity date of a derivative security. At time t (when the remaining time to maturity is $T-t$), assume the price of the derivative can be represented as $C(t, X(t))$.[1] Since C is a function of t and X, Itô's formula implies

---

[1] If the price of the derivative is a function of the asset price S and time, then we can always write it in this form as a function of the natural logarithm of S and time.

$$dC = \frac{\partial C}{\partial t}\, dt + \frac{\partial C}{\partial X}\, dX + \frac{1}{2}\frac{\partial^2 C}{\partial X^2}(dX)^2$$

$$= \frac{\partial C}{\partial t}\, dt + \frac{\partial C}{\partial X}\left(\nu\, dt + \sigma\, dB\right) + \frac{1}{2}\frac{\partial^2 C}{\partial X^2}\sigma^2\, dt \; . \qquad (10.1)$$

On the other hand, under the risk-neutral measure, the instantaneous expected rate of return on the derivative is the risk-free rate, so

$$\frac{dC}{C} = r\, dt + \text{something}\; dB \; .$$

where the "something" is the volatility of the derivative value. We can of course rearrange this as

$$dC = rC\, dt + \text{something}\; C\, dB \; . \qquad (10.2)$$

In order for both (10.1) and (10.2) to hold, the drifts on both right-hand sides must be equal.[2] This implies

$$rC = \frac{\partial C}{\partial t} + \nu\frac{\partial C}{\partial X} + \frac{1}{2}\sigma^2\frac{\partial^2 C}{\partial X^2} \; . \qquad (10.3)$$

This equation is the "fundamental pde." It is an equation that we want to solve for the function C. Every derivative written on S satisfies this same equation. Different derivatives have different values because of boundary conditions. The boundary conditions are the intrinsic value at maturity, optimality conditions for early exercise, barriers and the like.

To translate the terms in (10.3) into more familiar ones, notice that, because $S = e^X$, we have

$$\frac{\partial S}{\partial X} = e^X = S \; .$$

Therefore, by the chain rule of calculus,

$$\frac{\partial C}{\partial X} = \frac{\partial C}{\partial S}\frac{\partial S}{\partial X} = S\frac{\partial C}{\partial S} \; .$$

Thus the term $\partial C/\partial X$ is the delta of the derivative multiplied by the price of the underlying. Similarly, by ordinary calculus, the term $\partial^2 C/\partial X^2$ can be written in terms of the delta and the gamma of the derivative.

Sometimes one writes the derivative value as a function of time to maturity $(\tau = T - t)$ instead of t. The partial derivative of C with respect to τ is the negative of the partial derivative with respect to t, so the fundamental pde is

---

[2] Suppose a process X satisfies $dX = \alpha_1\, dt + \sigma_1\, dB = \alpha_2\, dt + \sigma_2\, dB$ for coefficients α_i and σ_i. This implies $(\alpha_1 - \alpha_2)\, dt = (\sigma_2 - \sigma_2)\, dB$. The right-hand side defines a (local) martingale and the left-hand side defines a continuous finite-variation process. As discussed in Sect. 2.2, the only continuous finite-variation martingales are constants, so the changes must be zero; i.e., $\alpha_1 = \alpha_2$ and $\sigma_1 = \sigma_2$.

the same except for a different sign on the first term of the right-hand side of (10.3). Rearranging a little, we have

$$\frac{\partial C}{\partial \tau} = -rC + \nu \frac{\partial C}{\partial X} + \frac{1}{2}\sigma^2 \frac{\partial^2 C}{\partial X^2} \,. \tag{10.3'}$$

In this form, the pde is similar to important equations in physics, in particular the equation for how heat propagates through a rod over time. In fact, it can be transformed exactly into the heat equation, which is how Black and Scholes originally solved the option valuation problem. The terminal condition for a call option, $C = \max(S - K, 0)$, can be viewed as defining C over the X dimension at $\tau = 0$, just as the temperature along the length of the rod might be specified at an initial date, and as τ increases C changes at each point X according to (10.3'), which is similar, as noted, to the equation for the change in temperature at a point on the rod as time passes.

10.2 Discretizing the PDE

To numerically solve the fundamental pde, we consider a discrete grid on the (t, x) space. We label the time points as $t_0, t_1, t_2, \ldots, t_N$, and the x points as $x_{-M}, x_{-M+1}, \ldots, x_0, x_1, \ldots, x_M$, with $t_0 = 0$, $t_N = T$, and $x_0 = \log S(0)$. The equation should hold for $-\infty < x < \infty$, but obviously we will have to bound this space, and we have denoted the upper and lower bounds by x_M and x_{-M} here. We take the points to be evenly spaced and set $\Delta t = t_i - t_{i-1}$ and $\Delta x = x_j - x_{j-1}$ for any i and j.

For specificity, we will consider a call option, though the discussion in this section applies to any derivative. We will compute a value for the call at each of the points on the grid. Then we return the value of the call at the point (t_0, x_0).

Consider a point (t_i, x_j). We could denote the estimated value of the call at this point by C_{ij} but for now we will just use the symbol C. Think of t being on the horizontal axis and x on the vertical axis. There are four points that can be reached from (t_i, x_j) by one step (an increase or decrease) in either t or x. Let's denote the estimated call value at $(t_i, x_j + \Delta x)$ as C_{up}, the value at $(t_i, x_j - \Delta x)$ as C_{down}, the value at $(t_i + \Delta t, x_j)$ as C_{right} and the value at $(t_i - \Delta t, x_j)$ as C_{left}.

We want to force (10.3) to hold on the grid. To estimate $\partial C/\partial X$ and $\partial^2 C/\partial X^2$, we make exactly the same calculations we made to estimate deltas and gammas in a binomial model. At the point (t_i, x_j), we estimate

$$\frac{\partial C}{\partial X} \approx \frac{C_{\text{up}} - C_{\text{down}}}{2\Delta x} \,. \tag{10.4a}$$

There are two other obvious estimates of this derivative:

$$\frac{C_{\text{up}} - C}{\Delta x} \quad \text{and} \quad \frac{C - C_{\text{down}}}{\Delta x} \,.$$

The first of these should be understood as an estimate at the midpoint of x_j and $x_j + \Delta x$ and the second as an estimate at the midpoint of x_j and $x_j - \Delta x$. The distance between these two midpoints is Δx, so the difference in these two estimates of $\partial C/\partial X$ divided by Δx is an estimate of the second derivative:

$$\frac{\partial^2 C}{\partial X^2} \approx \frac{C_{\text{up}} - 2C + C_{\text{down}}}{(\Delta x)^2} . \tag{10.4b}$$

The obvious estimate of $\partial C/\partial t$, which is analogous to the estimate of $\partial C/\partial X$, is

$$\frac{C_{\text{right}} - C_{\text{left}}}{2\Delta t} .$$

This is *not* the estimate we are going to use. The reason is that we want to solve for the call values on the grid in much the same way that we solved the binomial model—starting at the end and working backwards. If we use the above estimate of the time derivative, then at each point (t_i, x_j), equation (10.3) will link the call values at times t_{i-1}, t_j and t_{i+1}. This would substantially complicate the "backing up" process. However, in a sense, it is the right estimate, and the Crank-Nicolson method to be discussed below uses a similar idea.

The other two choices for estimating $\partial C/\partial t$ are analogous to the other two choices for estimating $\partial C/\partial X$. We can use either

$$\frac{\partial C}{\partial t} \approx \frac{C - C_{\text{left}}}{\Delta t} , \tag{10.4c}$$

or

$$\frac{\partial C}{\partial t} \approx \frac{C_{\text{right}} - C}{\Delta t} . \tag{10.4c$'$}$$

Using the first is called the "explicit" method of solving the pde, and using the second is called the "implicit" method. The reason for these names should become clear below.

10.3 Explicit and Implicit Methods

We first consider the explicit method. We set the value of the call at the final date t_N and each point x_j to be its intrinsic value, $\max(e^{x_j} - K, 0)$. Now consider calculating the value at date t_{N-1} and any point x_j. We do this by forcing the approximation to (10.3) based on (10.4a)–(10.4c) to hold at the point (t_N, x_j). Using the same notation as before, for $(t_i, x_j) = (t_N, x_j)$, implies

$$rC = \frac{C - C_{\text{left}}}{\Delta t} + \nu \left(\frac{C_{\text{up}} - C_{\text{down}}}{2\Delta x} \right) + \frac{1}{2}\sigma^2 \left(\frac{C_{\text{up}} + C_{\text{down}} - 2C}{(\Delta x)^2} \right) . \tag{10.5}$$

Given that t_i is the final date t_N, the values C, C_{up} and C_{down} have already been calculated as the intrinsic value of the call at maturity. The only unknown is C_{left}, which is the value of the call at (t_{N-1}, x_j). We can solve this *explicitly* for C_{left}, whence the name of the algorithm. We do this at each point x_j at date t_{N-1} (except for the top and bottom points, which we will discuss below) and then we follow the same procedure to back up sequentially to the initial date, as in the binomial model.

Equation (10.5) cannot be used to find C_{left} at the bottom point x_{-M}, because at this point there is no C_{down} at date t_N. Similarly, we cannot use it to find C_{left} at the top point x_M, because at that point there is no C_{up}. We have to define the values along the top and bottom of the grid in some other fashion. We do this using conditions the derivative is known to satisfy as the stock price approaches $+\infty$ or 0. For example, for a European call option, we use the conditions that $\partial C/\partial S \to 1$ as $S \to \infty$ and $\partial C/\partial S \to 0$ as $S \to 0$. We will explain this in more detail in the following section.

The solution of (10.5) for C_{left} can be written as

$$C_{\text{left}} = (1 - r\Delta t)\left(p_u C_{\text{up}} + pC + p_d C_{\text{down}}\right), \qquad (10.5')$$

where

$$p_u = \frac{\sigma^2 \Delta t + \nu \Delta t \Delta x}{2(1 - r\Delta t)(\Delta x)^2},$$

$$p_d = \frac{\sigma^2 \Delta t - \nu \Delta t \Delta x}{2(1 - r\Delta t)(\Delta x)^2},$$

and

$$p = 1 - p_u - p_d.$$

This can be interpreted as discounting the probability-weighted values of the call at the next date, where we consider that starting at the grid point (t_i, x_j), the logarithm of the stock price takes three possible values ($x_j - \Delta x$, x_j, and $x_j + \Delta x$) at the next date t_{i+1}, and where we use $1 - r\Delta t$ as the discount factor. Thus, it is essentially a trinomial model. This relationship was first noted by Brennan and Schwartz [11].

Actually, for this to be a sensible trinomial model, the "probabilities" p_u, p and p_d should be nonnegative. Assuming $1 - r\Delta t > 0$, this will be the case if and only if

$$\Delta x \leq \frac{\sigma^2}{|\nu|} \quad \text{and} \quad \Delta t \leq \frac{(\Delta x)^2}{\sigma^2 + r(\Delta x)^2}.$$

The first of these conditions characterizes p_u and p_d being nonnegative. The second is derived from $p_u + p_d \leq 1$. It is interesting to examine these conditions in terms of the number N of time periods and the number of steps in the x

dimension, which is $2M$. To simplify the notation in the following somewhat, denote the distance of the upper x boundary from x_0 by D (i.e., $D = x_M - x_0$). Then $\Delta t = T/N$ and $\Delta x = D/M$. The probabilities are nonnegative if and only if

$$M \geq \frac{|\nu|D}{\sigma^2} \quad \text{and} \quad N \geq rT + \left(\frac{\sigma^2 T}{D^2}\right) M^2 .$$

Consider fixing D and increasing the number of time periods and space steps (i.e., steps along the x dimension). To maintain positive probabilities, the above shows that the number of time periods must increase as the square of the number of space steps: increasing M by a factor of 10 requires increasing N by a factor of 100. The upshot is it can be computationally expensive to use a large number of space steps, if we want to maintain nonnegative probabilities.

One can reasonably ask whether this is important, because we can certainly solve (10.5) to estimate the call values even when the "probabilities" are negative. The answer is that it is important, but for a reason we have not yet discussed. In a numerical algorithm for solving a partial differential equation (or for solving many other types of problems) there are two types of errors: discretization error and roundoff error. If we increase N and M sufficiently, we should reduce the discretization error. However, each calculation on the computer introduces roundoff error. An algorithm is said to be "stable" if the roundoff errors stay small and bounded as the discretization error is reduced. An unfortunate fact about the explicit method is that it is stable only if the number of time steps increases with the square of the number of space steps. In the absence of this condition, the roundoff errors can accumulate and prevent one from reaching a solution of the desired accuracy.

The implicit method is known to be fully stable, so it is is to be preferred to the explicit method. We will discuss briefly how to implement this method, before moving in the next section to the "Crank-Nicolson" method, which is also fully stable and known to be more efficient than the implicit method.

The implicit method uses the approximation (10.4c$'$) for $\partial C/\partial t$. As before, the call values are defined at the final date as the intrinsic value. Backing up a period, consider a grid point (t_{N-1}, x_j). We will try to estimate the call value at this date by forcing (10.3) to hold at this point. This means

$$rC = \frac{C_{\text{right}} - C}{\Delta t} + \nu \left(\frac{C_{\text{up}} - C_{\text{down}}}{2\Delta x}\right) + \frac{1}{2}\sigma^2 \left(\frac{C_{\text{up}} + C_{\text{down}} - 2C}{(\Delta x)^2}\right) . \quad (10.6)$$

We know C_{right}, because it is the intrinsic value at (t_N, x_j). This equation links three unknowns (C, C_{up}, and C_{down}) to the known value C_{right}. We cannot solve it explicitly for these three unknowns. Instead, we need to solve a system of linear equations to simultaneously solve for all the call values at date t_{N-1}. There are $2M - 1$ equations of the form (10.6) plus conditions that we will impose at the upper and lower boundaries, and we need to solve these for the $2M + 1$ call values. This system of equations has the same form, and is solved in the same way, as the system of equations in the Crank-Nicolson method.

10.4 Crank-Nicolson

The estimate $(10.4c')$ of $\partial C/\partial t$ used in the implicit method is best understood as an estimate of $\partial C/\partial t$ at the midpoint of (t_i, x_j) and (t_{i+1}, x_j), i.e., at $(t_i + \Delta t/2, x_j)$. This is the basic idea of the Crank-Nicolson method. With this method, we continue to estimate the call values at the grid points, but we do so by forcing (10.3) to hold at midpoints of this type. To do this, we also need estimates of C, $\partial C/\partial X$ and $\partial^2 C/\partial X^2$ at the midpoints, but these are easy to obtain.

Let's modify the previous notation somewhat, writing C' for C_{right} and C'_{up} and C'_{down} for the values to the right and one step up and down, i.e., at the grid points $(t_i + \Delta t, x_j + \Delta x)$ and $(t_i + \Delta t, x_j - \Delta x)$ respectively. The obvious estimate of the call value at the midpoint $(t_i + \Delta t/2, x_j)$ is the average of C and C', so set

$$C^{\text{mid}} = \frac{C + C'}{2}.$$

Analogously, define

$$C^{\text{mid}}_{\text{up}} = \frac{C_{\text{up}} + C'_{\text{up}}}{2}, \qquad \text{and} \qquad C^{\text{mid}}_{\text{down}} = \frac{C_{\text{down}} + C'_{\text{down}}}{2}. \qquad (10.7)$$

The formulas (10.7) give us estimates of the call value at the midpoints one space step up and one space step down from x_j—i.e., at $(t_i + \Delta t/2, x_{j+1})$ and $(t_i + \Delta t/2, x_{j-1})$. We can now estimate $\partial C/\partial X$ and $\partial^2 C/\partial X^2$ at the midpoint $(t_i + \Delta t/2, x_j)$ exactly as before:

$$\frac{\partial C}{\partial X} \approx \frac{C^{\text{mid}}_{\text{up}} - C^{\text{mid}}_{\text{down}}}{2\Delta x},$$

and

$$\frac{\partial^2 C}{\partial X^2} \approx \frac{C^{\text{mid}}_{\text{up}} + C^{\text{mid}}_{\text{down}} - 2C^{\text{mid}}}{(\Delta x)^2}.$$

Now, (10.3) becomes

$$rC^{\text{mid}} = \frac{C' - C}{\Delta t} + \nu\left(\frac{C^{\text{mid}}_{\text{up}} - C^{\text{mid}}_{\text{down}}}{2\Delta x}\right) + \frac{1}{2}\sigma^2\left(\frac{C^{\text{mid}}_{\text{up}} + C^{\text{mid}}_{\text{down}} - 2C^{\text{mid}}}{(\Delta x)^2}\right).$$

$$(10.8)$$

Substituting from the formulas for C^{mid}, $C^{\text{mid}}_{\text{up}}$, and $C^{\text{mid}}_{\text{down}}$, we can re-write (10.8) as

$$\left(\frac{r}{2} + \frac{1}{\Delta t} + \frac{\sigma^2}{2(\Delta x)^2}\right)C - \left(\frac{\sigma^2}{4(\Delta x)^2} + \frac{\nu}{4\Delta x}\right)C_{\text{up}}$$

$$- \left(\frac{\sigma^2}{4(\Delta x)^2} - \frac{\nu}{4\Delta x}\right)C_{\text{down}} = \left(\frac{1}{\Delta t} - \frac{r}{2} - \frac{\sigma^2}{2(\Delta x)^2}\right)C'$$

$$+ \left(\frac{\sigma^2}{4(\Delta x)^2} + \frac{\nu}{4\Delta x}\right)C'_{\text{up}} + \left(\frac{\sigma^2}{4(\Delta x)^2} - \frac{\nu}{4\Delta x}\right)C'_{\text{down}} \quad (10.8')$$

We can also write this as

$$a_1 C - a_2 C_{\text{up}} - a_3 C_{\text{down}} = a_4 C' + a_2 C'_{\text{up}} + a_3 C'_{\text{down}} , \qquad (10.8'')$$

where the constants a_i are the factors in parentheses in (10.8').

As before, we start at the final date t_N and define the call value at that date by its intrinsic value. Consider a grid point (t_{N-1}, x_j). Forcing (10.3) to hold at the midpoint $(t_{N-1} + \Delta t/2, x_j)$ leads us to (10.8''). In this equation, C', C'_{up} and C'_{down} are known from the intrinsic value at maturity, and we need to solve for C, C_{up} and C_{down}. There are $2M - 1$ linear equations of this type and we will add linear equations at the upper and lower boundaries of the grid and solve the resulting system of $2M + 1$ linear equations for the $2M + 1$ call values. After finding the call values at date t_{N-1}, we then repeat the calculation at t_{N-2} and continue backing up in this way until we reach the initial date.

Notice that the Crank-Nicolson equations (10.8') are similar to the equations (10.6) in the implicit method, but more information is used in each step of the Crank-Nicolson method than is used in each step of the implicit method. Equation (10.8') links the call values C, C_{up} and C_{down} to the previously calculated C', C'_{up} and C'_{down}, whereas in the implicit method they were linked only to C' (which we called C_{right}).

10.5 European Options

To value a European option, one simply defines the values at the final date as the intrinsic value and then backs up to the initial date, using any of the methods described (explicit, implicit, or Crank-Nicolson). The value that should be returned is the value at the middle node at the initial date, which corresponds to the initial price of the underlying.

The boundary conditions normally used at the bottom and top of the grid are conditions that the first derivative $\partial C / \partial S$ of the option value are known to satisfy as $S \to 0$ and $S \to \infty$. These are conditions of the form

$$\lim_{S \to \infty} \frac{\partial C}{\partial S} = \lambda_0 , \qquad \text{and} \qquad \lim_{S \to 0} \frac{\partial C}{\partial S} = \lambda_\infty , \qquad (10.9)$$

for constants λ_0 and λ_∞. In the case of a call option, we have $\lambda_0 = 0$ and $\lambda_\infty = 1$. For a put option, we have $\lambda_0 = -1$ and $\lambda_\infty = 0$.

These conditions are implemented on the grid by forcing each value C at a point (t_i, x_{-M}) on the bottom of the grid to satisfy

$$C - C_{\text{up}} = \lambda_0 (S - S_{\text{up}}) \qquad (10.10a)$$

and by forcing each value C at a point (t_i, x_M) on the top of the grid to satisfy

$$C - C_{\text{down}} = \lambda_\infty (S - S_{\text{down}}) . \qquad (10.10b)$$

These two linear equations in the values at time t_i augment the $2M - 1$ equations already described to form a system of $2M + 1$ linear equations to be solved for the derivative values at the $2M + 1$ grid points at time t_i.

10.6 American Options

The explicit method is easily adapted to American options. As in a binomial model, we compute the option value at each node as the larger of its discounted expected value and its intrinsic value. To be somewhat more precise, we replace the "trinomial" value (10.5′) with

$$C_{\text{left}} = \max\left((1 - r\Delta t)(p_u C_{\text{up}} + pC + p_d C_{\text{down}}), \text{intrinsic value}\right) .$$

In the Crank-Nicolson method, one can in similar fashion compute the value of the derivative at each space node at any date by solving the system of equations (10.8″) and then replace the computed values by the intrinsic value when that is higher. However, because the values at the different space nodes are linked (i.e., the method is an implicit-type method), this one-at-a-time replacement of values by intrinsic values is not the most efficient method. See Wilmott [63] for more details (and for VBA code implementing the "projected successive over-relaxation" method).

10.7 Barrier Options

Finite-difference methods work well for valuing discretely-sampled barrier options. For a down-and-out option, one should place the bottom of the grid at the knock-out boundary. For an up-and-out option, one should place the top of the grid at the knock-out boundary. As discussed in Chap. 8, knock-in options can be valued as standard options minus knock-out options.

For barrier options, the boundary information (10.9) can be replaced by assigning a value of zero at the knock-out boundary. For example, for a down-and-out option, the condition (10.10a) can be replaced by $C = 0$. If the contract specifies that a rebate is to be paid to the buyer when the option is knocked out, then condition (10.10a) should be replaced by $C = $ Rebate.

10.8 Calculations in VBA

Crank-Nicolson

We will create a program that solves a system of equations of the form (10.8″), (10.10a) and (10.10b). We input the vector a of coefficients, a vector y of dimension $2M + 1$ containing the estimated values of the derivative at any date t_{i+1}, and an integer L from which M is defined as $M = (L-1)/2$ (i.e., $L = 2M+1$). The function will return the vector of values at date t_i.

We will write the boundary conditions (10.10a) and (10.10b), respectively, in the more general forms

$$C = z_1 + b_1 C_{\text{up}} , \qquad (10.11a)$$

and

$$C = z_L + b_L C_{\text{down}} , \qquad (10.11b)$$

where z_1, b_1, z_L and b_L are numbers to be calculated or input by the user. The equations (10.10a) and (10.10b) are the special cases in which $z_1 = \lambda_0(S - S_{\text{up}})$, $b_1 = 1$, $z_L = \lambda_\infty(S - S_{\text{down}})$, and $b_L = 1$. The additional generality in allowing b_1 and b_L to be different from one is important for many purposes, and we will see an example of it in the valuation of barrier options.

The system of equations that we want to solve is therefore

$$
\begin{pmatrix}
1 & -b_1 & 0 & 0 & 0 & \cdots & 0 & 0 & 0 & 0 & 0 \\
-a_3 & a_1 & -a_2 & 0 & 0 & \cdots & 0 & 0 & 0 & 0 & 0 \\
0 & -a_3 & a_1 & -a_2 & 0 & \cdots & 0 & 0 & 0 & 0 & 0 \\
\vdots & \vdots & \vdots & \vdots & \vdots & \vdots & \vdots & \vdots & \vdots & \vdots & \vdots \\
0 & 0 & 0 & 0 & 0 & \cdots & 0 & -a_3 & a_1 & -a_2 & 0 \\
0 & 0 & 0 & 0 & 0 & \cdots & 0 & 0 & -a_3 & a_1 & -a_2 \\
0 & 0 & 0 & 0 & 0 & \cdots & 0 & 0 & 0 & -b_L & 1
\end{pmatrix}
\begin{pmatrix}
C_1 \\ C_2 \\ C_3 \\ \vdots \\ C_{L-2} \\ C_{L-1} \\ C_L
\end{pmatrix}
=
\begin{pmatrix}
z_1 \\ z_2 \\ z_3 \\ \vdots \\ z_{L-2} \\ z_{L-1} \\ z_L
\end{pmatrix}
$$

where we are denoting the derivative values to be determined at date t_i across the $L\ (= 2M+1)$ space nodes as $C_1, \ldots, C_L$. The coefficients a_i are defined in (10.8''). The numbers $z_2, \ldots, z_{L-1}$ are the right-hand sides of (10.8'') and are determined by the coefficients a_i and the derivative values $y_1, \ldots, y_L$ at date t_{i+1}. The system of equations that must be solved to implement the implicit method is of this same form.

The first equation in this array (equation (10.11a)) can be written as

$$C_1 = u_1 + b_1 C_2 , \qquad (10.12a)$$

where $u_1 = z_1$. By induction, we will see that we can write, for each $j = 2, \ldots, L$,

$$C_{j-1} = u_{j-1} + b_{j-1} C_j \qquad (10.12b)$$

for some coefficients u_{j-1} and b_{j-1} to be determined. The j-th equation ($j = 2, \ldots, L-1$) in the array (equation (10.8'')) is

$$-a_3 C_{j-1} + a_1 C_j - a_2 C_{j+1} = z_j .$$

Supposing (10.12b) holds and using it to substitute for C_{j-1}, we have

$$-a_3 \left(u_{j-1} + b_{j-1} C_j \right) + a_1 C_j - a_2 C_{j+1} = z_j \quad \Longleftrightarrow \quad C_j = u_j + b_j C_{j+1} ,$$

where

$$u_j = \frac{z_j + a_3 u_{j-1}}{a_1 - a_3 b_{j-1}},$$

$$b_j = \frac{a_2}{a_1 - a_3 b_{j-1}}.$$

This establishes that (10.12b) holds for each $j = 2, \ldots, L$.

The last equation in the array (equation (10.11b)) is

$$C_L = z_L + b_L C_{L-1}. \tag{10.12c}$$

Our induction argument gives us

$$C_{L-1} = u_{L-1} + b_{L-1} C_L,$$

and when we combine these we have two equations in two unknowns and can solve for C_L as

$$C_L = \frac{z_L + b_L u_{L-1}}{1 - b_L b_{L-1}}.$$

We then successively obtain $C_{L-1}, C_{L-2}, \ldots, C_1$ from (10.12b).

```
Function CrankNicolson(a, y, L, z1, b1, zL, bL)
'
' Inputs are a = 4-vector of coefficients
'            y = L-vector of function values at a point in time
'            L = number of space points in the grid
'            z1 = constant for bottom boundary condition
'            b1 = coefficient for bottom boundary condition
'            zL = constant for top boundary condition
'            bL = coefficient for top boundary condition
'
' This returns a row vector of function values
'
Dim c(), b(), u(), z(), j
ReDim c(1 To L), b(1 To L), u(1 To L), z(1 To L)
u(1) = z1
b(1) = b1
For j = 2 To L - 1
    z(j) = a(4) * y(j) + a(2) * y(j + 1) + a(3) * y(j - 1)
    u(j) = (a(3) * u(j - 1) + z(j)) / (a(1) - a(3) * b(j - 1))
    b(j) = a(2) / (a(1) - a(3) * b(j - 1))
Next j
c(L) = (zL + bL * u(L - 1)) / (1 - bL * b(L - 1))
For j = L - 1 To 1 Step -1
    c(j) = u(j) + b(j) * c(j + 1)
Next j
CrankNicolson = c
End Function
```

Crank-Nicolson for European Options

We will demonstrate the Crank-Nicolson method by valuing a European call. Any other path-independent European derivative is valued in the same way, by appropriately redefining the value of the derivative at the final date and redefining the constants z_1 and z_L in the boundary conditions (10.11) at the bottom and top of the grid.

As elsewhere in this chapter, N denotes the number of time periods, and $2M + 1$ will be the number of x values on the grid. We use the symbol D to denote the distance of the top (or bottom) of the grid from $\log S(0)$. In other words, $D = x_M - x_0$. A reasonable value for D would be three standard deviations for $\log S$, which would mean $D = |\nu|T + 3\sigma\sqrt{T}$. For example, for a one-year option on a stock with a volatility of 30%, it should suffice to input $D = 1$.

As should be clear, the program is conceptually very similar to a binomial model. The difference is that the "backing up" procedure, which involves node-by-node discounting in a binomial model, here is accomplished via the Crank-Nicolson algorithm.[3]

```
Function European_Call_CrankNicolson(S0,K,r,sigma,q,T,N,M,Dist)
'
' Inputs are S0 = initial stock price
'            K = strike price
'            r = risk-free rate
'            sigma = volatility
'            q = dividend yield
'            T = time to maturity
'            N = number of time periods
'            M = number of space points above (or below) log(S0)
'            Dist = distance of boundary of grid from log(S0)
'
Dim dt, dx, dx2, u, sig2, nu, St, Sb, S, z1, b1, zL, bL, i, j, L
Dim CallV, y(), a(1 To 4)
L = 2 * M + 1          ' number of space points in the grid
ReDim y(1 To L)
'
' First we define parameters
'
dt = T / N             ' size of each time step
dx = Dist / M          ' size of each space step
dx2 = dx * dx
u = Exp(dx)            ' up parameter, same as in binomial model
```

---

[3] We use a different variable (y) for the call values at the final date—and consequently need to separate the first step of backing up (to the penultimate date) and the other steps of backing up (to date zero)—because element-by-element definition of an array and assignment of an array to a variable require different variable types in VBA. See Appendix A for more discussion.

```
sig2 = sigma * sigma
nu = r - q - sig2 / 2    ' risk-neutral drift
St = S0 * Exp(Dist)      ' stock price at the top nodes
Sb = S0 * Exp(-Dist)     ' stock price at the bottom nodes
a(1) = r / 2 + 1 / dt + sig2 / (2 * dx2)
a(2) = sig2 / (4 * dx2) + nu / (4 * dx)
a(3) = a(2) - nu / (2 * dx)
a(4) = -a(1) + 2 / dt
'
' Now we compute the call value at the final date
'
S = Sb
y(1) = Application.Max(S - K, 0)
For j = 2 To L           ' loop over nodes at last date
    S = S * u
    y(j) = Application.Max(S - K, 0)
Next j
'
'  ' Now we calculate the call value at the penultimate date
'
z1 = 0                   ' constant for bottom boundary condition
b1 = 1                   ' coefficient for bottom boundary condition
zL = St - St / u         ' constant for top boundary condition
bL = 1                   ' coefficient for top boundary condition
CallV = CrankNicolson(a, y, L, z1, b1, zL, bL)
'
' Now we back up to date 0
'
For i = (N - 2) To 0 Step -1
    CallV = CrankNicolson(a, CallV, L, z1, b1, zL, bL)
Next i
European_Call_CrankNicolson = CallV(M + 1)  ' value at middle node
End Function
```

Crank-Nicolson for Barrier Options

To price a down-and-out (or up-and-out option), we put the bottom (or top) of the grid at the boundary. The boundary condition that we use is that the value at the boundary is zero. We will consider the example of a down-and-out call option. In this case, the boundary condition at the bottom of the grid is (10.11a) with $z_1 = 0$ and $b_1 = 0$. The boundary condition at the top is the same as for an ordinary call. We can easily handle a rebate paid when the option is knocked out by inputting the value of the rebate as z_1.

The main new issue that we encounter in valuing barriers is locating the boundary of the grid at the barrier. For the down-and-out, we will input the value of the stock price at which the option is knocked out as Bar. We want the bottom of the grid to lie at the natural logarithm of this number. This will influence our choice of the space step Δx, because we want to have an integer number of steps between the bottom of the grid and $\log S(0)$. We assume that the value M input by the user represents the desired number of space steps above $\log S(0)$. We start with $\Delta x = D/M$ as an initial estimate of the size of the space step. We then decrease it, if necessary, to ensure that the distance between Bar and $\log S(0)$ is an integer multiple of Δx. We then increase M, if necessary, to ensure that the top of the grid will still be at or above $D + \log S(0)$. Finally, we define the top of the grid to be at $\log S(0) + M \cdot \Delta x$.

```
Function Down_And_Out_Call_CN(S0,K,r,sigma,q,T,N,M,Dist,Bar)
'
' Inputs are S0 = initial stock price
'            K = strike price
'            r = risk-free rate
'            sigma = volatility
'            q = dividend yield
'            T = time to maturity
'            N = number of time periods
'            M = number of space points above log(S0)
'            Dist = distance of top of grid from log(S0)
'            Bar = knock-out barrier < S0
'
Dim dx, DistBot, DistTop, dt, dx2, u, sig2, nu, St, S, z1, b1
Dim zL, bL, CallV, y(), a(4), i, j, L, NumBotSteps, NumTopSteps
dx = Dist / M                    ' first guess at size of space step
DistBot = Log(S0) - Log(Bar)     ' dist of log(S0) from bottom of grid
NumBotSteps = Application.Ceiling(DistBot / dx, 1)
dx = DistBot / NumBotSteps       ' new (smaller) space step
NumTopSteps = Application.Ceiling(Dist / dx, 1)
DistTop = NumTopSteps * dx       ' dist of log S(0) from top of grid
L = NumBotSteps + NumTopSteps + 1 ' number of space points
ReDim y(L)
dt = T / N                       ' size of time step
dx2 = dx * dx
u = Exp(dx)                      ' up parameter, as in binomial
sig2 = sigma * sigma
nu = r - q - sig2 / 2
St = S0 * Exp(DistTop)           ' stock price at top node
a(1) = r / 2 + 1 / dt + sig2 / (2 * dx2)
a(2) = sig2 / (4 * dx2) + nu / (4 * dx)
a(3) = a(2) - nu / (2 * dx)
a(4) = -a(1) + 2 / dt
S = Bar                          ' stock price at bottom node
y(1) = Application.Max(S - K, 0)
For j = 2 To L                   ' loop over nodes at last date
    S = S * u
    y(j) = Application.Max(S - K, 0)
Next j
z1 = 0                           ' constant for bottom boundary
```

```
b1 = 0                              ' coefficient for bottom boundary
zL = St - St / u                    ' constant for top boundary
bL = 1                              ' coefficient for top boundary
CallV = CrankNicolson(a, y, L, z1, b1, zL, bL)
For i = N - 2 To 0 Step -1          ' back up to date 0
    CallV = CrankNicolson(a, CallV, L, z1, b1, zL, bL)
Next i
Down_And_Out_Call_CN = CallV(NumBotSteps + 1)
End Function
```

Problems

10.1. Create an Excel worksheet to compare the estimates of the value of a discretely sampled barrier option given by the functions `Down_And_Out_Call_MC` created in Prob. 9.11 and the function `Down_And_Out_Call_CN`. Allow the user to input S, K, r, σ, q, the knock-out barrier, the number of Monte Carlo simulations, and the number of space steps above $\log S(0)$ in the Crank-Nicolson algorithm.

10.2. Create a VBA function `Up_And_Out_Put_CN` to value an up-and-out put option by the Crank-Nicolson method.

10.3. Create a VBA function `European_Call_Explicit` that uses the explicit method (10.5′) to value a European call option.

10.4. Write the system of equations (10.6) for the implicit method, together with boundary conditions of the form (10.11) as a matrix system and solve for u_j and b_j in (10.12), as in the subsection that defines the function `CrankNicolson`.

10.5. Create a VBA function `Implicit` that solves the system of equations in the preceding exercise.

10.6. Create a VBA function `European_Call_Implicit` that uses the implicit method to value a European call option.

Part III

Fixed Income

11

Fixed Income Concepts

In this part of the book, we will study the pricing and hedging of fixed-income derivatives, that is, derivatives that can be viewed as written on bonds or on interest rates. The complexities of this subject stem from the fact that the underlying bonds or rates will have time-varying and generally random volatilities and have volatilities that must be linked in some way to each other. We are not free to arbitrarily specify the volatilities and correlations, as we can, for example, with basket equity options, because such a specification may imply there is an arbitrage opportunity available from trading the underlying assets, and, of course, one could not trust a derivative value or a hedging strategy derived from such a model. There are many books that provide a more comprehensive and advanced treatment than we will be able to give here, among which Rebonato [55, 56], James and Webber [41] and Brigo and Mercurio [12] seem particularly useful.

We will focus on derivatives and underlyings that have very little default risk and ignore the pricing of default (credit) risk. Credit risk and credit derivatives are booming areas in both theory and practice. For this topic, one can consult the recent books of Bielecki and Rutkowski [2], Duffie and Singleton [25], Schönbucher [57], and Tavakoli [59]. Another important topic that will not be covered is mortgages.

In the first section, we introduce a fundamental construct: the yield curve, by which we mean the yields of (possibly theoretical) zero-coupon bonds of various maturities. The last two sections of the chapter (on principal components) are optional—nothing else in the book builds upon them.

11.1 The Yield Curve

Given prices of discount (zero-coupon) bonds of all maturities, any coupon paying bond can be priced as a portfolio of discount bonds. The relationship between time to maturity and the price of the corresponding discount bond with $1 face value is sometimes called the discount function. An equivalent concept is the yield curve, which is the relationship between time to maturity and the yield of the corresponding discount bond. Yields can be quoted according to different compounding conventions (semi-annual being the most common), but we will continue to use continuous compounding.[1] With this convention, the yield of a zero-coupon bond with \$1 face value having τ years to maturity and price P is defined to be the rate y such that $P = e^{-y\tau}$. Denoting this yield y at maturity τ by $y(\tau)$, the yield curve is the collection of yields $\{y(\tau)|\tau > 0\}$, conventionally plotted with maturity τ on the horizontal axis and the yields $y(\tau)$ on the vertical. Usually (but certainly not always) this is an upward sloping curve, meaning that rates at longer maturities are higher than rates at shorter maturities.

Some amount of estimation is necessary to compute the yield curve. We would like to know yields at arbitrary maturities, but there are not enough actively traded zero-coupon bonds to provide this information. Thus, we have to "fill in" the missing maturities. We may also use coupon-paying bonds (or swap rates, as we will discuss later) to estimate the yields.

The most popular method of estimating the yield curve from bond prices is to fit a "cubic spline," using the prices of a finite set of actively traded bonds. Given the set of bonds, let P_i denote the price of bond i, N_i the number of dates at which bond i pays a coupon or its face value, and $\{\tau_{ij}|j = 1, \ldots, N_i\}$ the dates at which bond i pays a coupon or its face value. Finally, let C_{ij} denote the cash flow paid by bond i at date τ_{ij} for $j = 1, \ldots, N_i$. Then for each i it "should" be the case that

$$P_i = \sum_{j=1}^{N_i} e^{-y(\tau_{ij})\tau_{ij}} C_{ij} . \tag{11.1}$$

This simply says that the price should be the present value of the cash flows. However, in practice, we will typically be unable to find yields $y(\tau_{ij})$ such that (11.1) holds exactly for each bond i. This is due to "measurement errors" in the form of bid-ask spreads and nonsynchronous pricing. Furthermore, even if we can find such yields, we still face the issue of estimating the yields at other maturities $\tau \notin \{\tau_{ij}\}$. The cubic spline is one way to address these issues.

A cubic spline consists of several cubic polynomials "spliced" together at a set of "knot points." Specifically, it consists of maturities $\tau_1, \ldots, \tau_n$ (the "knot points") and coefficients (a_i, b_i, c_i, d_i) for $i = 0, \ldots, n$, and the yield curve is modeled as

$$y(\tau) = \begin{cases} a_0\tau^3 + b_0\tau^2 + c_0\tau + d_0 & \text{for } 0 < \tau \le \tau_1 , \\ a_1\tau^3 + b_1\tau^2 + c_1\tau + d_1 & \text{for } \tau_1 < \tau \le \tau_2 , \\ \cdots & \cdots \\ a_n\tau^3 + b_n\tau^2 + c_n\tau + d_n & \text{for } \tau_n < \tau \le T, \end{cases}$$

---

[1] The relationship between the annually compounded rate y_a, the semi-annually compounded rate y_s and the continuously compounded rate y_c is $e^{y_c} = 1 + y_a = (1 + y_s/2)^2$.

where T is the maximum maturity considered. By changing the constants $d_1, \ldots, d_n$, we can write this in an equivalent and more convenient form:

$$y(\tau) = \begin{cases} a_0\tau^3 + b_0\tau^2 + c_0\tau + d_0 & \text{for } 0 < \tau \leq \tau_1, \\ a_1(\tau - \tau_1)^3 + b_1(\tau - \tau_1)^2 + c_1(\tau - \tau_1) + d_1 & \text{for } \tau_1 < \tau \leq \tau_2, \\ \ldots & \ldots \\ a_n(\tau - \tau_n)^3 + b_n(\tau - \tau_n)^2 + c_n(\tau - \tau_n) + d_n & \text{for } \tau_n < \tau \leq T. \end{cases}$$

$$(11.2)$$

To "splice" these polynomials together at the knot points means to choose the coefficients so that the two polynomials that meet at a knot point have the same value and the same first and second derivatives at the knot point. For example, at the first knot point τ_1, we want the adjacent polynomials to satisfy

Equality of yields: $\qquad\qquad a_0\tau_1^3 + b_0\tau_1^2 + c_0\tau_1 + d_0 = d_1$,
Equality of first derivatives: $\qquad\qquad 3a_0\tau_1^2 + 2b_0\tau_1 + c_0 = c_1$,
Equality of second derivatives: $\qquad\qquad 6a_0\tau_1 + 2b_0 = 2b_1$.

Thus, given the coefficients a_0, b_0, c_0, d_0, the only free coefficient for the second polynomial is a_1. Likewise, at the second knot point τ_2, we want the adjacent polynomials to agree with regard to:

Yields: $\qquad a_1(\tau_2 - \tau_1)^3 + b_1(\tau_2 - \tau_1)^2 + c_1(\tau_2 - \tau_1) + d_1 = d_2$,
First derivatives: $\qquad 3a_1(\tau_2 - \tau_1)^2 + 2b_1(\tau_2 - \tau_1) + c_1 = c_2$,
Second derivatives: $\qquad 6a_1(\tau_2 - \tau_1) + 2b_1 = 2b_2$.

Thus, the only free coefficient for the third polynomial is a_2. Continuing in this way, we see that the cubic spline is defined by the knot points and the coefficients $a_0, b_0, c_0, d_0, a_1, a_2, \ldots, a_n$. One wants to choose the knot points (hopefully, not too many) and these coefficients so that the relations (11.1) hold as closely as possible in some sense. For more on this subject, a good reference is James and Webber [41].

11.2 LIBOR

Many fixed-income instruments have cash flows tied to interest rate indices that are quoted as simple (i.e., noncompounded) interest rates. If you deposit \$1 at an annualized simple interest rate of $\mathcal{R}$ for a period of time Δt, then at the end of the period you will have $1 + \mathcal{R}\,\Delta t$ dollars.[2] The most common interest rate index is LIBOR (London Inter-Bank Offered Rate) which is an average rate in the London inter-bank market for loans for a specific term.

---

[2] The calligraphic symbol $\mathcal{R}$ will be used only for simple interest rates and hopefully will not be confused with the symbol R used for the accumulation factor $R(t) = e^{\int_0^t r(s)\,ds}$ and the corresponding risk-neutral expectation E^R.

For example, if $\mathcal{R}$ denotes six-month LIBOR, a \$1 million deposit at this rate will grow in six months to \$1 million times $1 + \mathcal{R}\,\Delta t$, where $\Delta t = 1/2$. We will use "LIBOR" as a generic term for such indices.

We will frequently find it convenient to express LIBOR rates in terms of equivalent bond prices. As before, let $P(t, u)$ denote the price at date t of a zero-coupon bond with a face value of \$1 maturing at date u, having $\Delta t = u - t$ years to maturity. Previously, we used this notation only for default-free bonds such as Treasuries. However, there is a small amount of credit risk in LIBOR rates because of the possibility of a bank failure. In discussing derivatives linked to LIBOR rates, such as swaps, caps, and floors, we will use the notation $P(t, u)$ for the price of a bond having the same default risk as a LIBOR deposit, but our models will ignore the possibility of default. An investment of \$1 at date t in the bond will purchase $1/P(t, u)$ units of the bond, which, in the absence of default, will be worth $1/P(t, u)$ dollars at maturity. We will assume

$$\frac{1}{P(t, u)} = 1 + \mathcal{R}\,\Delta t\,. \tag{11.3}$$

When necessary for clarity, we call the rate $\mathcal{R}$ a "spot rate" (at date t for the time period Δt), to distinguish it from "forward rates" to be defined later. The spot rate is also called a "floating rate," because it changes with market conditions.

11.3 Swaps

A "plain vanilla" interest rate swap involves the swap of a fixed interest rate for a floating interest rate on a given "notional principal." Let $\bar{\mathcal{R}}$ denote the fixed rate on a swap. The floating rate will be LIBOR (or some other interest rate index). In addition to $\bar{\mathcal{R}}$ and the floating rate index, the swap is defined by "payment dates," which we will denote by $t_1, \ldots, t_N$, with $t_{i+1} - t_i = \Delta t$. In the most common form, the "reset dates" are $t_0, \ldots, t_{N-1}$, with $t_1 - t_0 = \Delta t$ also.

At each reset date t_i, the simple interest rate $\mathcal{R}_i$ for period Δt is observed. This rate determines a payment at the following date t_{i+1}. In terms of bond prices, $\mathcal{R}_i$ is defined in accord with (11.3), substituting date t_i for date t and date t_{i+1} for date u; i.e.,

$$\frac{1}{P(t_i, t_{i+1})} = 1 + \mathcal{R}_i\,\Delta t\,. \tag{11.4}$$

One can enter a swap as the fixed-rate payer, normally called simply the "payer" or as the fixed-rate receiver, normally called the "receiver." The payer pays the fixed rate $\bar{\mathcal{R}}$ and receives the spot rate $\mathcal{R}_i$ at each payment date t_{i+1}. Only the net payment is exchanged. If $\bar{\mathcal{R}} > \mathcal{R}_i$ then the amount $(\bar{\mathcal{R}} - \mathcal{R}_i)\,\Delta t$ is paid at date t_{i+1} by the payer to the receiver for each \$1 of notional principal. If $\bar{\mathcal{R}} < \mathcal{R}_i$, then the payer receives $(\mathcal{R}_i - \bar{\mathcal{R}})\,\Delta t$ from the receiver for each \$1 of notional principal at date t_{i+1}. To state this more simply, the cash flow to the payer is $(\mathcal{R}_i - \bar{\mathcal{R}})\,\Delta t$ and the cash flow to the receiver is $(\bar{\mathcal{R}} - \mathcal{R}_i)\,\Delta t$, with the usual understanding that a negative cash flow is an outflow. Note that there is no exchange of principal at initiation, and there is no return of principal at the end of the swap. The principal is "notional" because it is used only to define the interest payments.

The value of a swap to the fixed-rate payer at any date $t \le t_0$ is

$$P(t, t_0) - P(t, t_N) - \bar{\mathcal{R}}\,\Delta t \sum_{i=1}^{N} P(t, t_i)\,. \tag{11.5}$$

To see this, note that $P(t, t_0)$ is the cost at date t of receiving \$1 at t_0. This \$1 can be invested at t_0 at the rate $\mathcal{R}_0$ and the amount $\mathcal{R}_0\,\Delta t$ withdrawn at t_1 with the \$1 principal rolled over at the new rate $\mathcal{R}_1$. Continuing in this way, one obtains the cash flow $\mathcal{R}_i\,\Delta t$ at each payment date t_{i+1} and the recovery of the \$1 principal at date t_N. The value of the \$1 principal at date t_N is negated in expression (11.5) by the term $-P(t, t_N)$. Thus, $P(t, t_0) - P(t, t_N)$ is the value of the floating rate payments on the notional principal. On the other hand, $\bar{\mathcal{R}}\,\Delta t \sum_{i=1}^{N} P(t, t_i)$ is the value of the fixed-rate payments. Therefore, expression (11.5) is the difference in the values of the floating and fixed-rate legs.

As with a forward price, the swap rate is usually set so that the value of the swap is zero at initiation. A swap initiated at date $t \le t_0$ has zero value at initiation if the fixed rate $\bar{\mathcal{R}}$ equates the expression (11.5) to zero. This means that $\bar{\mathcal{R}} = \mathcal{R}(t)$, where $\mathcal{R}(t)$ is defined by

$$P(t, t_0) = P(t, t_N) + \mathcal{R}(t)\,\Delta t \sum_{i=1}^{N} P(t, t_i)\,. \tag{11.6}$$

If $t = t_0$ the rate $\mathcal{R}(t)$ is a "spot swap rate," and if $t < t_0$ the rate $\mathcal{R}(t)$ is a "forward swap rate." The concept of forward swap rates will be important in the discussion of swaptions in Sect. 12.5. Of course, there are many spot and forward swap rates at any date, corresponding to swaps with different maturities and different payment (and reset) dates.

The "swap yield curve" or simply "swap curve" is the relation between time-to-maturity and the yields of discount bonds, where the discount bond prices and yields are inferred from market swap rates. To explain this in a manner consistent with Sect. 11.1, consider date $t = 0$ and consider swaps with $t_0 = 0$ (i.e., spot swaps). In the notation of Sect. 11.1, and noting that $P(0,0) = 1$, equation (11.6) can be written in terms of yields as

$$1 = e^{-y(t_N)t_N} + \sum_{i=1}^{N} e^{-y(t_i)t_i}\mathcal{R}(0)\,\Delta t\,. \tag{11.7}$$

Consider for example a collection of nineteen swaps at date 0 with semi-annual payments and maturity dates $t_N = 1.0$, $t_N = 1.5, \ldots, t_N = 10.0$. Each market swap rate (a different rate $\mathcal{R}(0)$ for each maturity) can be considered to satisfy (11.7). In these equations we have the twenty yields $y(0.5)$, $y(1.0)$, $y(10.0)$. The yield $y(0.5)$ will be given by six-month LIBOR according to (11.3). The other nineteen yields can be obtained by simultaneously solving the system of nineteen equations of the form (11.7), given the nineteen market swap rates. In practice, there are missing maturities and the swap curve is estimated using a cubic spline or some other technique, as discussed in Sect. 11.1.

11.4 Yield to Maturity, Duration, and Convexity

Consider a bond with cash flows $C_1, \ldots, C_N$ at dates $u_1 < \cdots < u_N$ and price P at date t, where $t < u_1$. Write $\tau_j = u_j - t$ as the time remaining until the j-th cash flow is paid. The (continuously compounded) yield to maturity of the bond at date t is defined to be the rate $\mathbf{y}$ such that

$$P = \sum_{j=1}^{N} e^{-\mathbf{y}\tau_j} C_j . \tag{11.8}$$

The bold character $\mathbf{y}$ is meant to distinguish this from the yield y of a discount bond. Viewing the right-hand side of (11.8) as a function of $\mathbf{y}$, we can express the first derivative in differential form as

$$dP = -\sum_{j=1}^{N} \tau_j e^{-\mathbf{y}\tau_j} C_j \, d\mathbf{y} ,$$

or, equivalently,

$$\frac{dP}{P} = -\sum_{j=1}^{N} \frac{e^{-\mathbf{y}\tau_j} C_j}{P} \tau_j \, d\mathbf{y} .$$

The factor

$$\sum_{j=1}^{N} \frac{e^{-\mathbf{y}\tau_j} C_j}{P} \tau_j$$

is called the "Macaulay duration" of the bond, and we will simplify this to "duration." It is a weighted average of the times to maturity τ_j of the cash flows, the weight on each time τ_j being the fraction of the bond value that the cash flow constitutes (using the same rate $\mathbf{y}$ to discount all of the cash flows). Thus, we have

$$\frac{dP}{P} = -\text{Duration} \times d\mathbf{y} . \tag{11.9}$$

Given the initial yield $\mathbf{y}$ and a change in the yield to $\mathbf{y}'$, this equation suggests the following approximation for the return $\Delta P/P$:

$$\frac{\Delta P}{P} \approx -\text{Duration} \times \Delta \mathbf{y} , \tag{11.10}$$

where $\Delta \mathbf{y} = \mathbf{y}' - \mathbf{y}$.

The relationship (11.10) is the foundation for "duration hedging." For example, to duration hedge a liability with a present value of x_L dollars and a duration of D_L years, one needs an asset with a value of x_A dollars and a duration of D_A satisfying $x_A D_A = x_L D_L$. If the change in the yields to maturity of the asset and liability are the same number $\Delta \mathbf{y}$, then by (11.10) the change in the value of the liability will be approximately $-D_L \Delta \mathbf{y}$ per dollar of initial value, for a total change in value of $-x_L D_L \Delta \mathbf{y}$ dollars. The change in the value of the asset will approximately offset the change in the value of the liability.

Actually, because of the convexity of the bond price (11.8) as a function of the yield $\mathbf{y}$, the approximation (11.10) will overstate the loss on an asset when the yield rises and understate the gain when the yield falls. Given a change in yield $\Delta \mathbf{y}$, the change in price $\Delta P = P' - P$ would actually satisfy

$$\Delta P > -\text{Duration} \times P \times \Delta \mathbf{y} .$$

Thus, if a liability is duration hedged, and the asset value is a "more convex" function of its yield than is the liability value, then an equal change $\Delta \mathbf{y}$ in their yields will lead to a net gain, the asset value falling less than the liability if $\Delta \mathbf{y} > 0$ and gaining more than the liability if $\Delta \mathbf{y} < 0$. The value of convexity (as a function of the yield to maturity) in a bond portfolio is the same as the value of convexity (as a function of the price of the underlying) in option hedging—cf. Sect. 3.5.

In general, the changes in the yields of an asset and a liability (or two different coupon bonds) will not be equal. To understand how the changes will be related, note that the definition (11.8) of the yield to maturity and the formula (11.1) relating a bond price to the yields of discount bonds imply

$$\sum_{j=1}^{N} e^{-\mathbf{y}\tau_j} C_j = \sum_{j=1}^{N} e^{-y(\tau_j)\tau_j} C_j .$$

Taking differentials of both sides, we have

$$dP = -P \times \text{Duration} \times dy = -\sum_{j=1}^{N} \tau_j e^{-y(\tau_j)\tau_j} C_j \, dy(\tau_j) .$$

If we suppose that the changes $dy(\tau_j)$ in the yields of the discount bonds are equal, to, say, dy, then we have

$$dP = -P \times \text{Duration} \times dy = -P \times \text{Duration}' \times dy,$$

where we define

$$\text{Duration}' = \sum_{j=1}^{N} \frac{e^{-y(\tau_j)\tau_j} C_j}{P} \tau_j. \tag{11.11}$$

This new definition of duration differs from the previous by using the yields of discount bonds to define the fraction of the bond value that each cash flow contributes, rather than the yield to maturity. If the changes in the yields of discount bonds are equal, then we say that there has been a "parallel shift" in the yield curve—it has moved up or down with the new curve being at each point the same distance from the old. Our previous discussion shows that duration hedging works if we use the new definition (11.11) of duration and if the yield curve shifts in a parallel fashion. Of course, parallel shifts in the yield curve are not the only, or even most common, types of shifts. In the next two sections, we discuss hedging against more general types of shifts in the yield curve.

Something very similar to duration hedging works if we can continuously rebalance the hedge and there is only a single factor determining the yield curve (meaning a single Brownian motion driving all yields). To understand this, we must first note that the expressions given in this section for dP are not the Itô differential, which explains how the bond price evolves over time and would include a second-derivative term. For example, in (11.9) we are simply asking how different the price would be if the yield to maturity had been different at a given point in time. To define the Itô differential, let now $\mathbf{y}(t)$ denote the yield to maturity of the bond at date t. Equation (11.8) can be restated as

$$P(t) = f(t, \mathbf{y}(t)) = \sum_{j=1}^{N} e^{\mathbf{y}(t)(u_j - t)} C_j . \tag{11.12}$$

Note that even if the yield to maturity were constant over time, the bond price would change with t as a result of the changes in the times to maturity $u_j - t$ of the cash flows. This creates the dependence on t in the function $f(t, \mathbf{y})$.

From Itô's formula, we have

$$dP = \frac{\partial f}{\partial t} \, dt + \frac{\partial f}{\partial \mathbf{y}} \, d\mathbf{y} + \frac{1}{2} \frac{\partial^2 f}{\partial \mathbf{y}^2} (d\mathbf{y})^2 .$$

As explained above, the factor $\partial f/\partial \mathbf{y}$ equals $-\text{Duration} \times P$. The value of convexity appears here in the last term, the derivative $\partial^2 f/\partial \mathbf{y}^2$ being positive as a result of convexity and analogous to the gamma of an option. Assuming $d\mathbf{y}(t) = \alpha(t) \, dt + \sigma(t) \, dB(t)$ for a Brownian motion B and some α and σ, we have

$$\frac{dP}{P} = \frac{1}{P} \left(\frac{\partial f}{\partial t} + \frac{\partial f}{\partial \mathbf{y}} \alpha + \frac{1}{2} \frac{\partial^2 f}{\partial \mathbf{y}^2} \sigma^2 \right) dt - \text{Duration} \times \sigma \, dB .$$

If the yields of an asset and liability are driven by the same Brownian motion and the duration hedge is adjusted for the relative volatilities of the asset and liability (holding more of the asset if its yield volatility is lower), then a duration hedge will hedge the risky part of the change in the liability value. If adjusted continuously, it will provide a perfect hedge, exactly analogous to a delta hedge of an option position. The Vasicek model we will discuss in Chap. 13 and the Cox-Ingersoll-Ross model we will discuss in Chap. 14 are examples of continuous-time models that assume a single Brownian motion driving all yields (i.e., they are single-factor models).The Ho-Lee, Black-Derman-Toy, and Black-Karasinski binomial models that will be discussed in Chap. 14 are also single-factor models and have the same implication for hedging. In the following section, we will discuss the fact that, empirically, there appears to be more than one factor determining the yield curve.

11.5 Principal Components

This section will describe a popular statistical method for determining the factors that have the most impact on the yield curve. We consider yields at fixed maturities $\tau_1, \ldots, \tau_N$, the yield for maturity τ_j at date t being denoted $y(t, \tau_j)$. We assume that we have a sample of past yields at dates $t_0, \ldots, t_M$ at equally spaced dates. Thus, $t_i - t_{i-1} = \Delta t$ for some Δt and each i. We compute the changes in yields:

$$\Delta_{ij} = y(t_i, \tau_j) - y(t_{i-1}, \tau_j) \ .$$

Thus we are looking at the changes in the yield curve over time periods of length Δt, focusing on N points on the yield curve defined by the maturities τ_j. Let V denote the sample covariance matrix of the changes in yields: the element in row j and column k of V is the sample covariance of the changes in yields at maturities τ_j and τ_k; thus, the diagonal elements are the sample variances.[3]

The method of principal components is to compute the "eigenvectors" and "eigenvalues" of the estimated covariance matrix V. An eigenvector is a vector x for which there corresponds a number λ such that $Vx = \lambda x$. The number λ is called the eigenvalue corresponding to the eigenvector x. Given the $N \times N$ symmetric matrix V, we can construct an $N \times N$ matrix C whose columns are eigenvectors of V and an $N \times N$ diagonal matrix D containing the eigenvalues of V on the diagonal. The eigenvectors can be normalized to have unit length and to be mutually orthogonal, which means that the matrix C of eigenvectors has the property that $C^{-1} = C^\top$, where C^{-1} denotes the inverse of C and $C^\top$ its transpose. The property $Vx = \lambda x$ for the columns x of C implies that $VC = CD$. Hence $C^\top VC = C^\top CD = D$.

---

[3] The sample covariance matrix V is an estimate of the "unconditional" covariances. It is a common finding that variances and covariances change over time. Thus, we could (and probably should) use methods such as those described in Chap. 4 to estimate the covariance matrix. The following applies equally well to other estimates V of the covariance matrix.

This can be understood as a factor model for the changes in yields, where there are as many factors as maturities. At date t_i, the vector of factor realizations z_{ij} is computed as

$$\begin{pmatrix} z_{i1} \\ \vdots \\ z_{iN} \end{pmatrix} = C^\top \begin{pmatrix} \Delta_{i1} \\ \vdots \\ \Delta_{iN} \end{pmatrix}. \tag{11.13}$$

Thus,

$$\begin{pmatrix} \Delta_{i1} \\ \vdots \\ \Delta_{iN} \end{pmatrix} = C \begin{pmatrix} z_{i1} \\ \vdots \\ z_{iN} \end{pmatrix}. \tag{11.14}$$

Given any random vector ξ with covariance matrix Σ and a linear transformation $\xi' = L\xi$, the covariance matrix of ξ' is $L\Sigma L^\top$. Therefore, (11.13) implies that the covariance matrix of the factors is $C^\top V C$, and we observed in the previous paragraph that $C^\top V C = D$. Therefore, the factors are uncorrelated, and the factor variances are the eigenvalues of V.

Let $\beta_{k\ell}$ denote the (k, ℓ)–th element of C. Then we can write (11.14) as

$$\begin{aligned} \Delta_{i1} &= \beta_{11} z_{i1} + \cdots + \beta_{1N} z_{iN}, \\ \vdots \ \ \vdots & \qquad\qquad \vdots \\ \Delta_{iN} &= \beta_{N1} z_{i1} + \cdots + \beta_{NN} z_{iN}. \end{aligned} \tag{11.15}$$

As in any factor model, the factors are common to all of the maturities. Each factor is random, taking a different value at each date t_i. The β's represent the "loadings" of the yield changes on the factors, $\beta_{k\ell}$ being the loading of the change in the yield at maturity τ_k on the ℓ-th factor.

In a normal factor model, there are fewer factors than variables being explained. It serves no point to have a factor model with as many factors as there are variables to be explained (in our case, as many factors as maturities). We can improve the usefulness of the above by omitting some factors. We will omit the factors that are least important in explaining the changes in yields. For example, if we omit all but the first three factors, we will have

$$\begin{aligned} \Delta_{i1} &= \beta_{11} z_{i1} + \beta_{12} z_{i2} + \beta_{13} z_{i3} + \varepsilon_{i1}, \\ \vdots \ \ \vdots & \qquad\qquad\qquad \vdots \\ \Delta_{iN} &= \beta_{N1} z_{i1} + \beta_{N2} z_{i2} + \beta_{N3} z_{i3} + \varepsilon_{iN}, \end{aligned} \tag{11.16}$$

where

$$\varepsilon_{ij} = \beta_{j4} z_{i4} + \cdots + \beta_{jN} z_{iN} \tag{11.17}$$

is interpreted as the "residual" part[4] of Δ_{ij}.

---

[4] Frequently, the definition of "factor model" requires the residuals to be uncorrelated, in which case they are called "idiosyncratic risks." Rather than producing uncorrelated residuals, the principal components methods identifies factors such that the residuals are "small."

The importance of a factor depends on the factor loadings and the variance of the factor. The loadings on the j-th factor are the elements in the j-th column of C, which is the j-th eigenvector of V. Because the eigenvectors all have unit length and are mutually orthogonal, each vector of loadings has the same importance as any other for explaining the changes in yields. Thus, the importance of a factor in this model depends on the variance of the factor. The variance of the j-th factor is the j-th element on the diagonal of D, which is the eigenvalue corresponding to the j-th eigenvector. The factors that we should omit are clearly those with small eigenvalues.

As an example, an analysis of monthly changes in (continuously compounded) U.S. Treasury yields from 1992 through 2002 at maturities of 1 month, 3 months, 1 year, 2 years, 3 years, 4 years, and 5 years[5] produces seven eigenvalues (corresponding to the seven maturities) that sum to 5.526×10^{-5}. This sum is the total variance of the seven factors. The largest eigenvalue is 74% of the total, the two largest eigenvalues constitute 94% of the total, and the three largest constitute 98% of the total. Thus, three factors contribute 98% of the total factor variance for this data set, so a factor model with three (or even two) factors explains a very high percentage of the changes in yields for this data set.

The factors can be interpreted by examining the corresponding eigenvectors. The eigenvectors corresponding to the three largest eigenvalues in this data set are the columns below (the first column corresponding to the largest eigenvalue, etc.):

$$
\begin{array}{rrr}
0.1967 & -0.8512 & 0.4782 \\
0.2234 & -0.3740 & -0.6389 \\
0.3775 & -0.1077 & -0.4783 \\
0.4415 & 0.0855 & -0.0530 \\
0.4528 & 0.1532 & 0.0980 \\
0.4428 & 0.2013 & 0.2043 \\
0.4158 & 0.2294 & 0.2834 \\
\end{array}
$$

We can interpret these as follows. A positive value in a given month for the factor with the highest variance will lead to an increase in all of the yields, because all of the elements in the first column are positive (it will also lead to a slight increase in the slope due to the loadings at longer maturities being generally slightly larger than the loadings at smaller maturities). A positive value for the next factor will decrease yields at short maturities and increase the yields at longer maturities, thus leading to an increase in the slope of the yield curve. A positive value for the third factor will lead to an increase in yields at short and long maturities and a decrease in yields at intermediate maturities, thus affecting the curvature of the yield curve. Results of this sort are common for data sets containing longer maturities also, leading to the conclusion that the most important factor is the level of the yield curve, the

---

[5] These were computed from discount bond price and yield data from the Center for Research in Security Prices (CRSP) at the University of Chicago.

second most important factor is the slope of the yield curve, and the third most important factor is the curvature of the yield curve, with three factors explaining nearly all of the variations in yields.

11.6 Hedging Principal Components

Consider again a bond with cash flows $C_1, \ldots, C_N$ at dates $\tau_1 < \cdots < \tau_N$ and price P and recall that the price at date $0 < \tau_1$ should be

$$P = \sum_{j=1}^{N} e^{-y_j \tau_j} C_j \, ,$$

where for convenience we are writing y_j for the yield $y(0, \tau_j)$ of the discount bond maturing at τ_j. Viewing the price as a function of $y_1, \ldots, y_N$, we can write the differential as

$$dP = -\sum_{j=1}^{N} \tau_j e^{-y_j \tau_j} C_j \, dy_j \, .$$

Equivalently,

$$\frac{dP}{P} = -\sum_{j=1}^{N} \frac{e^{-y_j \tau_j} C_j \tau_j}{P} \, dy_j \, .$$

Given discrete changes Δy_j in the yields, this implies the following approximation for the return:

$$\frac{\Delta P}{P} \approx -\sum_{j=1}^{N} \frac{e^{-y_j \tau_j} C_j \tau_j}{P} \, \Delta y_j \, . \tag{11.18}$$

As in Sect. 11.4, because of convexity, the approximation understates the new price $P' = P + \Delta P$. In Sect. 11.4 we considered this equation assuming equal changes in the yields (a parallel shift in the yield curve). Here, we will discuss the more general case.

The approximation (11.18) suggests how we can hedge against the factors identified in the previous section. For example, let $\beta_{1\ell}, \ldots, \beta_{N\ell}$ denote the loadings of the yields at maturities $\tau_1, \ldots, \tau_N$ on the factor with the ℓ-th greatest variance, for $\ell = 1, 2, 3$.[6] Then the factor model suggests

$$\Delta y_j \approx \beta_{j1} z_1 + \beta_{j2} z_2 + \beta_{j3} z_3 \, ,$$

---

[6] It is unlikely that these specific maturities would have been used in the principal components algorithm, so the loadings would have to be estimated by interpolating or fitting some type of curve to the loadings of the maturities that were used.

where the z_k denote the realizations of the factors. Combining this with (11.18) results in

$$\frac{\Delta P}{P} \approx -\sum_{j=1}^{N} \left(\frac{e^{-y_j \tau_j} C_j \tau_j}{P} \sum_{k=1}^{3} \beta_{jk} z_k \right)$$

$$= -\sum_{k=1}^{3} \left(\sum_{j=1}^{N} \frac{e^{-y_j \tau_j} C_j \tau_j \beta_{jk}}{P} \right) z_k.$$

This means that the bond return is approximately a linear combination of the factors, with the coefficient (loading) on the k–th factor being

$$-\sum_{j=1}^{N} \frac{e^{-y_j \tau_j} C_j \tau_j \beta_{jk}}{P}. \tag{11.19}$$

To hedge a liability against the factor, we want this coefficient for the asset multiplied by the dollar value of the asset to equal the corresponding coefficient for the liability multiplied by its dollar value. There are three conditions of this type in a three-factor model, which means that a portfolio of three distinct assets is necessary to hedge a liability. A similar application is in bond portfolio management. If we want to avoid exposing ourselves to factor risk relative to a benchmark portfolio, then we should match the loadings (11.19) for our portfolio with the corresponding loadings of the benchmark.

Problems

Assume the following discount bond prices for each of the following exercises:

$$P(0, 0.5) = 0.995$$
$$P(0, 1.0) = 0.988$$
$$P(0, 1.5) = 0.978$$
$$P(0, 2.0) = 0.966$$
$$P(0, 2.5) = 0.951$$
$$P(0, 3.0) = 0.935$$
$$P(0, 3.5) = 0.916$$
$$P(0, 4.0) = 0.896$$
$$P(0, 4.5) = 0.874$$
$$P(0, 5.0) = 0.850$$

11.1. Compute the six-month and one-year LIBOR rates.

11.2. Compute the swap rate for a two-year swap with semi-annual payments.

11.3. Compute the forward swap rate for a two-year swap with semi-annual payments beginning in

(a) one year,
(b) two years,
(c) three years.

11.4. Compute and plot the continuously-compounded discount-bond yields at maturities 0.5, 1.0, 1.5, ..., 5.0.

11.5. Consider a cubic spline as defined in (11.2) with knot points at two and four years and the following coefficients:

$$a_0 = -0.00163$$
$$b_0 = 0.00812$$
$$c_0 = -0.00676$$
$$d_0 = 0.01184$$
$$a_1 = 0.00052$$
$$b_1 = -0.00169$$
$$c_1 = 0.00609$$
$$d_1 = 0.01770$$
$$a_2 = -0.00175$$
$$b_2 = 0.00141$$
$$c_2 = 0.00552$$
$$d_2 = 0.02725$$

(a) Plot the cubic spline and compare it to the plot from the previous exercise.
(b) Confirm that the adjacent polynomials have the same values, first derivatives, and second derivatives at each of the two knot points.

11.6. Create a VBA function `DiscountBondPrice` that takes the time τ to maturity as an input and returns an estimated price for a discount bond maturing at τ, using the cubic spline formula (11.2), knot points at 2 and 4 years, and the coefficients in the previous exercise. Confirm that it gives approximately the same discount bond prices as those at the beginning of this set of exercises. *A Warning about Extrapolating to Longer Maturities*: What price does the function give for a discount bond maturing in ten years?

11.7. Consider a two-year coupon bond with $1 face value and a semi-annual coupon of $0.03, with the first coupon being six months away.

(a) Compute the price of the bond.
(b) Compute the yield to maturity of the bond using the Excel solver tool.
(c) Compute the duration of the bond.

11.8. Repeat the previous exercise for a one-year coupon bond with $1 face value and a semi-annual coupon of $0.04.

11.9. Repeat the previous exercise for a five-year coupon bond with $1 face value and a semi-annual coupon of $0.02.

11.10. Suppose you have shorted $100 million of the two-year bond with the semi-annual coupon of $0.03. How much of the five-year bond with the semi-annual coupon of $0.02 should you hold to duration hedge the short position?

11.11. Suppose you have shorted $100 million of the two-year bond with the semi-annual coupon of $0.03. Using the data from the principal components example in Sect. 11.6, find a portfolio of the one-year bond with a coupon of $0.04 and the five-year bond with a coupon of $0.02 that will hedge the first two principal components of the short position. Assume the loadings of the six-month yield on the two factors are the averages of the loadings of the one-month and one-year yields, the loadings of the 1.5 year yield on the two factors are the averages of the loadings of the 1.0 and 2.0 years, the loadings of the 2.5 year yield are the averages of the loadings of the 2.0 and 3.0 year yields, etc.

12

Introduction to Fixed Income Derivatives

In this chapter, we will introduce some fundamental fixed-income derivatives (caps, floors and swaptions) and explain the "market model" approach to valuation. We will also explain the relation between caps, floors and swaptions on the one hand and discount and coupon bond options on the other. This leads to other approaches for valuing caps, floors and swaptions, which will be developed in the following chapters.

12.1 Caps and Floors

Caps and floors have a structure very similar to that of swaps, as described in Sect. 11.3. At each reset date t_i, the simple interest rate $\mathcal{R}_i$ for period Δt is observed. This rate determines a payment at the following date t_{i+1}. As discussed in the preceding chapter, the cash flow to the payer in a swap is $(\mathcal{R}_i - \bar{\mathcal{R}})\Delta t$ and the cash flow to the receiver is $(\bar{\mathcal{R}} - \mathcal{R}_i)\Delta t$, for each \$1 of notional principal. A swap is really a series of forward contracts, in which both parties have obligations. On the other hand, caps and floors are series of options. A premium is paid up-front by the buyer of a cap or floor to the seller and all future cash flows are paid by the seller to the buyer. The owner of a cap with cap rate $\bar{\mathcal{R}}$ receives $\max(0, \mathcal{R}_i - \bar{\mathcal{R}})\Delta t$ at date t_{i+1} for each \$1 of notional principal, and the owner of a floor receives $\max(0, \bar{\mathcal{R}} - \mathcal{R}_i)\Delta t$ at date t_{i+1} for each \$1 of notional principal.

Caps and floors are used in conjunction with hedging floating rate obligations or for speculative purposes. Portfolios of caps and floors have properties analogous to option portfolios. For example, the combination of a long cap and a short floor at the same rate $\bar{\mathcal{R}}$ creates the payer side of a swap, in the same way that a long call and short put create a synthetic long forward (and a short cap and a long floor at the same rate $\bar{\mathcal{R}}$ creates the receiver side of a swap just as a short call and long put create a synthetic short forward contract). A long cap at rate $\bar{\mathcal{R}}_c$ and a short floor at rate $\bar{\mathcal{R}}_f < \bar{\mathcal{R}}_c$ creates a collar (for an underlying floating rate obligation), etc.

The individual payments on a cap are called "caplets," and a cap is simply a portfolio of caplets. Similarly, the individual payments on a floor are called "floorlets," and the values of caplets and floorlets are linked by put-call parity, as we will see below.

A caplet can be viewed as a call option on the spot rate with strike equal to the fixed rate. Thus, it is a "bet" on higher interest rates. Because interest rates and bond prices are inversely related, it can also be viewed as a bet on lower bond prices. In this regard, it is similar to a put option on bond prices. In fact, we will see in Sect. 12.7 that a caplet is exactly equivalent to a put option on a discount bond. Likewise, a floorlet can be viewed either as a put option on the spot rate or a call option on a discount bond.

12.2 Forward Rates

Suppose we wish to borrow money at date u for a period of Δt years, and we want to lock in the rate on the loan at date $t < u$. To do this, we can buy the discount bond maturing at u and finance the purchase by shorting $P(t, u)/P(t, u + \Delta t)$ units of the bond maturing at $u + \Delta t$. This generates a cash flow of \$1 at date u and $-P(t, u)/P(t, u + \Delta t)$ dollars at date $u + \Delta t$. This implies a simple interest rate of $\mathcal{R}$ defined as

$$\frac{P(t, u)}{P(t, u + \Delta t)} = 1 + \mathcal{R}\,\Delta t . \tag{12.1}$$

This rate is called a "forward rate."

Forward rates will be important for loans at the reset dates maturing at the subsequent payment dates. We will denote the forward rate at date $t \leq t_i$ for a loan between t_i and t_{i+1} as $\mathcal{R}_i(t)$. This rate is defined in accord with (12.1), substituting the date t_i for date u; i.e.,

$$\frac{P(t, t_i)}{P(t, t_{i+1})} = 1 + \mathcal{R}_i(t)\,\Delta t . \tag{12.2}$$

Note that when $t = t_i$, $P(t, t_i) = 1$, so $\mathcal{R}_i(t_i) = \mathcal{R}_i$ defined in (11.4)—i.e., the forward rate equals the spot rate at t_i.

12.3 Portfolios that Pay Spot Rates

One way to value caps and floors is to view them as portfolios of options on rates, as we will see later. In order to apply the option pricing formulas derived earlier, we need to know that each rate is the value of some asset, so the option can be viewed as an option on an asset. This is very straightforward.

To obtain the spot rate $\mathcal{R}_i$ at date t_{i+1}, one needs \$1 to invest at date t_i. This can be arranged at date $t < t_i$ by buying one unit of the bond maturing at t_i. Investing the dollar paid by the bond at the spot rate at t_i will generate $1 + \mathcal{R}_i \Delta t$ dollars at date t_{i+1}. The "extra" dollar can be eliminated by being short one unit of the bond maturing at t_{i+1}, leaving $\mathcal{R}_i \Delta t$ dollars. Thus, the portfolio that pays the spot rate multiplied by the period length Δt consists of being long one unit of the bond maturing at t_i and short one unit of the bond maturing at t_{i+1}. This implies that the value at date $t < t_i$ of receiving $\mathcal{R}_i \Delta t$ dollars at date t_{i+1} is $P(t, t_i) - P(t, t_{i+1})$.

At date t_i the spot rate $\mathcal{R}_i$ becomes known. Between t_i and t_{i+1}, the value of receiving $\mathcal{R}_i \Delta t$ dollars at date t_{i+1} is the present value of this known cash flow, which is $\mathcal{R}_i \Delta t \, P(t, t_{i+1})$. To summarize, the value of receiving $\mathcal{R}_i \Delta t$ dollars at date t_{i+1} is

$$S_i(t) = \begin{cases} P(t, t_i) - P(t, t_{i+1}) & \text{if } t < t_i \,, \\ \mathcal{R}_i \, \Delta t \, P(t, t_{i+1}) & \text{if } t_i \leq t \leq t_{i+1} \,. \end{cases} \tag{12.3}$$

Actually, we will view a caplet as an option on a forward contract and apply Black's formula. To do this, we need to know the forward price of the asset with price $S_i(t)$ for a contract maturing at t_{i+1}. We denote this forward price by $F_i(t)$. The synthetic forward argument presented in Sect. 7.3 shows that for any non-dividend paying asset with price S, the forward price for a contract maturing at T is $S(t)/P(t, T)$. So, F_i is given by

$$F_i(t) = \begin{cases} \frac{P(t, t_i)}{P(t, t_{i+1})} - 1 & \text{if } t < t_i \,, \\ \mathcal{R}_i \, \Delta t & \text{if } t_i \leq t \leq t_{i+1} \,. \end{cases}$$

$$= \begin{cases} \mathcal{R}_i(t) \, \Delta t & \text{if } t < t_i \,, \\ \mathcal{R}_i \, \Delta t & \text{if } t_i \leq t \leq t_{i+1} \,, \end{cases} \tag{12.4}$$

where $\mathcal{R}_i(t)$ is the forward rate defined in (12.2). Thus, the asset with price $S_i(t)$ pays the spot rate at date t_i times the period length at date t_{i+1}, and the forward price of this asset is the forward rate times the period length.

12.4 The Market Model for Caps and Floors

The valuation we will describe here is standard market practice for valuing caps and floors or at least for quoting the prices of caps and floors. Specifically, it is standard to quote prices in terms of implied volatilities, where the volatility is to be input into Black's formula for options on forwards. This model is sometimes called the "market model."

We can apply Black's formula to the forward contract with price F_i described in the previous section. We view the caplet as a call option maturing at t_i on this forward contract that matures at t_{i+1}. As explained in Chapter 7, the value at the maturity date T of a call option with strike K on a forward contract with price F maturing at $T' \geq T$ is $\max(0, F(T) - K)P(T, T')$.

Therefore, the value at maturity of a call option maturing at t_i with strike $\bar{\mathcal{R}}\,\Delta t$ on the forward contract with price F_i is

$$\max(0, F_i(t_i) - \bar{\mathcal{R}}\,\Delta t)P(t_i, t_{i+1}) \,.$$

Since $F_i(t_i) = \mathcal{R}_i\,\Delta t$, this equals

$$\max(0, \mathcal{R}_i - \bar{\mathcal{R}})\,\Delta t\, P(t_i, t_{i+1}) \,. \tag{12.5}$$

This is also the value of the caplet at date t_i.

It follows that the value of the caplet at any date $t \le t_i$ is the value of the call option on the forward contract. To apply Black's formula, we need the forward price to have a constant (or at least non-randomly varying) volatility. As noted earlier, at dates $t \le t_i$, $F_i(t) = \mathcal{R}_i(t)\,\Delta t$, where $\mathcal{R}_i(t)$ is the forward rate, so the volatility of the forward price is the volatility of the forward rate. Black's formula yields:

Assuming the forward rate $\mathcal{R}_i(t)$ has a constant volatility σ, the value at date $0 < t_i$ of a caplet with reset date t_i and payment date t_{i+1} is

$$P(0, t_{i+1})\mathcal{R}_i(0)\,\Delta t\, N(d_1) - P(0, t_{i+1})\bar{\mathcal{R}}\,\Delta t\, N(d_2) \,, \tag{12.6a}$$

and the value at date $0 < t_i$ of a floorlet with reset date t_i and payment date t_{i+1} is

$$P(0, t_{i+1})\bar{\mathcal{R}}\,\Delta t\, N(-d_2) - P(0, t_{i+1})\mathcal{R}_i(0)\,\Delta t\, N(-d_1) \,, \tag{12.6b}$$

where

$$d_1 = \frac{\log\left(\mathcal{R}_i(0)/\bar{\mathcal{R}}\right) + \frac{1}{2}\sigma^2 t_i}{\sigma\sqrt{t_i}} \,, \tag{12.6c}$$

$$d_2 = d_1 - \sigma\sqrt{t_i} \,. \tag{12.6d}$$

The put-call parity relationship for caplets and floorlets can be seen as follows. If we add $\bar{\mathcal{R}}\Delta t$ to the caplet payment, we obtain

$$\bar{\mathcal{R}}\,\Delta t + \max(0, \mathcal{R}_i - \bar{\mathcal{R}})\,\Delta t = \max(\bar{\mathcal{R}}, \mathcal{R}_i)\,\Delta t \,.$$

On the other hand, if we add $\mathcal{R}_i\,\Delta t$ to the floorlet payment, we obtain the same thing:

$$\mathcal{R}_i\,\Delta t + \max(0, \bar{\mathcal{R}} - \mathcal{R}_i)\,\Delta t = \max(\mathcal{R}_i, \bar{\mathcal{R}})\,\Delta t \,.$$

Hence, the value of a caplet plus the value of receiving $\bar{\mathcal{R}}\,\Delta t$ at date t_{i+1} must equal the value of a floorlet plus the value of receiving $\mathcal{R}_i\,\Delta t$ at date t_{i+1}. The value at any date $t \le t_i$ of receiving $\bar{\mathcal{R}}\,\Delta t$ at date t_{i+1} is the value of $\bar{\mathcal{R}} \Delta t$ discount bonds maturing at date t_{i+1}, which we are denoting by $\bar{\mathcal{R}} \Delta t P(t, t_{i+1})$. The value at at any date $t \leq t_i$ of obtaining $\mathcal{R}_i \Delta t$ dollars at date t_{i+1} is $S_i(t) = P(t, t_i) - P(t, t_{i+1})$. We conclude that

Value of Caplet $+ \bar{\mathcal{R}} \Delta t P(t, t_{i+1}) =$ Value of Floorlet $+ P(t, t_i) - P(t, t_{i+1})$.

12.5 The Market Model for European Swaptions

The owner of a European swaption has an option to enter into a swap at the maturity date of the swaption. A payer swaption gives the owner of the option the right to enter into a swap as a fixed-rate payer, for a given swap rate (not necessarily a rate that makes the swap have zero value at any date). The owner of a receiver swaption has the right to enter into the swap as a fixed-rate receiver. The values of payer swaptions and receiver swaptions are linked by put-call parity.

A payer swaption has similarities to a cap. The owner of a cap has the right to receive the floating rate and pay the fixed rate and will do so in each period in which the floating rate is higher. Similarly, the owner of a payer swaption has the right to receive floating and pay fixed. However, the owner of a cap chooses each period whether to exercise his option, whereas the owner of a swaption makes a once-and-for-all decision whether to exercise, at the maturity of the swaption. A cap is therefore a portfolio of options, whereas a swaption is an option on a portfolio. In general of course, other things being equal, a portfolio of options is worth more than an option on a portfolio.

Consider a payer swaption with maturity date T and swap rate $\bar{\mathcal{R}}$, where the underlying swap has payment dates $t_1, \ldots, t_N$ with the first reset date of the swap being $t_0 = t_1 - \Delta t \geq T$. We assume the notional principal of the swap is \$1. Expression (11.5) in Sect. 11.3 gives the value at date T to the payer in the swap as

$$S(T) - Z(T) ,$$

where we define

$$S(t) = P(t, t_0) - P(t, t_N) , \tag{12.7}$$

$$Z(t) = \bar{\mathcal{R}} \Delta t \sum_{i=1}^{N} P(t, t_i) . \tag{12.8}$$

As explained in Sect. 11.3, $S(t)$ is the value of the floating-rate payments in the swap and $Z(t)$ is the value of the fixed-rate payments. The value of a payer swaption at its maturity T is therefore

$$\max(0, S(T) - Z(T)) .$$

We can value the swaption using Margrabe's formula for exchange options provided the ratio of prices S/Z has a constant (or non-randomly varying) volatility. The volatility of the ratio is the same as the volatility of the forward swap rate. To see this, recall that in Sect. 11.3 the forward swap rate $\mathcal{R}(t)$ was defined to be the rate such that

$$P(t, t_0) - P(t, t_N) = \mathcal{R}(t)\,\Delta t \sum_{i=1}^{N} P(t, t_i)\,,$$

which means that the swap would have zero value if initiated at date t at the rate $\mathcal{R}(t)$—cf. equation (11.6). Thus,

$$\frac{S(t)}{Z(t)} = \frac{P(t, t_0) - P(t, t_N)}{\bar{\mathcal{R}}\,\Delta t \sum_{i=1}^{N} P(t, t_i)} = \frac{\mathcal{R}(t)\,\Delta t \sum_{i=1}^{N} P(t, t_i)}{\bar{\mathcal{R}}\,\Delta t \sum_{i=1}^{N} P(t, t_i)} = \frac{\mathcal{R}(t)}{\bar{\mathcal{R}}}\,,$$

where $\mathcal{R}(t)$ is the forward swap rate. Margrabe's formula implies:[1]

Assuming the forward swap rate $\mathcal{R}(t)$ has a constant volatility σ, the date–0 value of a European payer swaption is

$$[P(0, t_0) - P(0, t_N)]\,N(d_1) - \left[\bar{\mathcal{R}}\,\Delta t \sum_{i=1}^{N} P(0, t_i)\right] N(d_2)\,, \qquad (12.9a)$$

and the date–0 value of a European receiver swaption is

$$\left[\bar{\mathcal{R}}\,\Delta t \sum_{i=1}^{N} P(0, t_i)\right] N(-d_2) - [P(0, t_0) - P(0, t_N)]\,N(-d_1)\,, \qquad (12.9b)$$

where

$$d_1 = \frac{\log\left(P(0, t_0) - P(0, t_N)\right) - \log\left(\bar{\mathcal{R}}\,\Delta t \sum_{i=1}^{N} P(0, t_i)\right) + \frac{1}{2}\sigma^2 T}{\sigma\sqrt{T}}\,,$$

$$\qquad (12.9c)$$

$$d_2 = d_1 - \sigma\sqrt{T}. \qquad (12.9d)$$

Put-call parity for swaptions is as follows: fixed-rate cash flows plus the option to exchange for floating is equivalent to floating-rate cash flows plus the option to exchange for fixed. In each side of this equivalence, one obtains, at the option maturity, the larger of the values of the fixed and floating-rate legs. More formally,

---

[1] To improve the clarity of the typesetting, we have written $\log(S) - \log(Z)$ in (12.9c) instead of our customary $\log(S/Z)$.

$$\bar{\mathcal{R}}\,\Delta t \sum_{i=1}^{N} P(T,t_i) + \max\left(0,\; P(T,t_0) - P(T,t_N) - \bar{\mathcal{R}}\,\Delta t \sum_{i=1}^{N} P(T,t_i)\right)$$

$$= P(T,t_0) - P(T,t_N) + \max\left(0,\; \bar{\mathcal{R}}\,\Delta t \sum_{i=1}^{N} P(T,t_i) - P(T,t_0) + P(T,t_N)\right).$$

Therefore, at any date $t \leq T$,

$$\bar{\mathcal{R}}\,\Delta t \sum_{i=1}^{N} P(t,t_i)\; +\; \text{Value of Payer Swaption}$$

$$= P(t,t_0) - P(t,t_N)\; +\; \text{Value of Receiver Swaption.} \quad (12.10)$$

12.6 A Comment on Consistency

It is well known that it is inconsistent to assume both that the forward rates $\mathcal{R}_i(t)$ and the forward swap rate $\mathcal{R}(t)$ have constant volatilities. We can obtain some intuition for this as follows. Recall that

$$\mathcal{R}_i(t)\,\Delta t = \frac{P(t,t_i) - P(t,t_{i+1})}{P(t,t_{i+1})}\,,$$

and

$$\mathcal{R}(t)\,\Delta t = \frac{P(t,t_0) - P(t,t_N)}{\sum_{i=0}^{N-1} P(t,t_{i+1})}\,.$$

The numerator in the last equation is the sum (over $i = 0,\ldots,N-1$) of the numerators in the previous equation; hence, it is the sum of the $\mathcal{R}_i(t)\,\Delta t\,P(t,t_{i+1})$. This implies that

$$\mathcal{R}(t) = \frac{\sum_{i=0}^{N-1} P(t,t_{i+1})\mathcal{R}_i(t)}{\sum_{i=1}^{N-1} P(t,t_{i+1})}\,,$$

which we can write as

$$\mathcal{R}(t) = \sum_{i=0}^{N-1} w_i(t)\mathcal{R}_i(t)\,,$$

where the weights $w_i(t)$ are defined as

$$w_i(t) = \frac{P(t,t_{i+1})}{\sum_{i=0}^{N-1} P(t,t_{i+1})}\,.$$

Therefore, the forward swap rate is a weighted average of the forward rates. A sum (or average) of lognormal variables is not lognormal, so if the forward rates have constant volatilities, then the forward swap rate will not (and vice versa), absent very peculiar assumptions about the weights $w_i(t)$. This means that one should not really simultaneously use Black's formula for valuing caps (or floors) and Margrabe's formula for valuing swaptions (though there is evidence that the error introduced by doing so may be small). In the following chapters, we will consider other models that do not suffer from this type of inconsistency.

12.7 Caplets as Puts on Discount Bonds

Previously, we considered a caplet as a call option on the forward rate and thus a "bet" on higher interest rates. This is equivalent to a bet on lower bond prices, and we will now show that a caplet with payment date t_{i+1} is equivalent to $1 + \bar{\mathcal{R}}\,\Delta t$ put options on the t_{i+1}–maturity discount bond. The put options mature at the reset date t_i of the caplet and have strike equal to $1/(1 + \bar{\mathcal{R}}\,\Delta t)$. To see this equivalence, note that the value of $1 + \bar{\mathcal{R}}\,\Delta t$ such options at their maturity date t_i is

$$
\begin{aligned}
[1 + \bar{\mathcal{R}}\,\Delta t] \max\left(0, \frac{1}{1 + \bar{\mathcal{R}}\,\Delta t} - P(t_i, t_{i+1})\right) \\
= \max\left(0, 1 - [1 + \bar{\mathcal{R}}\,\Delta t]\,P(t_i, t_{i+1})\right) \\
= P(t_i, t_{i+1}) \max\left(0, \tfrac{1}{P(t_i, t_{i+1})} - 1 - \bar{\mathcal{R}}\,\Delta t\right).
\end{aligned}
$$

Given that $1/P(t_i, t_{i+1}) = 1 + \mathcal{R}_i\,\Delta t$, this equals

$$
P(t_i, t_{i+1}) \max(0, \mathcal{R}_i - \bar{\mathcal{R}})\,\Delta t.
$$

This is the value at date t_i of the caplet with payment date t_{i+1} shown in expression (12.5). It follows that the caplet and the $1 + \bar{\mathcal{R}}\,\Delta t$ put options must have the same value at any date prior to t_i.

Similarly, a floorlet with payment date t_{i+1} is equivalent to $1 + \bar{\mathcal{R}}\,\Delta t$ call options on the t_{i+1}–maturity discount bond, with the call options maturing at date t_i and having strike equal to $1/(1 + \bar{\mathcal{R}}\,\Delta t)$. In the following chapters, we will describe models for valuing bond options. These models will also be applied to price caps, as portfolios of put options on discount bonds, and to price floors, as portfolios of calls.

12.8 Swaptions as Options on Coupon Bonds

As noted previously, the value at date T of the payer swaption, if exercised, is

$$
P(T, t_0) - P(T, t_N) - \bar{\mathcal{R}}\,\Delta t \sum_{i=1}^{N} P(T, t_i).
$$

In Sect. 12.5, we considered this as the difference of two pieces, the first piece being $S(T) = P(T, t_0) - P(T, t_N)$, which is the value of the floating-rate leg, and the second being $Z(T) = \bar{\mathcal{R}} \, \Delta t \sum_{i=1}^{N} P(T, t_i)$, the value of the fixed-rate leg. We can also separate it differently—the first part being $P(T, t_0)$, which is the value at the swaption maturity of the discount bond maturing at t_0, and the second part being $P(T, t_N) + \bar{\mathcal{R}} \, \Delta t \sum_{i=1}^{N} P(T, t_i)$, which is the value of a fixed-rate bond including the face value at maturity. Thus, a payer swaption is equivalent to an option to exchange a fixed-rate coupon bond for a discount bond. A receiver swaption is an option to engage in the reverse exchange.

Typically, $t_0 = T$ (the swap starts at the swaption maturity), in which case $P(T, t_0) = 1$ and the payer (receiver) swaption is a standard put (call) option on the coupon bond, with exercise price equal to 1. The models developed in later chapters for valuing options on coupon bonds can therefore also be applied to value swaptions.

12.9 Calculations in VBA

Valuing Caps with Black's Formula

We will compute the value of each caplet with the `Black_Call` function. In the `Black_Call` function, we input the forward price for the caplet with reset date t_i as

$$F_i(0) = \mathcal{R}_i(0) \, \Delta t = \frac{P(0, t_i)}{P(0, t_{i+1})} - 1 \,, \tag{12.11}$$

as given in equations (12.2) and (12.4). The exercise price of each caplet is $K = \bar{\mathcal{R}} \, \Delta t$. We also need to input the discounting factor $P(0, t_{i+1})$. The discount bond prices $P(0, t_1), \ldots, P(0, t_N)$ are input as a vector P. The discount bond price $P(0, t_0)$ is input as PO.[2] We will assume the same volatility for each forward rate.

```
Function MarketModel_Cap(P0, P, rbar, sigma, N, t0, dt)
'
' Inputs are P0 = price of discount bond maturing at t0
'            P = N-vector of discount bond prices, from t1 to tN
'            rbar = fixed rate in the cap
'            sigma = volatility of the forward LIBOR rates
'            N = number of reset (or payment) dates
'            t0 = time until first reset date
'            dt = time between reset (or payment) dates
'
Dim K, F, i
K = rbar * dt         ' strike prices of caplets
```

---

[2] The reason for separating the bond maturing at t_0 is that vectors input as arrays from Excel worksheets are automatically indexed beginning with the index 1. To keep the indexing consistent with our notation, we want the first element of the input vector to be $P(0, t_1)$, the second element to be $P(0, t_2)$, etc. This consistency is convenient, though obviously not essential.

```
If t0 = 0 Then
'
'   if valuing at the reset date of the first caplet,
'   the caplet value is its intrinsic value
'
    MarketModel_Cap = P(1)*Application.Max(0,1/P(1)-1-rbar*dt)
Else                    ' if valuing prior to maturity of first caplet
    F = P0 / P(1) - 1
    MarketModel_Cap = Black_Call(F, K, P(1), sigma, t0)
End If
For i = 1 To N - 1   ' adds caplets with reset dates t1, ..., t(N-1)
    F = P(i) / P(i + 1) - 1
    MarketModel_Cap = MarketModel_Cap _
                      + Black_Call(F,K,P(i+1),sigma,t0+i*dt)
Next i
End Function
```

Valuing Swaptions with Margrabe's Formula

```
Function MarketModel_Payer_Swaption(P0, P, rbar, sigma, N, T, dt)
'
' Inputs are P0 = price of discount bond maturing at t0
'            P = N-vector of discount bond prices, from t1 to tN
'            rbar = fixed rate in the swap
'            sigma = volatility of the forward swap rate
'            N = number of swap cash flows
'            T = time to maturity
'            dt = time between swap cash flow dates
'
Dim Floating, Fixed, i
Floating = P0 - P(N)          ' value of the floating leg
Fixed = P(1)
For i = 2 To N
    Fixed = Fixed + P(i)
Next i
Fixed = rbar * dt * Fixed    ' value of the fixed leg
MarketModel_Payer_Swaption = Margrabe(Floating,Fixed, sigma,0,0,T)
End Function
```

Problems

12.1. Modify the function MarketModel_Cap so that rather than taking P0 and the vector P of discount bond prices as inputs, it "looks up" discount bond prices from a function DiscountBondPrice that returns a discount bond price for any maturity. For example, DiscountBondPrice might be based on a cubic spline fit to the yield curve as discussed in Sect. 11.1. To test the new function MarketModel_Cap, you will need to create a test function

DiscountBondPrice. For example, you could use the following, which corresponds to a rather steeply increasing yield curve, especially at the short end.

```
Function DiscountBondPrice(t)
DiscountBondPrice = Exp(-t * (0.01 + 0.0052 * t - 0.00012 * t ^ 2))
End Function
```

12.2. Create a function MarketModel_Cap_ImpliedVol that uses bisection to find the forward rate volatility (assume the same volatility for each forward rate) that equates the cap price given by MarketModel_Cap to a market cap price. The function should take the same inputs as MarketModel_Cap except that the forward rate volatility should be replaced by the market cap price.

12.3. Repeat Prob. 12.1 for the function MarketModel_Payer_Swaption.

12.4. Create a function MarketModel_Payer_Swaption_ImpliedVol that uses bisection to find the forward swap rate volatility that equates the swaption price given by MarketModel_Payer_Swaption to a market swaption price. The function should take the same inputs as MarketModel_Payer_Swaption except that the forward swap rate volatility should be replaced by the market swaption price.

12.5. Create a VBA function MarketModel_Floor to value a floor, assuming the forward rates have constant and equal volatilities. Write the function so that it looks up discount bond prices from the DiscountBondPrice function.

12.6. Create a VBA function MarketModel_Receiver_Swaption to value a receiver swaption, assuming the forward swap rate has a constant volatility. Write the function so that it looks up discount bond prices from the DiscountBondPrice function.

12.7. The following exercise is motivated by an example presented in one of my classes by David Eichhorn of NISA Investment Advisors.

(a) Using the DiscountBondPrice function above, calculate what the swap rate should be today for a 10-year swap with semiannual cash flows.
(b) Calculate the value of a 3 × 10 European receiver swaption (an option maturing in 3 years to enter into a 10-year swap as the receiver) with the underlying swap having semiannual cash flows and the fixed rate being equal to the spot swap rate calculated in part (a). Assume the forward swap rate has a constant volatility equal to 0.1.
(c) Consider a 3 × 10 European payer swaption with the underlying swap having semiannual cash flows. Calculate the fixed rate of the swap (using bisection or the Excel Solver tool) that makes the payer swaption have the same value as the receiver swaption calculated in part (b). Assume the forward swap rate has a constant volatility equal to 0.1.
(d) Calculate the forward swap rate for a 10-year swap with semiannual cash flows beginning in 3 years.

Commentary

A collar is an alternative to a forward contract. With standard options, as mentioned in Sect. 1.1, a collar consists of a long call and short put (or the reverse). If the strike of each equals the forward price, then the collar is equivalent to a forward and will have zero cost—by put-call parity, both options have the same price, so the purchase of the call can be financed by selling the put. One can also construct a zero-cost collar with the call and put having different strikes.

A swaption collar is an alternative to a forward swap contract. For example, consider a long receiver swaption with the underlying swap rate being $\bar{\mathcal{R}}_r$ and a short payer swaption with the underlying swap rate being $\bar{\mathcal{R}}_p > \bar{\mathcal{R}}_r$, with the swaptions having the same time to maturity T. If the market swap rate $\mathcal{R}(T)$ at date T is below $\bar{\mathcal{R}}_r$ then the receiver swaption will be in the money and will be exercised—one would rather receive $\bar{\mathcal{R}}_r$ than $\mathcal{R}(T)$ in this circumstance. Likewise, if $\mathcal{R}(T) > \bar{\mathcal{R}}_p$ then the payer swaption will be exercised—one would rather pay $\bar{\mathcal{R}}_p$ than $\mathcal{R}(T)$ in this circumstance. When $\bar{\mathcal{R}}_p > \mathcal{R}(T) > \bar{\mathcal{R}}_r$, neither swaption is in the money. Thus, we have the following for the investor who is long the receiver swaption and short the payer swaption:

- $\mathcal{R}(T) < \bar{\mathcal{R}}_r \implies$ receive $\bar{\mathcal{R}}_r$ and pay floating for the maturity of the swap contract,
- $\bar{\mathcal{R}}_r < \mathcal{R}(T) < \bar{\mathcal{R}}_p \implies$ neither swaption exercised,
- $\mathcal{R}(T) > \bar{\mathcal{R}}_p \implies$ receive $\bar{\mathcal{R}}_p$ and pay floating for the maturity of the swap contract.

Note that one can always pay floating just by borrowing short term; it is receiving fixed that is important here. In the second case above, the investor can engage in a swap at date T and receive the market swap rate $\mathcal{R}(T)$. Thus, the collar guarantees that he will receive at least $\bar{\mathcal{R}}_r$ and will receive no more than $\bar{\mathcal{R}}_p$, whatever might be the market swap rate at date T.

An institution such as a pension fund can use a receiver swap to help hedge its fixed-rate liabilities. As an alternative, it can use a collar as just described. The difference is that a forward receiver swap would guarantee a fixed rate to be received, whereas the collar leaves some residual risk—it guarantees only that the fixed rate will be between $\bar{\mathcal{R}}_r$ and $\bar{\mathcal{R}}_p$. This risk may appear attractive. In Prob. 12.7, the lower rate $\bar{\mathcal{R}}_r$ is the market swap rate, and the higher rate $\bar{\mathcal{R}}_p$ is significantly higher; thus, it may appear that the worst-case scenario is the same as what one can get in the market with a swap and the best case is significantly better. However, this is an illusion, because the appropriate comparison is with a forward swap (starting at the maturity of the swaptions) and the forward swap rate is substantially higher than the spot swap rate (due to the yield curve being steeply upward sloping). This "illusion" may create a good marketing opportunity for sellers of collars.

13

Valuing Derivatives in the Extended Vasicek Model

In the preceding chapter, we presented models for valuing swaptions, caps, and floors based on assumptions that certain forward rates have constant volatilities. As noted, the assumptions used to value swaptions on the one hand and caps and floors on the other are inconsistent. Because the values of fixed-income derivatives are derived from the yield curve, one can obtain a consistent model by developing a model of how the yield curve will evolve over time. In this chapter, we will give a fairly full account of one popular and relatively simple model. The next chapter contains much briefer descriptions of other models.

We will begin by describing the basic Vasicek model. The extended Vasicek model (of which there are several versions) includes time-dependent parameters, so that it can be fit to the yield curve at the time it is used and possibly also to other market variables, such as cap prices or yield volatilities.

13.1 The Short Rate and Discount Bond Prices

An assumption of the Vasicek model and related models discussed in the next chapter is that there is an instantaneously risk-free rate. Letting $r(s)$ denote this rate at date s, the meaning of this assumption, as discussed in Sect. 1.1, is that there is an asset with price process

$$R(t) = \exp\left(\int_0^t r(s)\,\mathrm{d}s\right) .$$

The instantaneous rate of return on this asset is

$$\frac{\mathrm{d}R(t)}{R(t)} = r(t)\,\mathrm{d}t .$$

There is no random term (of the form $\sigma\,\mathrm{d}B$) in this rate of return; thus, we view the return as known at date t, whence the name "instantaneously risk-free." If there were another instantaneously risk-free asset with rate of return

$\hat{r}(t)$ which differed from $r(t)$ on a non-negligible set of dates and states of the world, there would be an arbitrage opportunity. Hence, we will assume there is a unique instantaneously risk-free rate. We will call this rate the "short rate." The interpretation of the price $R(t)$ is that it is the amount that would be accumulated by date t, beginning with $1 at date 0 and continuously rolling over the investment in the instantaneously risk-free asset. It is conceptually similar to the net asset value of a money market fund, so R is sometimes called the price of a money market account. We will also call it an "accumulation factor."

The probability measure associated with R being the numeraire is called the risk-neutral measure, just as when the short rate is constant. Under the risk-neutral measure, the price $P(t,u)$ at date t of a discount bond maturing at u must be

$$P(t,u) = R(t)E_t^R\left[\frac{1}{R(u)}\right] = E_t^R\left[\exp\left(-\int_t^u r(s)\,ds\right)\right], \qquad (13.1)$$

the first equality being a result of our fundamental pricing formula (1.17) and the fact that the discount bond pays $1 at maturity and the second equality following from the definition of R. Thus, a model for the evolution of the short rate under the risk-neutral measure implies a model for discount bond prices and hence the yield curve.

13.2 The Vasicek Model

In the basic Vasicek [62] model, it is assumed that

$$dr(t) = \kappa\big[\theta - r(t)\big]\,dt + \sigma\,dB(t) \qquad (13.2)$$

for constants $\kappa \geq 0$, θ, and σ, where B is a Brownian motion under the risk-neutral measure. In this model, θ is the long-run mean of the short rate process and κ is interpreted as the rate of mean reversion. When $r(t) > \theta$, the drift term will be negative and push r down towards θ. Likewise, when $r(t) < \theta$, the drift term will be positive and push r up towards θ. The rate at which r drifts towards θ is obviously determined by κ. **We are going to depart from our convention and call σ the "volatility" of the short rate** even though $\sigma^2\,dt$ is the instantaneous variance of dr rather than dr/r.

The short rate is normally distributed in the Vasicek model. Given information at date t—i.e., knowledge of $r(t)$—then, if $\kappa > 0$, the rate $r(s)$ for $s > t$ is normally distributed with mean[1]

$$\theta + e^{-\kappa(s-t)}\big[r(t) - \theta\big] \qquad (13.3)$$

---

[1] These results on the mean and variance of $r(s)$ follow from the solution (13.5) of the Vasicek equation (13.2).

and variance

$$\frac{\sigma^2 \left(1 - e^{-2\kappa(s-t)}\right)}{2\kappa} . \tag{13.4}$$

For any $r(t)$, the mean converges to θ at long horizons (i.e., as $s \to \infty$), justifying the interpretation of θ as the long run mean. In fact, the mean converges exponentially to θ at rate κ. The variance at long horizons is $\sigma^2/2\kappa$. On the other hand, if $\kappa = 0$, then the short rate is a Brownian motion with volatility σ. The mean of $r(s)$ is $r(t)$ for $s > t$, and the variance of $r(s)$ is $(s-t)\sigma^2$. Thus, the uncertainty at long horizons, as measured by the variance, is unbounded when $\kappa = 0$. Whether κ is positive or not, an implication of the normal distribution is that there is a positive probability of the short rate $r(s)$ being negative at any date $s > t$, which should not be the case for (nominal) interest rates. However, the probability of negative rates over any given horizon will be small if the variance is sufficiently small. From the formula (13.4), we see that, when $\kappa > 0$, the variance will be small if either the volatility σ is small or the rate of mean reversion κ is large.

These facts about the distribution of the short rate demonstrate the importance of assuming mean reversion ($\kappa > 0$). If $\kappa = 0$, the short rate (because it is a Brownian motion) has the property that for any real number K, with probability one there will be some date $s > t$ such that $r(s) > K$, irrespective of the starting position $r(t)$. Likewise, there will be some date $s' > t$ such that $r(s') < -K$. Thus, the short rate "wanders" off in both directions in an unbounded way. This property may be reasonable for the logarithm of a stock price, because a stock price may become arbitrarily high or arbitrarily close to zero (implying that the logarithm is unbounded both above and below). However, it is not reasonable for an interest rate, which should exhibit more stability. The assumption of mean reversion ($\kappa > 0$) provides this stability, guaranteeing that the uncertainty at long horizons is bounded and that on average the short rate will converge back to a finite long-run mean. Nevertheless, we will give results in this chapter for the case $\kappa = 0$, because this is a particularly simple model and because it is the basic building block for what is called the "continuous-time Ho-Lee model." The actual Ho-Lee model, which is a binomial model, will be discussed in the following chapter.

The relative simplicity of the Vasicek model stems from the fact that under assumption (13.2) the accumulation factor R is lognormally distributed. Specifically, at any date t and given any date $u > t$, the random variable

$$\int_t^u r(s) \, ds$$

is normally distributed. Its mean and variance depend on the parameters κ, θ, and σ and the length of the time interval $u - t$. The mean also depends on the short rate $r(t)$ at date t. Therefore, the discount bond price depends on the same things.

We will now present an explicit formula for discount bond prices, based on the fundamental formula (13.1). By taking the Itô differential of $r(s)$ in the following, one can verify that it is the solution of the Vasicek model (13.2):[2]

$$r(s) = \theta - e^{-\kappa(s-t)}\left[\theta - r(t)\right] + \sigma \int_t^s e^{-\kappa(s-y)}\,dB(y).\tag{13.5}$$

If $\kappa = 0$, this simplifies to

$$r(s) = r(t) + \sigma \int_t^s dB(y) = r(t) + \sigma[B(s) - B(t)]\,.\tag{13.5$'$}$$

These equations imply the following.

In the Vasicek model, consider dates $t < u$ and define $\tau = u - t$. Given the short rate $r(t)$ at date t, the discount bond price has the form

$$P(t, u) = \exp\left(-a(\tau) - b(\tau)r(t)\right)\,,\tag{13.6a}$$

where, if $\kappa = 0$,

$$a(\tau) = -\sigma^2\tau^3/6\,,\tag{13.6b}$$
$$b(\tau) = \tau\,,\tag{13.6c}$$

and, if $\kappa > 0$,

$$a(\tau) = \theta\tau - \frac{\theta}{\kappa}\left(1 - e^{-\kappa\tau}\right) - \frac{\sigma^2}{4\kappa^3}\left(2\kappa\tau - e^{-2\kappa\tau} + 4e^{-\kappa\tau} - 3\right)\,,\tag{13.6b$'$}$$
$$b(\tau) = \frac{1}{\kappa}\left(1 - e^{-\kappa\tau}\right)\,.\tag{13.6c$'$}$$

We will end this section with a proof of (13.6) in the case $\kappa = 0$. The formula for $\kappa > 0$ can be established by the same reasoning, the calculations being only slightly more complicated. From (13.5$'$), we have

$$\int_t^u r(s)\,ds = \int_t^u \left[r(t) + \sigma \int_t^s dB(y)\right]ds$$
$$= \tau r(t) + \sigma \int_t^u \left\{\int_t^s dB(y)\right\}ds\,.$$

We can change the order of integration in the double integral above to obtain

---

[2] Equation (13.5) implies the distributional properties of $r(s)$ stated earlier. The integral $\int_t^s e^{-\kappa(s-y)}\,dB(y)$ is normally distributed with mean zero; therefore equation (13.5) shows that the mean of $r(s)$ in the case $\kappa > 0$ is as given in (13.3). The variance is computed, following the rules in Chap. 2, as $\int_t^s e^{-2\kappa(s-y)}\,dy$, which simplifies to the formula given in (13.4).

$$\int_t^u \left\{ \int_t^s \mathrm{d}B(y) \right\} \mathrm{d}s - \int_t^u \left\{ \int_y^u \mathrm{d}s \right\} \mathrm{d}B(y)$$

$$= \int_t^u (u - y) \, \mathrm{d}B(y) \, .$$

Given the information at date t, this is a normally distributed random variable with mean zero and variance equal to

$$\int_t^u (u - y)^2 \, \mathrm{d}y = \frac{\tau^3}{3} \, .$$

Therefore, $\int_t^u r(s) \, \mathrm{d}s$ is normally distributed with mean $\tau \, r(t)$ and variance $\sigma^2 \tau^3 / 3$. We now use the fact that if x is normally distributed with mean μ and variance σ^2, then the mean of e^x is $\mathrm{e}^{\mu + \sigma^2/2}$. Substituting this into (13.1) gives the result.

13.3 Estimating the Vasicek Model

One way to choose the parameters κ, θ and σ of the Vasicek model is to imply them from market bond prices and the formula (13.6). This is analogous to implying the volatility of a stock from market option prices and the Black-Scholes formula. This can also be viewed as an alternative to fitting a cubic spline, as discussed in Chap. 11. Rather than selecting the parameters of the cubic spline to provide the best approximation to market bond prices, one can choose the Vasicek parameters κ, θ and σ. The formula (13.1) will then give prices of discount bonds of all maturities.[3]

An alternative procedure is to estimate the parameters from historical data on the short rate. The short rate is a theoretical construct and one must choose some proxy to use for empirical work, for example, the Federal Funds rate, or the yield of a short-term (typically one-month or three-month) Treasury bill. One also needs to use a proxy for the short rate when implying the parameters from market bond prices as described in the previous paragraph (or one could view the short rate as one of the parameters to be implied).

An issue that arises when estimating the model from historical data is that the Vasicek equation (13.2) characterizes the evolution of the short rate relative to a Brownian motion under the risk-neutral measure, whereas the historical data is governed by the actual measure. Vasicek [62] actually assumed that (13.2) holds relative to a Brownian motion under the actual probability measure. We will write his assumption as

$$\mathrm{d}r(t) = \kappa^* [\theta^* - r(t)] \, \mathrm{d}t + \sigma^* \, \mathrm{d}B^*(t) \, , \tag{13.7}$$

---

[3] Of course, the fit to market bond prices will typically be poorer with fewer parameters; therefore, the Vasicek yield curve will typically not fit as well as a cubic spline.

where κ^*, θ^*, σ^* are constants and B^* is a Brownian motion under the actual probability measure.[4] Vasicek then assumed a constant "market price of risk" λ, from which it follows that[5]

$$dr(t) = \kappa^*[\theta^* - r(t)]\,dt + \sigma^*\lambda\,dt + \sigma^*\,dB(t)\,, \qquad (13.2')$$

where B is a Brownian motion under the risk-neutral measure. This is the same as (13.2) when we define $\kappa = \kappa^*$, $\sigma = \sigma^*$, and $\theta = \theta^* + \sigma^*\lambda/\kappa^*$. This means that the mean-reversion and volatility parameters are the same under the actual and risk-neutral measures, whereas the long-run mean parameters θ and θ^* are related via the market price of risk. Thus, to estimate κ, θ and σ in (13.2), we could estimate $\kappa^* = \kappa$, $\sigma^* = \sigma$, and θ^* from (13.7) and historical data and then choose θ to best fit market bond prices.

We can estimate the parameters of (13.7) by linear regression. Suppose we have data on a proxy for the short rate at dates $t_0, t_1, \ldots, t_N$, with $t_i - t_{i-1} = \Delta t$ for each i. The solution (13.5) of the Vasicek equation (13.2) for the short rate, adapted to (13.7), implies the following equation for the changes in the short rate:

$$r(t_i) - r(t_{i-1}) = \left(1 - e^{-\kappa^*\Delta t}\right)\theta^* - \left(1 - e^{-\kappa^*\Delta t}\right)r(t_{i-1})$$
$$+ \sigma^*\int_{t_{i-1}}^{t_i} e^{-\kappa^*(t_i - a)}\,dB^*(a)\,.$$

We can write this as

$$r(t_i) - r(t_{i-1}) = a + b\,r(t_{i-1}) + \varepsilon$$

where ε is a normally distributed random variable, independent of $r(t_{i-1})$, with mean zero and variance

$$\sigma^{*2}\int_{t_{i-1}}^{t_i} e^{-2\kappa^*(t_i - a)}\,da = \frac{\sigma^{*2}\left(1 - e^{-2\kappa^*\Delta t}\right)}{2\kappa^*}\,.$$

We can estimate a, b and the variance of ε by linear regression and then obtain κ^*, θ^* and σ^* from the equations

$$a = \left(1 - e^{-\kappa^*\Delta t}\right)\theta^*\,,$$
$$b = -\left(1 - e^{-\kappa^*\Delta t}\right)\,,$$
$$\mathrm{var}(\varepsilon) = \frac{\sigma^{*2}\left(1 - e^{-2\kappa^*\Delta t}\right)}{2\kappa^*}\,.$$

---

[4] Note that our notation is the reverse of what many people use: here B denotes a Brownian motion under the risk-neutral measure (under which we will primarily operate) and B^* denotes a Brownian motion under the actual measure.

[5] The price of risk is the new drift of B^* when we change from the actual measure to the risk-neutral measure. See Appendix B.1.

13.4 Hedging in the Vasicek Model

Because the Brownian motion driving the short rate also drives all the discount bond prices, the bond prices are (instantaneously) perfectly correlated in the Vasicek model (and also in the extensions we will consider next). This is true in any model in which there is a single factor, such as the short rate, that determines all bond prices. The volatilities of the bond returns determine the hedge ratios for hedging one bond with another.

Equation (13.6a) and Itô's formula (the rule that, if $P = e^X$, then $dP/P = dX + (dX)^2/2$) imply that for any date t and any fixed maturity date $u > t$, we have

$$\frac{dP(t,u)}{P(t,u)} = r(t)\,dt - \sigma b(u-t)\,dB(t) \; . \tag{13.8}$$

Because B is a Brownian motion under the risk-neutral measure, this equation implies that the expected rate of return on a discount bond under the risk-neutral measure is the short rate. This is not a surprise, because it is true for every asset under the risk-neutral measure. What is important in this equation is that the volatility is a non-random function of the date t, given the maturity u.

As an example, consider hedging a short two-year bond with a long position in a one-year bond at some date t. According to the formula (13.8), the change in the value of the short position in the two-year bond will be

$$-P(t,t+2)r(t)\,dt - P(t,t+2)\sigma b(2)\,dB(t)$$

Suppose we hold x units of the one-year bond as a hedge and borrow

$$xP(t,t+1) - P(t,t+2)$$

dollars at the instantaneously risk-free rate to finance the hedge. Then the change in the value of the portfolio will be

$$
\begin{aligned}
xP(t,t+1)r(t)\,dt &+ xP(t,t+1)\sigma b(1)\,dB(t) \\
&- P(t,t+2)r(t)\,dt - P(t,t+2)\sigma b(2)\,dB(t) \\
&- [xP(t,t+1) - P(t,t+2)]\,r\,dt \\
&= xP(t,t+1)\sigma b(1)\,dB(t) - P(t,t+2)\sigma b(2)\,dB(t) \; ,
\end{aligned}
$$

so a perfect hedge is obtained by setting

$$x = \frac{b(2)P(t,t+2)}{b(1)P(t,t+1)} \; .$$

In the case $\kappa = 0$, this simplifies to $x = 2P(t,t+2)/P(t,t+1)$, which means that the dollar value $xP(t,t+1)$ of the holding in the one-year bond is twice the dollar value of the short position in the two-year bond. One holds twice as much of the one-year bond, because it is half as volatile as the two-year bond in this model.

By this same reasoning, one can compute a perfect hedge for any discount bond by using another discount bond of any maturity. In fact, one can hedge any term structure derivative in this model by investing the right amount in a bond of any maturity. This is clearly overly simplistic, and such arbitrary hedges would not be used in practice.

There is a close connection between hedging in this model and duration hedging, as discussed in Sect. 11.4. To see this, consider a coupon bond that at date t has remaining cash flows $C_1, \ldots, C_n$ at dates $t + \tau_1 < \cdots < t + \tau_n$ and price $P(t)$. Its price should satisfy

$$P(t) = \sum_{j=1}^{n} C_j P(t, t + \tau_j) .$$

In the Vasicek model, the random part of the change in the price of the coupon bond is

$$-\sigma \left(\sum_{j=1}^{n} b(\tau_j) C_j P(t, t + \tau_j) \right) \, dB(t) ,$$

and the random part of the return is

$$-\sigma \left(\sum_{j=1}^{n} \frac{b(\tau_j) C_j P(t, t + \tau_j)}{P(t)} \right) \, dB(t) . \tag{13.9}$$

Consider the case $\kappa = 0$. Then $b(\tau_j) = \tau_j$, so the random part of the return of the coupon bond is

$$-\sigma \left(\sum_{j=1}^{n} \frac{\tau_j C_j P(t, t + \tau_j)}{P(t)} \right) \, dB(t) .$$

The factor in parentheses is Duration' defined in (11.11). Denoting it now as $D(t)$ and recognizing that the expected rate of return of the bond must be the short rate, we have, in the Vasicek model with $\kappa = 0$,

$$\frac{dP(t)}{P(t)} = r(t) \, dt - \sigma D(t) \, dB(t) .$$

It follows that in this model one can hedge any fixed-income liability by holding enough of any coupon bond such that the dollar value multiplied by its duration equals the dollar value of the liability multiplied by its duration. In Sect. 11.4, we noted that duration matching works for parallel shifts in the yield curve. Later, we will see that in the Vasicek model with $\kappa = 0$ only parallel shifts in the yield curve are possible. Therefore, it should not be surprising that duration matching works in this model.

In the case $\kappa > 0$, one can interpret the volatility of the coupon bond similarly. The analogue to duration for this model is the weighted average of the function $b(\tau_j)$ of the times to maturity, as the formula (13.9) shows. The weights again are the fractions of the bond value that each cash flow contributes.

13.5 Extensions of the Vasicek Model

To fit the model to current market conditions at the time the model is being used, the parameters κ, θ and σ can be taken to be time-dependent. The model with time-dependent parameters is studied in Hull and White [38] and is usually called the Hull-White model. It is convenient to denote by $\theta(t)$ the time-dependent function replacing the constant $\kappa\theta$ in the definition of the Vasicek model (13.2); that is, we redefine as θ what was previously $\kappa\theta$.[6] As usual, we will let date 0 be the date at which we are using the model. The model is then, for $t > 0$,

$$dr(t) = \theta(t)\,dt - \kappa(t)r(t)\,dt + \sigma(t)\,dB(t)\,. \tag{13.10}$$

We will focus on the simplest case, in which κ and σ are constants, deferring discussion of the general case to the last section of this chapter. The general case is quite similar, with the formulas being only slightly more complicated. So, we assume now that

$$dr(t) = \theta(t)\,dt - \kappa r(t)\,dt + \sigma\,dB(t)\,. \tag{13.10$'$}$$

for a non-random function θ. We will call this model with $\kappa > 0$ the Hull-White model. The model with $\kappa = 0$ (i.e., in the absence of mean reversion) is called the continuous-time Ho-Lee model. We will refer to the general case (13.10) as the general Hull-White model, and we will discuss it in Sect. 13.11. In the Hull-White model, we can interpret $\theta(t)/\kappa$ as a time-varying long-run mean of the short rate process, because we have

$$dr(t) = \kappa\left[\frac{\theta(t)}{\kappa} - r(t)\right]\,dt + \sigma\,dB(t)\,.$$

---

[6] Some authors (including Hull and White) do not make this definition—i.e., they denote by $\kappa(t)\theta(t)$ what we are denoting by $\theta(t)$—so be careful when combining results from different sources.

One can show by the following by directly differentiating.

In the Hull-White model,

$$r(s) = \phi(s) + \hat{r}(s) \,, \tag{13.11a}$$

where

$$\phi(s) = \int_0^s e^{-\kappa(s-y)} \theta(y) \, dy \,, \tag{13.11b}$$

and

$$\hat{r}(s) = e^{-\kappa s} r(0) + \sigma \int_0^s e^{-\kappa(s-y)} \, dB(y) \,. \tag{13.11c}$$

When $\kappa = 0$ (the continuous-time Ho-Lee model), these formulas simplify to

$$\phi(s) = \int_0^s \theta(y) \, dy \,. \tag{13.11b'}$$

and

$$\hat{r}(s) = r(0) + \sigma B(s) \,. \tag{13.11c'}$$

Note that (13.11c) and (13.11c') imply

$$d\hat{r}(t) = -\kappa \hat{r}(t) \, dt + \sigma \, dB(t) \,, \tag{13.12}$$

where $\kappa = 0$ for (13.11c'). Thus, $\hat{r}$ is a Vasicek short-rate process having a long-run mean of zero.

The virtue of the expression for r given in (13.11) is that the basic bond pricing equation (13.1) now gives us

$$
\begin{aligned}
P(t, u) &= E_t^R \left[\exp\left(-\int_t^u r(s) \, ds \right) \right] \\
&= E_t^R \left[\exp\left(-\int_t^u \phi(s) + \hat{r}(s) \, ds \right) \right] \\
&= E_t^R \left[\exp\left(-\int_t^u \phi(s) \, ds \right) \exp\left(-\int_t^u \hat{r}(s) \, ds \right) \right] \\
&= \exp\left(-\int_t^u \phi(s) \, ds \right) E_t^R \left[\exp\left(-\int_t^u \hat{r}(s) \, ds \right) \right] \,.
\end{aligned}
\tag{13.13}
$$

Thus, the non-random part $\phi(s)$ of the short-rate process "pulls out" of the expectation in the pricing formula. Furthermore, we have already calculated the expectation in (13.13) because it is the discount bond price in the Vasicek model with a long-run mean of zero. This yields the following.

Consider dates $t < u$ and define $\tau = u - t$. The price at date t of a discount bond maturing at date u in the extended Vasicek model is

$$P(t, u) = \exp\left(-\int_t^u \phi(s)\,\mathrm{d}s - a(\tau) - b(\tau)\hat{r}(t)\right) , \qquad (13.14a)$$

where, in the continuous-time Ho-Lee model,

$$a(\tau) = -\sigma^2 \tau^3/6 , \qquad (13.14b)$$
$$b(\tau) = \tau , \qquad (13.14c)$$

and, in the Hull-White model,

$$a(\tau) = -\frac{\sigma^2}{4\kappa^3}\left(2\kappa\tau - \mathrm{e}^{-2\kappa\tau} + 4\mathrm{e}^{-\kappa\tau} - 3\right) , \qquad (13.14b')$$

$$b(\tau) = \frac{1}{\kappa}\left(1 - \mathrm{e}^{-\kappa\tau}\right) . \qquad (13.14c')$$

Using the fact that the expected rate of return on a discount bond must be the short rate under the risk-neutral measure, we obtain from (13.14a) and (13.12) that, for each fixed maturity date u,

$$\frac{\mathrm{d}P(t, u)}{P(t, u)} = r(t)\,\mathrm{d}t - \sigma b(\tau)\,\mathrm{d}B(t) . \qquad (13.15)$$

Thus, the volatilities are determined by the function b, just as in the basic Vasicek model. The computation of hedge ratios is therefore analogous to the computations presented in Sect. 13.4 for the basic Vasicek model. In fact, the volatilities of discount bond returns in the Hull-White model are the same as in the Vasicek model with $\kappa > 0$, and the volatilities in the continuous-time Ho-Lee model are the same as in the Vasicek model with $\kappa = 0$.

13.6 Fitting Discount Bond Prices and Forward Rates

It is simple to fit the Hull-White model and continuous-time Ho-Lee model to market discount bond prices by choosing the function ϕ. We will use the superscript "mkt" to denote market prices. Letting date 0 denote the date at which we are fitting the model, we want to have

$$P^{\mathrm{mkt}}(0, u) = P(0, u)$$

for all maturities u, where $P(0, u)$ denotes the model prices given in (13.14a) as

$$P(0, u) = \exp\left(-\int_0^u \phi(s)\,\mathrm{d}s - a(u) - b(u)r(0)\right).$$

The functions a and b depend on κ and σ, but we regard them now as having already been chosen. Therefore, to match the model prices to market prices, we simply have to set

$$\exp\left(-\int_0^u \phi(s)\right) = \exp\left(a(u) + b(u)r(0)\right) P^{\mathrm{mkt}}(0, u). \qquad (13.16)$$

Usually we will only need to solve (13.16) for a finite number of maturities u in order to calibrate the model sufficiently. We will illustrate this in the analysis of coupon bond options in Sect. 13.8. However, the equation can be solved in principle for each maturity u as follows: take the natural logarithm of both sides, multiply by minus 1, and differentiate with respect to u to obtain

$$\phi(u) = -\frac{\partial a(u)}{\partial u} - \frac{\partial b(u)}{\partial u}r(0) - \frac{\partial \log P^{\mathrm{mkt}}(0, u)}{\partial u}. \qquad (13.17)$$

Equation (13.17) gives the solution of the model for $\phi(u)$, and it also has an important interpretation. We can obviously rearrange it as

$$\phi(u) + \frac{\partial a(u)}{\partial u} + \frac{\partial b(u)}{\partial u}r(0) = -\frac{\partial \log P^{\mathrm{mkt}}(0, u)}{\partial u}. \qquad (13.17')$$

The expression

$$-\frac{\partial \log P^{\mathrm{mkt}}(0, u)}{\partial u}$$

is the *market* instantaneous forward rate at date 0 for maturity u. The left-hand side of (13.17') is equal to

$$-\frac{\partial \log P(0, u)}{\partial u},$$

and hence is the *model* instantaneous forward rate at date 0 for maturity u. Therefore, fitting the model to the yield curve is equivalent to fitting model forward rates to market forward rates.[7]

---

[7] To understand the interpretation of the derivatives of the log bond prices as forward rates, consider the market prices $P^{\mathrm{mkt}}(0, u)$. Recall from our discussion in Sect. 12.2 that to lock in at date 0 the rate of interest on a \$1 loan from dates u to u', one can buy a unit of the discount bond maturing at u and raise the funds $P^{\mathrm{mkt}}(0, u)$ required by short-selling $P^{\mathrm{mkt}}(0, u)/P^{\mathrm{mkt}}(0, u')$ units of the bond maturing at u', which leads to an obligation of $P^{\mathrm{mkt}}(0, u)/P^{\mathrm{mkt}}(0, u')$ dollars at date u'. The continuously compounded forward rate for the loan between dates u and u' is therefore x defined by

$$\mathrm{e}^{(u'-u)x} = \frac{P^{\mathrm{mkt}}(0, u)}{P^{\mathrm{mkt}}(0, u')}.$$

It is very important to note that if we choose the function $\phi(t)$ at date 0 to match the market, then when we want to recalibrate the model at some later date to match the market at that date, we will have to select a different ϕ function at the later date. In other words, we use a model today that we know we will discard as incorrect tomorrow. The ϕ function (as well as the κ and σ functions in the general Hull-White model) is continually discarded and refit to match the market. This is an unpleasant reality, but it is not really too different from using implied volatilities in the Black-Scholes formula. We know that the implied volatility curve changes over time, so the volatility we use tomorrow may well be different from the volatility we use today. No model is every a literally correct description of the real world. As has been said, the test of the pudding is in the tasting. The test of a model is whether it generates reasonably correct values and hedges. This is an empirical question, and it is not equivalent to the question of whether the assumptions of the model are correct.

13.7 Discount Bond Options, Caps and Floors

We can use Black's formula to value discount bond options in the Vasicek model and in its extensions. As explained in the previous chapter, caps and floors are portfolios of discount bond options, so the values of caps (floors) can be computed by summing the values of the individual caplets (floorlets). The reason we can apply Black's formula is that the volatility of the forward price of a discount bond is non-random in the Vasicek model and its extensions.

Consider valuing at date 0 an option maturing at date T on a discount bond maturing at $u > T$. Because forward must equal spot at maturity, an option written on a forward contract maturing at T is equivalent.[8] The forward price of the discount bond at date $t \leq T$ for a forward contract maturing at T is given by

$$F(t) = \frac{P(t, u)}{P(t, T)}.$$

From Itô's formula and the equation (13.15) for the discount bond returns, we have

---

Equivalently,

$$x = \frac{\log P^{\mathrm{mkt}}(0, u) - \log P^{\mathrm{mkt}}(0, u')}{u' - u} = -\frac{\log P^{\mathrm{mkt}}(0, u') - \log P^{\mathrm{mkt}}(0, u)}{u' - u}.$$

As we make the maturity of the loan shorter, with $u' \to u$, the limit of the above (the forward rate for an instantaneous loan) is by definition the derivative (the usual calculus derivative) of $-\log P^{\mathrm{mkt}}(0, u)$.

[8] This is the same reasoning we used to derive Merton's formulas from Black's formulas in Sect. 7.3.

$$\frac{\mathrm{d}F(t)}{F(t)} = \text{something}\,\mathrm{d}t + \frac{\mathrm{d}P(t,u)}{P(t,u)} - \frac{\mathrm{d}P(t,T)}{P(t,T)}$$

$$= \text{something}\,\mathrm{d}t - \sigma[b(u-t) - b(T-t)]\,\mathrm{d}B(t)\,.$$

Thus, the volatility depends on the non-random function b.

In the continuous-time Ho-Lee model, we have

$$b(u-t) - b(T-t) = u - T\,.$$

Therefore, the forward price has a constant volatility equal to $(u-T)\sigma$. In the Hull-White model, we have

$$b(u-t) - b(T-t) = \frac{1}{\kappa}\left(\mathrm{e}^{-\kappa(T-t)} - \mathrm{e}^{-\kappa(u-t)}\right) = \frac{\left(\mathrm{e}^{-\kappa T} - \mathrm{e}^{-\kappa u}\right)\mathrm{e}^{\kappa t}}{\kappa}\,.$$

Thus, the volatility is time varying (it depends on t), so we compute the average volatility as in Sects. 3.8 and 7.9. Specifically,

$$\begin{aligned}
\sigma_{\mathrm{avg}} &= \frac{\sigma\left(\mathrm{e}^{-\kappa T} - \mathrm{e}^{-\kappa u}\right)}{\kappa}\sqrt{\frac{1}{T}\int_0^T \mathrm{e}^{2\kappa t}\,\mathrm{d}t} \\
&= \frac{\sigma\left(\mathrm{e}^{-\kappa T} - \mathrm{e}^{-\kappa u}\right)}{\kappa}\sqrt{\frac{\mathrm{e}^{2\kappa T} - 1}{2\kappa T}}\,.
\end{aligned} \tag{13.18}$$

Substituting $P^{\mathrm{mkt}}(0,u)/P^{\mathrm{mkt}}(0,T)$ as the forward price of the discount bond maturing at u in Black's formula gives the following.

Consider an option with exercise price K maturing at date T on a discount bond maturing at date $u > T$. In the extended Vasicek model, the values at date 0 of such options are

$$\text{Call Price} = P^{\mathrm{mkt}}(0,u)\,\mathrm{N}(d_1) - P^{\mathrm{mkt}}(0,T)K\,\mathrm{N}(d_2)\,, \tag{13.19a}$$

$$\text{Put Price} = P^{\mathrm{mkt}}(0,T)K\,\mathrm{N}(-d_2) - P^{\mathrm{mkt}}(0,u)\,\mathrm{N}(-d_1)\,, \tag{13.19b}$$

where

$$d_1 = \frac{\log P^{\mathrm{mkt}}(0,u) - \log\left(P^{\mathrm{mkt}}(0,T)K\right) + \frac{1}{2}\sigma_{\mathrm{avg}}^2 T}{\sigma_{\mathrm{avg}}\sqrt{T}}\,, \tag{13.19c}$$

$$d_2 = d_1 - \sigma_{\mathrm{avg}}\sqrt{T}\,. \tag{13.19d}$$

The volatility σ_{avg} is defined as $\sigma_{\mathrm{avg}} = (u-T)\sigma$ in the continuous-time Ho-Lee model and according to the formula (13.18) in the Hull-White model.

As our notation indicates, the discount bond prices appearing in these formulas should be taken to be the *market* prices of the bonds at the date the options are valued, rather than the *model* prices. Of course, if we have fit the model to the market prices of discount bonds, there is no distinction. Using model prices that are different from market prices would be similar to inputting a stock price in the Black-Scholes formula obtained from a discounted-cash-flow analysis rather than using the market price of the stock. This is not something that one would normally do.

Because the discount bond volatilities are the same in the continuous-time Ho-Lee model as in the basic Vasicek model with $\kappa = 0$ and are the same in the Hull-White model as in the basic Vasicek model with $\kappa > 0$, the formulas for the values of discount bond options are the same in the corresponding models. These extensions of the basic Vasicek model permit one to match model bond prices to market bond prices, but they have no effect on the values of discount bond options, provided we remember to use the market prices in (13.19).

As noted before, these option pricing formulas can be used to value caplets and floorlets. A caplet with reset date t_i and payment date t_{i+1} is equivalent, as discussed in Sect. 12.7, to $1 + \bar{R}\Delta$ put options maturing at t_i on the discount bond maturing at t_{i+1}, with the exercise price of each option being $1/(1 + \bar{R}\Delta t)$. When we make these substitutions in the above formulas (and the same for floorlets) we obtain the following.

Consider a caplet and a floorlet with reset date t_i and payment date t_{i+1}. Define $\Delta t = t_{i+1} - t_i$. Let $\bar{R}$ denote the cap and floor rate. In the extended Vasicek model, the values at date 0 of the caplet and floorlet are as follows.

$$\text{Floorlet Price} = (1 + \bar{R}\Delta t)P^{\text{mkt}}(0, t_{i+1})\,\text{N}(d_1) - P^{\text{mkt}}(0, t_i)\,\text{N}(d_2)\,, \tag{13.20a}$$

$$\text{Caplet Price} = P^{\text{mkt}}(0, t_i)\,\text{N}(-d_2) - (1 + \bar{R}\Delta t)P^{\text{mkt}}(0, t_{i+1})\,\text{N}(-d_1)\,, \tag{13.20b}$$

where

$$d_1 = \frac{\log\left((1 + \bar{R}\Delta t)P^{\text{mkt}}(0, t_{i+1})\right) - \log P^{\text{mkt}}(0, t_i) + \frac{1}{2}\sigma_{\text{avg}}^2 t_i}{\sigma_{\text{avg}}\sqrt{t_i}}\,, \tag{13.20c}$$

$$d_2 = d_1 - \sigma_{\text{avg}}\sqrt{t_i}\,, \tag{13.20d}$$

The volatility σ_{avg} is defined as $\sigma\Delta t$ in the continuous-time Ho-Lee model and as

$$\sigma_{\text{avg}} = \frac{\sigma e^{-\kappa t_i}\left(1 - e^{-\kappa \Delta t}\right)}{\kappa}\sqrt{\frac{e^{2\kappa t_i} - 1}{2\kappa t_i}} \tag{13.20e}$$

in the Hull-White model.

A reasonable way to choose the parameter σ in the continuous-time Ho-Lee model and the parameters κ and σ in the Hull-White model would be to fit the model as well as possible to market prices of caps and/or floors. The function $\phi(t)$ can then be chosen as discussed in the preceding section.

We also used Black's formula to value caplets and floorlets in Sect. 12.4. However, the formulas in this section and the formulas in Sect. 12.4 give different values, because they are based on different assumptions. In the Vasicek model and its extensions, the forward price

$$F(t) = \frac{P(t, t_{i+1})}{P(t, t_i)}$$

of the discount bond corresponding to the caplet with payment date t_{i+1} has a non-random volatility, and this is the assumption on which (13.20) is based. In Sect. 12.4, we assumed the forward LIBOR rate had a non-random volatility. The definition (12.2) of the forward rate is

$$R_i(t) = \frac{P(t, t_i)}{P(t, t_{i+1})} - 1 = \frac{1}{F(t)} - 1 .$$

If the forward price has a non-random volatility, then the forward rate is the sum of a variable $(1/F(t))$ with a constant volatility and a constant (-1) and hence will have a random volatility. Likewise, if the forward rate has a non-random volatility, then the forward price will have a random volatility. Therefore, the assumptions of the two models are inconsistent.

13.8 Coupon Bond Options and Swaptions

In this section, we will discuss the valuation of coupon bond options in the extended Vasicek models. As discussed in Sect. 12.8, swaptions are equivalent to options on coupon bonds, so the valuation methods can also be applied to swaptions.

Consider a bond paying a coupon of $\$c$ at dates $t_1, t_2, \ldots, t_N$ and its face value of $\$1$ at date t_N. Consider a European call option on the bond maturing at date T. In the case of a European swaption, all of the coupon payment dates occur after the option maturity $(t_i > T$ for $i = 1, \ldots, N)$, and the coupon is taken to be $c = \bar{R}\,\Delta t$, where $\bar{R}$ is the swap rate. In the case of an option on a coupon bond, some of the coupons may occur before the option maturity, but naturally the value of the option depends only on the coupons that occur after the option matures, because those are the only coupons to which an exerciser of the option would be entitled. By focusing only on those coupons, we can take $t_i > T$ for $i = 1, \ldots, N$.

A coupon bond is a portfolio of discount bonds, so an option on a coupon bond is an option on a portfolio of discount bonds. In general, an option on a portfolio is worth less than a portfolio of options, so an option on a coupon bond should be worth less than a portfolio of discount bond options. However, in any "single-factor" model, it is possible to define the strike prices of the discount bond options so that the option on the coupon bond is worth exactly the same as a portfolio of discount bond options.[9] This reduces the valuation problem for coupon bond options to the problem of valuing discount bond options, which we have already solved for this model. We will use the same technique in Sect. 13.9 to value options on caps ("captions") and options on floors ("floortions").

As shown in (13.14a), the single factor that determines all discount bond prices in this model is the random variable $\hat{r}$. We let r^* denote the value of $\hat{r}(T)$ such that, according to the model, the coupon bond option will be at the money at maturity when $\hat{r}(T) = r^*$. Being at the money means of course that

$$\sum_{i=1}^{N} cP(T, t_i) + P(T, t_N) = K . \tag{13.21}$$

Thus, based on the formula (13.14a) for the discount bond prices, we define r^* by

$$\sum_{i=1}^{N} c \exp\left(-\int_T^{t_i} \phi(s)\, ds - a(t_i - T) - b(t_i - T)r^*\right)$$
$$+ \exp\left(-\int_T^{t_N} \phi(s)\, ds - a(t_N - T) - b(t_N - T)r^*\right) = K . \tag{13.21'}$$

According to the model, if $\hat{r}(T)$ is lower than r^*, then the bond prices will be higher and the call option will be in the money; conversely, if $\hat{r}(T)$ is higher than r^*, then the bond prices will be lower and the call option will be out of the money.

For $i = 1, \ldots, N$, define K_i to be the model value of the discount bond price $P(T, t_i)$ when $\hat{r}(T) = r^*$. In other words, define K_i as

$$K_i = \exp\left(-\int_T^{t_i} \phi(s)\, ds - a(t_i - T) - b(t_i - T)r^*\right) . \tag{13.22}$$

Note that equations (13.21') and (13.22) imply

$$\sum_{i=1}^{N} cK_i + K_N = K . \tag{13.23}$$

Consider for each i ($i = 1, \ldots, N$) a hypothetical call option maturing at T with the underlying for the option being the discount bond maturing at t_i. Let K_i be the exercise price of option i. According to the model, the value of the underlying at date T for option i will be

$$P(T, t_i) = \exp\left(-\int_T^{t_i} \phi(s)\, ds - a(t_i - T) - b(t_i - T)\hat{r}(T)\right) .$$

---

[9] This method was first described by Jamshidian [42].

By the definition of K_i, therefore, option i will be in the money if and only if $\hat{r}(T) < r^*$. Thus, all of the options are in (or out) of the money in the same circumstances.

Now consider the portfolio consisting of c units of option i for $i = 1, \ldots, N-1$ and $1+c$ units of option N. According to the model, the value of the portfolio will be zero at date T if $\hat{r}(T) \geq r^*$, and, if $\hat{r}(T) < r^*$, it follows from (13.23) that the value of the portfolio will be

$$\sum_{i=1}^{N-1} c\big[P(t,t_i) - K_i\big] + (1+c)\big[P(t,T_N) - K_N\big] = \sum_{i=1}^{N} cP(t,t_i) + P(t,T_N) - K,$$

which is the intrinsic value of the coupon bond option when $r(T) < r^*$. Therefore, according to the model, the coupon bond option is equivalent to this portfolio of discount bond options. We can value the discount bond options from the formulas in the previous section and then value the portfolio and hence the coupon bond option by summing. The same type of reasoning also allows us to value put options on coupon bonds.

The only issue in implementing this method is that we need to compute the exercise prices K_i. Of course, we define them by (13.22), given r^*. To do this requires that we calculate the factor

$$\exp\left(-\int_T^{t_i} \phi(s)\,\mathrm{d}s\right), \tag{13.24}$$

by fitting the model to market bond prices, as discussed in Sect. 13.6. We will explain this more explicitly in the next paragraph. In order to apply equation (13.22), we need to compute r^*, which is given by equation (13.21'). We can solve (13.21') for r^* by bisection or some other root-finding method. Again, we need to know the factors (13.24) in order to solve this equation.

To compute the factors (13.24), we use (13.16). We repeat it here twice, for maturity dates t_i and T:

$$\exp\left(-\int_0^{t_i} \phi(s)\,\mathrm{d}s\right) = \exp\left(a(t_i) + b(t_i)r(0)\right) P^{\mathrm{mkt}}(0,t_i),$$

$$\exp\left(-\int_0^{T} \phi(s)\,\mathrm{d}s\right) = \exp\left(a(T) + b(T)r(0)\right) P^{\mathrm{mkt}}(0,T).$$

If we divide the top equation by the bottom, then on the left-hand side we obtain

$$\frac{\exp\left(-\int_0^{t_i} \phi(s)\,\mathrm{d}s\right)}{\exp\left(-\int_0^{T} \phi(s)\,\mathrm{d}s\right)} = \exp\left(-\int_0^{t_i} \phi(s)\,\mathrm{d}s + \int_0^{T} \phi(s)\,\mathrm{d}s\right)$$

$$= \exp\left(-\int_T^{t_i} \phi(s)\,\mathrm{d}s\right),$$

which is the number we want. Hence, we obtain by dividing the right-hand sides of the above equations:

$$\exp\left(-\int_T^{t_i} \phi(s)\,ds\right) = \exp\left\{a(t_i) - a(T) + [b(t_i) - b(T)]r(0)\right\} \frac{P^{\mathrm{mkt}}(0, t_i)}{P^{\mathrm{mkt}}(0, T)}.$$
$$(13.25)$$

13.9 Captions and Floortions

In this section, we will consider options on caps and floors, which are sometimes called "captions" and "floortions" respectively. For specificity, we will consider a call option on a cap. A cap is a portfolio of caplets, and each caplet is a call option on a spot rate. Therefore a call option on a cap is a call option on a portfolio of calls on spot rates and hence is a bet on (or insurance against) high spot rates.

Of course, high spot rates are equivalent to low bond prices, so a call on a cap can also be seen as a bet on low bond prices. We will actually take this latter approach and view a cap as a portfolio of put options on discount bonds, as in Sect. 13.7. This means that we will analyze a call option on a cap as a call on a portfolio of puts. We will use the trick of the previous section and reduce this call on a portfolio of puts to a portfolio of calls on puts. Each call on a put is a compound option and can be valued as in Chap. 8. The assumptions made in Chap. 8 for compound options are valid here, because the forward prices of the discount bonds (which are the underlyings for the puts) have non-random volatilities in the extended Vasicek models.

Consider a cap with reset dates $t_0, \ldots t_{N-1}$ and payment dates $t_1, \ldots, t_N$, with $t_{i+1} - t_i = \Delta t$ for each i. Let $\bar{R}$ denote the fixed rate on the cap. We consider a call option on the cap maturing at date $T \leq t_0$ and having exercise price K. In the model, the value of the cap at the option maturity will depend on the random variable $\hat{r}(T)$. As in the previous section, we will let r^* denote the value of $\hat{r}(T)$ such that the call option on the cap is at the money at date T when $\hat{r}(T) = r^*$. If $\hat{r}(T) > r^*$ then the cap will be more valuable, so the call will be in the money; conversely if $\hat{r}(T) < r^*$ the call will be out of the money. Also, similar again to the previous section, we will let K_i denote the value of the caplet with reset date t_i and payment date t_{i+1} $(i = 0, \ldots, N-1)$ when $\hat{r}(T) = r^*$. We consider hypothetical call options on the individual caplets with exercise prices K_i. If $\hat{r}(T) > r^*$, then, according to the model, all of the calls on the caplets will be in the money, and the sum of their values will equal the value of the call on the cap. On the other hand, if $\hat{r}(T) < r^*$ then all of the calls on the caplets will be out of the money, as will be the call on the cap. Thus, the value of the call on the cap is the sum of the values of the calls on the caplets. In order to apply the methods developed in Chap. 8 for valuing compound options to value the calls on the caplets (i.e., the calls on puts on discount bonds), all we need to do is to calculate the rate r^* and the exercise prices K_i. We will explain this only briefly.

Consider the caplet with payment date t_{i+1}. According to the valuation formula (13.20b), the value of the caplet at date T, when there is a remaining time to maturity of $t_i - T$, will be[10]

Caplet Value at Date $T = P(T, t_i)N(-d_2) - (1 + \bar{R}\Delta t)P(T, t_{i+1})N(-d_1)$,
$$(13.26a)$$

where

$$d_1 = \frac{\log\left((1 + \bar{R}\Delta t)P(T, t_{i+1})\right) - \log P(T, t_i) + \frac{1}{2}\sigma_{\text{avg}}^2(t_i - T)}{\sigma_{\text{avg}}\sqrt{t_i - T}}, \quad (13.26b)$$

$$d_2 = d_1 - \sigma_{\text{avg}}\sqrt{t_i - T}. \tag{13.26c}$$

Here σ_{avg} denotes the average volatility of the forward price of the discount bond between the option valuation date T and the option maturity date t_i. This shows that the values of the caplets and hence the value of the cap depend on the discount bond prices at date T, which, according to the model, depend on $\hat{r}(T)$. One can use bisection or some other root-finding method again to find the value of $\hat{r}(T)$ such that, according to the model, the value of the cap will be K at date T. Substituting this value of $\hat{r}(T)$ into the bond pricing formula (13.14) and then substituting the bond prices $P(T, t_i)$ and $P(T, t_{i+1})$ into (13.26), we define K_i as the caplet value in (13.26a).

13.10 Yields and Yield Volatilities

We are denoting the price at date t of a discount bond maturing at u, with remaining time to maturity of $\tau = u - t$, as $P(t, u)$. However, as in Sect. 11.1, we will denote the corresponding yield as $y(t, \tau)$. The yield is defined as

$$y(t, \tau) = \frac{-\log P(t, u)}{\tau}.$$

In the extended Vasicek model, the yield is given by the bond pricing formula (13.6a) as

$$y(t, \tau) = \frac{\int_t^{t+\tau} \phi(s)\,ds + a(\tau) + b(\tau)\hat{r}(t)}{\tau}.$$

Note that the component

$$\frac{a(\tau) + b(\tau)\hat{r}(t)}{\tau}$$

of the yield $y(t, t + \tau)$ is independent of t, except for its dependence on $\hat{r}(t)$. For example, in the continuous-time Ho-Lee model, this component equals

---

[10] We use now the model prices for the discount bonds at date T. Obviously, the market prices at the caption maturity date T are unknown at date 0, the date at which we are valuing the caption.

$$\frac{\sigma^2 \tau^2}{6} + \hat{r}(t) \; .$$

This is a decreasing function of the time to maturity τ with intercept equal to $\hat{r}(t)$. At a different date t, this component of the yield curve would be the same, except for having a different intercept.[11] This is the reason for the statement in Sect. 13.4 that only parallel shifts in the yield curve are possible in this model (and hence duration hedging works in the model).

The volatility at date t of the yield at a fixed maturity τ is[12] $\sigma b(\tau)/\tau$. In the continuous-time Ho-Lee model, this is σ and in the Hull-White model it is $\sigma \left(1 - e^{-\kappa\tau}\right)/(\kappa\tau)$. In Sect. 13.7 we mentioned that one might choose the parameter σ in the Ho-Lee model and the parameters σ and κ in the Hull-White model to fit cap or floor prices as well as possible. An alternative would be to choose them to match estimated yield volatilities. Of course the fit to either cap prices or yield volatilities will be of limited quality, given that there is only one or two parameters in these models. On the other hand, an exact fit can be obtained with the general Hull-White model.

13.11 The General Hull-White Model

We will briefly discuss the general Hull-White model, in which

$$\mathrm{d}r(t) = \theta(t)\,\mathrm{d}t - \kappa(t)r(t)\,\mathrm{d}t + \sigma(t)\,\mathrm{d}B(t) \; . \qquad (13.27)$$

By differentiating the following, one can see that the short rate r in the general Hull-White model satisfies

$$r(s) = \phi(s) + \hat{r}(s) \; , \qquad (13.28a)$$

where

$$\phi(s) = \int_0^s \exp\left(-\int_y^s \kappa(x)\,\mathrm{d}x\right) \theta(y)\,\mathrm{d}y \; . \qquad (13.28b)$$

and

---

[11] Note that this part decreases to $-\infty$ at long maturities ($\tau \to \infty$), which is very strange for yields, which should be nonnegative. To compensate for this strange behavior, one needs the part

$$\frac{1}{\tau}\int_t^{t+\tau} \phi(s)\,\mathrm{d}s \; ,$$

which is the average of $\phi(s)$ between t and $t + \tau$ to be growing sufficiently with τ. This phenomenon is a consequence of the absence of mean reversion.

[12] By "volatility" here we mean the instantaneous standard deviation of $\mathrm{d}y(t,\tau)$ not the instantaneous standard deviation of $\mathrm{d}y(t,\tau)/y(t,\tau)$.

$$\hat{r}(s) = \exp\left(-\int_0^s \kappa(x)\,\mathrm{d}x\right) r(0) + \int_0^s \exp\left(-\int_y^s \kappa(x)\,\mathrm{d}x\right) \sigma(y)\,\mathrm{d}B(y) \,.$$
(13.28c)

As before, discount bond prices are

$$P(t,u) = \exp\left(-\int_t^u \phi(s)\,\mathrm{d}s\right) E_t^R\left[\exp\left(-\int_t^u \hat{r}(s)\,\mathrm{d}s\right)\right] \,.$$
(13.29)

Moreover, (13.28c) implies, for $s > t$,

$$\hat{r}(s) = \exp\left(-\int_t^s \kappa(x)\,\mathrm{d}x\right) \hat{r}(t) + \int_t^s \exp\left(-\int_y^s \kappa(x)\,\mathrm{d}x\right) \sigma(y)\,\mathrm{d}B(y) \,.$$

Using this fact, the expectation in (13.29) can be calculated as before as the expectation of the exponential of a normally distributed random variable, leading to the result

$$P(t,u) = \exp\left(-\int_t^u \phi(s)\,\mathrm{d}s - a(t,u) - b(t,u)\hat{r}(t)\right) \,,$$
(13.30a)

where

$$b(t,u) = \int_t^u \exp\left(-\int_t^s \kappa(x)\,\mathrm{d}x\right) \mathrm{d}s \,,$$
(13.30b)

$$a(t,u) = -\frac{1}{2}\int_t^u b(y,u)^2 \sigma(y)^2\,\mathrm{d}y \,.$$
(13.30c)

Note that the functions a and b now depend on the date t of valuation and the date u of maturity rather than being determined entirely by the time to maturity $u - t$. Note also that we can write

$$b(t,u) = \exp\left(\int_0^t \kappa(x)\,\mathrm{d}x\right) \int_t^u \exp\left(-\int_0^s \kappa(x)\,\mathrm{d}x\right) \mathrm{d}s$$

$$= \exp\left(\int_0^t \kappa(x)\,\mathrm{d}x\right)$$
(13.31)

$$\times \left[\int_0^u \exp\left(-\int_0^s \kappa(x)\,\mathrm{d}x\right) \mathrm{d}s - \int_0^t \exp\left(-\int_0^s \kappa(x)\,\mathrm{d}x\right) \mathrm{d}s\right]$$

$$= \exp\left(\int_0^t \kappa(x)\,\mathrm{d}x\right) (b(0,u) - b(0,t)) \,.$$
(13.32)

Therefore, we can recover the functions $b(t,u)$ and $a(t,u)$ from the functions $b(0,t)$, $\exp\left(\int_0^t \kappa(x)\,\mathrm{d}x\right)$, and $\sigma(t)$.

The returns of discount bonds satisfy

$$\frac{\mathrm{d}P(t,u)}{P(t,u)} = r(t)\,\mathrm{d}t - \sigma(t)b(t,u)\,\mathrm{d}B(t) . \tag{13.33}$$

Hence, hedge ratios, which depend on relative volatilities, are determined by the function b as before. Given the functions κ and σ, the model can be calibrated to market discount bond prices by choosing the function ϕ exactly as discussed in Sect. 13.6.

The pricing of fixed-income derivatives discussed in Sects. 13.7–13.9 was based on forward prices of discount bonds having non-random volatilities. This is true also in the general Hull-White model. Given dates $t < T < u$, the forward price at date t of the discount bond maturing at date u, when the forward contract matures at T, is, as before, $F(t) = P(t,u)/P(t,T)$. In the general Hull-White model, we have

$$\frac{\mathrm{d}F(t)}{F(t)} = \text{something}\,\mathrm{d}t + \frac{\mathrm{d}P(t,u)}{P(t,u)} - \frac{\mathrm{d}P(t,T)}{P(t,T)}$$

$$= \text{something}\,\mathrm{d}t - \sigma(t)[b(t,u) - b(t,T)]\,\mathrm{d}B(t)$$

$$= \text{something}\,\mathrm{d}t - \sigma(t)\exp\left(\int_0^t \kappa(x)\,\mathrm{d}x\right)[b(0,u) - b(0,T)]\,\mathrm{d}B(t) .$$

Thus, the average volatility, from date 0 to date T, is

$$\sigma_{\text{avg}} = [b(0,u) - b(0,T)]\sqrt{\frac{1}{T}\int_0^T \sigma(t)^2 \exp\left(2\int_0^t \kappa(x)\,\mathrm{d}x\right)\mathrm{d}t} .$$

With this substitution, the pricing of fixed-income derivatives is the same as in the basic Hull-White model discussed in Sects. 13.7–13.9.

The advantage of the general Hull-White model is that it allows more flexibility in fitting the model to current market conditions. This may (though there is certainly no guarantee) provide better pricing and hedging of interest-rate derivatives. Hull and White suggest choosing the volatility function σ to fit anticipated future volatilities of the short rate and choosing the mean-reversion function κ to fit market cap/floor prices or yield volatilities. We will assume that the volatility function σ has been chosen, and we will describe how to choose the mean-reversion function to fit the model to estimated yield volatilities or cap prices. As usual, we let date 0 denote the date at which we are fitting the model.

The simplest approach is to take the function $\kappa(t)$ to be piecewise constant. Consider dates $0 = t_0 < t_1 < \cdots t_M$ with $t_i - t_{i-1} = \Delta t$ for each i. For valuing a swaption, for example, we would want t_M to be the sum of the time to maturity of the swaption and the length of the swap. Thus, we are fitting the model until the end of the swap underlying the swaption. We want to find numbers $\kappa_1, \ldots, \kappa_M$ and will set the function κ to equal κ_i for t between t_{i-1} and t_i.

Given this definition, we have

$$\exp\left(-\int_0^{t_1} \kappa(x)\,dx\right) = e^{-\kappa_1\,\Delta t} ,$$

$$\exp\left(-\int_0^{t_2} \kappa(x)\,dx\right) = e^{-\kappa_1\,\Delta t}e^{-\kappa_2\,\Delta t} ,$$

$$\exp\left(-\int_0^{t_3} \kappa(x)\,dx\right) = e^{-\kappa_1\,\Delta t}e^{-\kappa_2\,\Delta t}e^{-\kappa_3\,\Delta t} ,$$

etc. Furthermore,

$$b(0,t_1) = e^{-\kappa_1\,\Delta t}\Delta t ,$$

$$b(0,t_2) = e^{-\kappa_1\,\Delta t}\,\Delta t + e^{-\kappa_1\,\Delta t}e^{-\kappa_2\,\Delta t}\,\Delta t ,$$

$$b(0,t_3) = e^{-\kappa_1\,\Delta t}\,\Delta t + e^{-\kappa_1\,\Delta t}e^{-\kappa_2\,\Delta t}\,\Delta t + e^{-\kappa_1\,\Delta t}e^{-\kappa_2\,\Delta t}e^{-\kappa_3\,\Delta t}\,\Delta t ,$$

etc. This yields the following recursive structure for the $b(0,t_i)$.

$$b(0,t_1) = e^{-\kappa_1\,\Delta t}\Delta t , \tag{13.34a}$$

$$b(0,t_2) = b(0,t_1) + b(0,t_1)e^{-\kappa_2\,\Delta t} , \tag{13.34b}$$

$$(\forall i \geq 3) \quad b(0,t_i) = b(0,t_{i-1}) + [b(0,t_{i-1}) - b(0,t_{i-2})]e^{-\kappa_i\,\Delta t} . \tag{13.34c}$$

Matching the yield volatilities in the model to estimated market volatilities is extremely simple. The same analysis presented in Sect. 13.10 for the basic Hull-White model shows that the volatility at date 0 of the yield at maturity t_i in the general Hull-White model is

$$\frac{\sigma(0)b(0,t_i)}{t_i} .$$

Given the estimated yield volatility, we can define $b(0,t_i)$ by equating the model volatility to the estimated volatility. Using (13.34), the κ_i can be computed from the $b(0,t_i)$ for $i = 1, 2, \ldots$.

It is equally easy to match the model to market cap or floor prices. Suppose we have estimated implied volatilities for caplets using the pricing formula (13.20). Suppose the first caplet has reset date t_1 and payment date t_2, the second has reset date t_2 and payment date t_3, etc. According to the model, the average volatility of the forward price of the discount bond underlying the caplet with reset date t_i and payment date t_{i+1} is

$$[b(0,t_{i+1}) - b(0,t_i)]\sqrt{\frac{1}{t_i}\int_0^{t_i} \sigma(t)^2 \exp\left(2\int_0^t \kappa(x)\,dx\right)dt} .$$

Notice that everything in this expression, except $b(0,t_{i+1})$, is determined by $\kappa_1, \ldots, \kappa_i$ and the σ function. Therefore, we can select the number $b(0,t_{i+1})$ to match the model volatility to the implied volatility for this caplet. This defines κ_{i+1} from (13.34). Continuing in this way, we can successively define $\kappa_1, \kappa_2, \ldots$.

13.12 Calculations in VBA

Discount Bond Options

Rather than calculating σ_{avg} from the volatility σ of the short rate and the constant κ or function $\kappa(t)$, we will treat σ_{avg} as an input into the option pricing formulas. This is necessary if one wants to use (i.e., invert) the following functions to imply σ_{avg} from market prices and then used the implied σ_{avg} to calibrate the model.

```
Function Vasicek_Discount_Bond_Call(Underlying,K,MatDisc,sigavg,T)
'
' Inputs are Underlying = price of underlying discount bond
'            K = strike price
'            MatDisc = price of bond maturing when option matures
'            sigavg = average volatility of the forward bond price
'            T = time to maturity
'
Dim F
F = Underlying / MatDisc   ' forward price of the underlying
Vasicek_MatDisc_Bond_Call = Black_Call(F, K, MatDisc, sigavg, T)
End Function


Function Vasicek_Discount_Bond_Put(Underlying,K,MatDisc,sigavg,T)
'
' Inputs are Underlying = price of underlying discount bond
'            K = strike price
'            MatDisc = price of bond maturing when option matures
'            sigavg = average volatility of the forward bond price
'            T = time to maturity
'
Dim F
F = Underlying / MatDisc   ' forward price of the underlying
Vasicek_Discount_Bond_Put = Black_Put(F, K, MatDisc, sigavg, T)
End Function
```

Caps

We input the first reset date t_0, the time between payments Δt, the total number of payments N and the cap rate $\bar{\mathcal{R}}$. As in Sect. 12.9, we input the price of the discount bond maturing at t_0 as P0, and we input a vector P containing the N discount bond prices $P(0, t_1), \ldots, P(0, t_N)$. We will allow different volatilities for the different caplets. The input sigavg should have elements 1 through N, with sigavg(i) being the average volatility of the forward price of the discount bond maturing at t_i, where the forward contract matures at t_{i-1}, and the average volatility is computed from date 0 to date

t_{i-1}, according to the formulas in Sect. 13.7. Specifically, sigavg(i) should be $\sigma \Delta t$ for the continuous-time Ho-Lee model, and sigavg(i) should be

$$\frac{\sigma e^{-\kappa t_{i-1}} \left(1 - e^{-\kappa \Delta t}\right)}{\kappa} \sqrt{\frac{e^{2\kappa t_{i-1}} - 1}{2\kappa t_{i-1}}} \tag{13.35}$$

for the Hull-White model.

```
Function Vasicek_Cap(P0, P, rbar, sigavg, N, t0, dt)
'
' Inputs are P0 = price of discount bond maturing at t0
'            P = N-vector of discount bond prices, from t1 to tN
'            rbar = fixed rate in the cap
'            sigavg = N-vector of average vols of forward prices
'            N = number of reset (or payment) dates
'            t0 = time until first reset date
'            dt = time between reset (or payment) dates
'
Dim K, x, MatDisc, Underlying, mat, i
K = 1 / (1 + rbar * dt) ' exercise price of each caplet
If t0 = 0 Then
'   if valuing at the reset date of the first caplet,
'   the value of the first caplet is its intrinsic value
    x = P(1) * Application.Max(0, 1 / P(1) - 1 - rbar * dt)
Else        ' if valuing before maturity of first caplet
    MatDisc = P0
    Underlying = P(1)
    x = _
    Vasicek_Discount_Bond_Put(Underlying,K,MatDisc,sigavg(1),t0)
End If
For i = 1 To N - 1
    MatDisc = P(i)          ' price of bondmaturing at reset date
    Underlying = P(i + 1)   ' price of bondmaturing at payment date
    mat = t0 + i * dt       ' reset date of i-th caplet
    x = x + _
    Vasicek_Discount_Bond_Put(Underlying,K,MatDisc,sigavg(i+1),mat)
Next i
Vasicek_Cap = (1 + rbar * dt) * x ' each caplet is (1+rbar*dt) puts
End Function
```

Coupon Bond Options in the Hull-White Model

We will value a European call option on a coupon bond. First we create functions to calculate $a(\tau)$ defined in (13.14b'), $b(\tau)$ defined in (13.14c'), and σ_{avg} defined in (13.18).

```
Function hwa(sigma, kappa, tau)
hwa = -sigma ^ 2 * (2 * kappa * tau - Exp(-2 * kappa * tau) _
```

```
        + 4 * Exp(-kappa * tau) - 3) / (4 * kappa ^ 3)
End Function

Function hwb(kappa, tau)
hwb = (1 - Exp(-kappa * tau)) / kappa
End Function

Function hwsigavg(sigma, kappa, T, u)
hwsigavg = (sigma * (Exp(-kappa * T) - Exp(-kappa * u)) / kappa) _
          * Sqr((Exp(2 * kappa * T) - 1) / (2 * kappa * T))
End Function
```

We calculate the model value of the coupon bond at date T (given by the left-hand side of (13.21′) for $r = r^*$) in the function HW_Coup_Bond. The vector Cal contains N elements, and Cal(i) represents the calibration factor $e^{-\int_T^{t_i} \phi(s)\,ds}$, which will be calculated from (13.25) in the main function HW_Coup_Bond_Call.

```
Function HW_Coup_Bond(Coup,N,t1,dt,Cal,r,sigma,kappa,T)
Dim x, tau, a, b, i
x = 0
For i = 1 To N - 1                     ' coupon dates before maturity
    tau = t1 + (i - 1) * dt - T                ' tau = ti  - T
    a = hwa(sigma, kappa, tau)                 ' a(ti - T)
    b = hwb(kappa, tau)                        ' b(ti - T)
    x = x + Coup * Cal(i) * Exp(-a - b * r)
Next i
tau = t1 + (N - 1) * dt - T                    ' tau = tN - T
a = hwa(sigma, kappa, tau)                     ' a(tN - T)
b = hwb(kappa, tau)                            ' b(tN - T)
x = x+(1+Coup)*Cal(N)*Exp(-a-b*r)   ' face and coupon at maturity
HW_Coup_Bond = x
End Function
```

In the main function HW_Coupon_Bond_Call, we input the coupon paid by the bond, the number N of coupon payments, the amount of time t1 before the first coupon payment, the amount of time dt between coupon payments, the strike price K of the option, the price MatDisc of the discount bond maturing at the option maturity date T, a vector P containing the N market prices of the discount bonds maturing at the coupon payment dates $t_1,\ldots,t_N$, the current short rate r, the parameters sigma and kappa of the Vasicek model, and the amount of time T until the option matures.

```
Function HW_Coup_Bond_Call(Coup,N,t1,dt,K,MatDisc,P,r,sigma,kappa,T)
'
' Inputs are Coup = amount of each coupon
'               N = number of coupons
'               t1 = time until first coupon
'               dt = time between coupon payment dates
```

```
'              K = strike price
'              MatDisc = price of bond maturing when option matures
'              P = N-vector of bond prices from t1 to tN
'              r = initial value of the short rate
'              sigma = volatility of the short rate
'              kappa = mean-reversion of the short rate
'              T = time to maturity
'
Dim tol, guess, fguess, rstar, x, tau, lower, flower, upper, fupper
Dim a, b, aT, bT, strike, sigavg, Cal(), i
ReDim Cal(1 To N)
'
' First we calibrate the model to market bond prices
'
aT = hwa(sigma, kappa, T)
bT = hwb(kappa, T)
For i = 1 To N
    tau = t1 + (i - 1) * dt                      ' tau = ti
    a = hwa(sigma, kappa, tau)                   ' a(ti)
    b = hwb(kappa, tau)                          ' b(ti)
    Cal(i) = Exp(a - aT + (b - bT) * r) * P(i) / MatDisc
Next i
'
'  We find upper and lower bounds for rstar
'
lower = 0                        ' first try at lower bound
flower = HW_Coup_Bond(Coup,N,t1,dt,Cal,lower,sigma,kappa,T)-K
Do While flower < 0          ' reduce lower for a lower bound
    lower = lower - 1
    flower = _
    HW_Coup_Bond(Coup,N,t1,dt,Cal,lower,sigma,kappa,T)-K
Loop
upper = 1                        ' first try at upper bound
fupper = HW_Coup_Bond(Coup,N,t1,dt,Cal,upper,sigma,kappa,T)-K
Do While fupper > 0          ' increase upper for an upper bound
    upper = upper + 1
    fupper = _
    HW_Coup_Bond(Coup,N,t1,dt,Cal,upper,sigma,kappa,T)-K
Loop
'
' Now we do the bisection to find rstar
'
tol = 10 ^ -8
guess = 0.5 * lower + 0.5 * upper
fguess = HW_Coup_Bond(Coup,N,t1,dt,Cal,guess,sigma,kappa,T)-K
Do While upper - lower > tol
    If fupper * fguess < 0 Then
        lower = guess
        flower = fguess
```

```
            guess = 0.5 * lower + 0.5 * upper
            fguess = _
            HW_Coup_Bond(Coup,N,t1,dt,Cal,guess,sigma,kappa,T)-K
        Else
            upper = guess
            fupper = fguess
            guess = 0.5 * lower + 0.5 * upper
            fguess = _
            HW_Coup_Bond(Coup,N,t1,dt,Cal,guess,sigma,kappa,T)-K
        End If
Loop
rstar = guess
'
' Now we calculate the exercise prices and
' sum the values of the discount bond options
'
For i = 1 To N - 1
    tau = t1 + (i - 1) * dt - T            ' tau = ti - T
    a = hwa(sigma, kappa, tau)             ' a(ti - T)
    b = hwb(kappa, tau)                    ' b(ti - T)
    strike = Cal(i) * Exp(-a - b * rstar)  ' Ki
    sigavg = hwsigavg(sigma,kappa,T,T+tau) ' sigavg for option i
    x = x + _
    Coup*Vasicek_Discount_Bond_Call(P(i),strike,MatDisc,sigavg,T)
Next i
tau = t1 + (N - 1) * dt - T                ' tau = tN
a = hwa(sigma, kappa, tau)                 ' a(tN - T)
b = hwb(kappa, tau)                        ' b(tN - T)
strike = Cal(N) * Exp(-a - b * rstar)      ' KN
sigavg = hwsigavg(sigma,kappa,T,T+tau)     ' sigavg for option N
x = x + _
(1+Coup)*Vasicek_Discount_Bond_Call(P(N),strike,MatDisc,sigavg,T)
HW_Coup_Bond_Call = x
End Function
```

Problems

13.1. Modify the function Vasicek_Cap so that rather than taking the vector P of discount bond prices as an input, it "looks up" discount bond prices from a function DiscountBondPrice that returns a discount bond price for any maturity, as in Exercise 12.12.1.

13.2. Create a function ContinuousHoLee_Cap by modifying the function Vasicek_Cap so that the average volatilities are computed as $\sigma \Delta t$. In other words, input σ rather than the vector σ_{avg}. Write the function so that it looks up discount bond prices as in the previous exercise.

13.3. Create a function `ContinuousHoLee_ImpliedVol` using bisection that takes the same inputs as the previous function, except taking a cap price as input rather than σ, and which returns the volatility σ that is consistent with the cap price.

13.4. Create a function `HW_Cap` by modifying the function `Vasicek_Cap` so that the average volatilities are calculated from equation (13.35). Thus, σ and κ should be input rather than the vector σ_{avg}. Write the function so that it looks up discount bond prices as in the previous exercises.

13.5. Create an Excel worksheet in which the user inputs $\bar{r}$, σ and κ. Compute the values of caps using the function `HW_Cap` for caps of length $N = 1, \ldots 20$. Take $t_0 = 0$ and $\Delta t = 0.5$ and look up discount bond prices. For each N, use the Excel solver tool or the function created in Exercise 13.3 to compute the volatility σ for which the `ContinuousHoLee_Cap` function gives the same cap price. In other words, compute the implied Ho-Lee volatilities, given the cap prices. What is the pattern in implied volatilities and why?

13.6. Create a VBA function `Vasicek_Floor` to value a floor in the extended Vasicek model, looking up discount bond prices from the `DiscountBondPrice` function.

13.7. Modify the function `HW_Coup_Bond_Call` so that it looks up discount bond prices from the `DiscountBondPrice` function.

13.8. Create a VBA function `HW_Coup_Bond_Put` to value a put option on a coupon bond in the basic Hull-White Model, looking up discount bond prices from the `DiscountBondPrice` function.

13.9. Create a VBA function `HW_Payer_Swaption` to value a payer swaption in the basic Hull-White Model, looking up discount bond prices from the `DiscountBondPrice` function.

13.10. Create a VBA function `HW_Receiver_Swaption` to value a receiver swaption in the basic Hull-White Model, looking up discount bond prices as in previous exercises.

13.11. Repeat Prob. 12.7, assuming the basic Hull-White Model, for various values of σ and κ.

A Brief Survey of Term Structure Models

This chapter presents very brief descriptions of several important models. The list of models is certainly not exhaustive, and our descriptions will be far from complete. As mentioned previously, there are many good references for this material in book form already, and our goal here is merely to provide a short introduction.

14.1 Ho-Lee

The Ho-Lee [36] model is a binomial version of the Vasicek model without mean reversion, in which the one-period interest rate is assumed to have a deterministic drift. This was the first widely-used model that enabled one to fit the current yield curve.

Consider discrete times $0 = t_0 < t_1 < \cdots t_N$ with $t_i - t_{i-1} = \Delta t$ for each i. We denote the one-period interest rate from date t_i to t_{i+1} by $r(t_i)$. We express the rate as an annualized continuously compounded rate, so the one-period discount factor from date t_i to t_{i+1} is $e^{-r(t_i)\Delta t}$. We could put this in a continuous-time framework by assuming the short rate is constant $(= r(t_i))$ during each time interval (t_i, t_{i+1}). As always, the risk-neutral measure is the probability measure corresponding to the numeraire

$$R(t) = \exp\left(\int_0^t r(s)\,\mathrm{d}s\right). \tag{14.1}$$

However, we will only be doing valuation at the discrete dates $t_0, \ldots, t_N$ and at date t_i the accumulation factor (14.1) is

$$R(t_i) = \exp\left(\sum_{j=0}^{i-1} r(t_j)\,\Delta t\right).$$

Thus, the continuous-time framework is not necessary.

The model assumes that over each time period $t_i - t_{i-1}$ the change $\Delta r(t_i) = r(t_i) - r(t_{i-1})$ in the one-period rate is

$$\Delta r(t_i) = \theta(t_i)\,\Delta t \pm \sigma\sqrt{\Delta t}\,,$$

where the risk-neutral probability of "+" and "−" is $1/2$ each and θ is a non-random function. As in the discussion of the extended Vasicek model, it is convenient to define a random process $\hat{r}$ by $\hat{r}(0) = r(0)$ and

$$\Delta\hat{r}(t_i) = \pm\sigma\sqrt{\Delta t}$$

for each i. Also, define $\phi(t_i) = \sum_{j=1}^{i}\theta(t_i)$ for $i \geq 1$. Then we have, for $i > 0$,

$$r(t_i) = \phi(t_i) + \hat{r}(t_i)\,.$$

The following illustrates a three-period tree with initial one-period rate $r_0 = 5\%$, $\Delta t = 1$, $\theta(t_i) = 0$ for all i, and $\sigma = 1\%$.

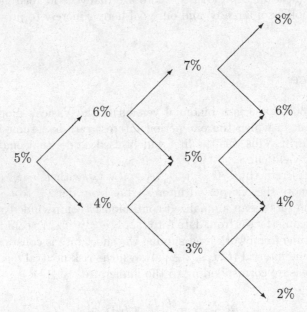

To value a fixed-income derivative, we discount the terminal value backwards through the tree as in Chaps. 5 and 9. The new feature is that the discount rate is changing over time. So, we now have three trees to consider: the tree for the underlying, the tree for the one-period interest rate, and the tree for the derivative value. However, the tree for the underlying can be created from the interest-rate tree, so the interest-rate tree is the basic input instead of the tree for the underlying.

To clarify this, we will start with the simplest example: valuing a discount bond. Consider the above interest-rate tree and a discount bond that matures at date 3. The value of the bond is 1 at maturity. Because we are using continuous compounding, the discount factor is $e^{-.07} = 0.932$ at the top node at date 2. This is also the value of the discount bond at that node. Likewise, the value of the bond at the middle node at date 2 is $e^{-.05} = 0.951$. This implies that the value of the bond at the top node at date 1 is

$$e^{-.06}\left(\frac{1}{2}\times 0.932 + \frac{1}{2}\times 0.951\right) = 0.887\,.$$

Continuing in this way, we derive the following tree for the discount bond, concluding that its price at date 0 should be \$0.861.

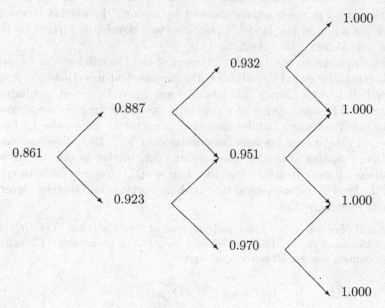

In general, the price at date 0 of a discount bond maturing at date t_n is

$$P(0,t_n) = E^R\left[\exp\left(-\sum_{i=0}^{n-1} r(t_i)\,\Delta t\right)\right]$$

$$= E^R\left[\exp\left(-r(t_0)\,\Delta t - \sum_{i=1}^{n-1}[\hat r(t_i) + \phi(t_i)]\,\Delta t\right)\right]$$

$$= \exp\left(-r(t_0)\,\Delta t - \sum_{i=1}^{n-1}\phi(t_i)\,\Delta t\right) E^R\left[\exp\left(-\sum_{i=1}^{n-1}\hat r(t_i)\,\Delta t\right)\right]\,.$$

$$(14.2)$$

Given the parameter σ, the expectation in the last line of the above can easily be computed.

The parameters $\phi(t_n)$ can be chosen to equate model prices of discount bonds maturing at $t_1, \ldots, t_N$ to market prices. This is done simply by choosing $\phi(t_n)$ to satisfy the following equation, which we will derive below:

$$
e^{-\phi(t_n)\,\Delta t} = \frac{2e^{r(0)\,\Delta t}}{e^{n\sigma\sqrt{\Delta t}\,\Delta t} + e^{-n\sigma\sqrt{\Delta t}\,\Delta t}} \times \frac{P^{\mathrm{mkt}}(0, t_{n+1})}{P^{\mathrm{mkt}}(0, t_n)}. \tag{14.3}
$$

Because the ratio of market prices $P^{\mathrm{mkt}}(0, t_{i+1})/P^{\mathrm{mkt}}(0, t_i)$ is the reciprocal of one plus the market forward rate at date 0 for loans from date t_i to t_{i+1}, this formula for the parameters $\phi(t_i)$ is equivalent to equating model forward rates to market forward rates, as was discussed for the Hull-White and continuous-time Ho-Lee models in Sect. 13.6. In fact, the fitting of the Ho-Lee model to market bond prices is often expressed by saying "the market forward rate curve is an input to the model." This idea was developed further by Heath, Jarrow and Morton [33]—see Sect. 14.6.

As for options on equities and currencies, the binomial model for interest rates is especially useful for valuing early exercise features. However, it should be noted that, even though this model is very easy to use, it has important limitations. The assumption of a constant volatility for the one-period rate and no mean reversion implies excessive uncertainty about the level of the one-period rate at long horizons, as discussed in Sect. 13.1. To offset this, one could use a smaller volatility when valuing long-maturity options. However, for options with early exercise features, this would imply too little uncertainty about the level of the one-period rate at short horizons and thereby undervalue the early exercise option.

We will conclude this section with a proof of formula (14.3). Let $\varepsilon(t_i)$ denote independent random variables that equal $\pm\sigma\sqrt{\Delta t}$ with probability $1/2$ each under the risk-neutral measure. Then we can write

$$
\hat{r}(t_i) = r(0) + \sum_{j=1}^{i} \varepsilon(t_j) \, .
$$

This implies that

$$
\sum_{i=0}^{n-1} \hat{r}(t_i) = (n-1)r(0) + \sum_{i=1}^{n-1}\sum_{j=1}^{i} \varepsilon(t_j)
$$
$$
= (n-1)r(0) + (n-1)\varepsilon(t_1) + (n-2)\varepsilon(t_2) + \cdots + \varepsilon(t_{n-1})
$$
$$
= (n-1)r(0) + \sum_{i=1}^{n-1}(n-i)\varepsilon(t_i) \, .
$$

Therefore, (14.2) gives us

$$
P(0, t_n) = \exp\left(-nr(0)\,\Delta t - \sum_{i=1}^{n-1}\phi(t_i)\,\Delta t\right) E^R\left[\exp\left(-\sum_{i=1}^{n-1}(n-i)\varepsilon(t_i)\Delta t\right)\right] \, .
$$

Moreover,

$$E^R \left[\exp\left(-\sum_{i=1}^{n-1} (n-i)\varepsilon(t_i)\,\Delta t \right) \right] = E^R \left[\prod_{i=1}^{n-1} \exp\left(-(n-i)\varepsilon(t_i)\,\Delta t \right) \right]$$

$$= \prod_{i=1}^{n-1} E^R \left[\exp\left(-(n-i)\varepsilon(t_i)\,\Delta t \right) \right]$$

$$= \prod_{i=1}^{n-1} \left(\frac{1}{2} \exp\left((n-i)\sigma\sqrt{\Delta t}\,\Delta t \right) + \frac{1}{2} \exp\left(-(n-i)\sigma\sqrt{\Delta t}\,\Delta t \right) \right).$$

Thus,

$$P(0,t_n) = \exp\left(-nr(0)\,\Delta t - \sum_{i=1}^{n-1} \phi(t_i)\,\Delta t \right)$$

$$\times \prod_{i=1}^{n-1} \left(\frac{1}{2} \exp\left((n-i)\sigma\sqrt{\Delta t}\,\Delta t \right) + \frac{1}{2} \exp\left(-(n-i)\sigma\sqrt{\Delta t}\,\Delta t \right) \right).$$

Likewise,

$$P(0,t_{n+1}) = \exp\left(-(n+1)r(0)\,\Delta t - \sum_{i=1}^{n} \phi(t_i)\,\Delta t \right)$$

$$\times \prod_{i=1}^{n} \left(\frac{1}{2} \exp\left((n+1-i)\sigma\sqrt{\Delta t}\,\Delta t \right) + \frac{1}{2} \exp\left(-(n+1-i)\sigma\sqrt{\Delta t}\,\Delta t \right) \right),$$

which we can write as

$$\exp\left(-(n+1)r(0)\,\Delta t - \sum_{i=1}^{n} \phi(t_i)\,\Delta t \right)$$

$$\times \prod_{i=0}^{n-1} \left(\frac{1}{2} \exp\left((n-i)\sigma\sqrt{\Delta t}\,\Delta t \right) + \frac{1}{2} \exp\left(-(n-i)\sigma\sqrt{\Delta t}\,\Delta t \right) \right)$$

$$= \exp\left(-r(0)\,\Delta t - \phi(t_n)\,\Delta t \right)$$

$$\times \left(\frac{1}{2} \exp\left(n\sigma\sqrt{\Delta t}\,\Delta t \right) + \frac{1}{2} \exp\left(-n\sigma\sqrt{\Delta t}\,\Delta t \right) \right) P(0,t_n).$$

Thus, the ratio of model prices is

$$\frac{P(0,t_{n+1})}{P(0,t_n)} = \exp\left(-r(0)\,\Delta t - \phi(t_n)\,\Delta t \right)$$

$$\times \left(\frac{1}{2} \exp\left(n\sigma\sqrt{\Delta t}\,\Delta t \right) + \frac{1}{2} \exp\left(-n\sigma\sqrt{\Delta t}\,\Delta t \right) \right).$$

Equating this to the ratio of market prices $P^{\mathrm{mkt}}(0,t_{n+1})/P^{\mathrm{mkt}}(0,t_n)$ gives the formula (14.3).

14.2 Black-Derman-Toy

The Black-Derman-Toy [4] model is, like the Ho-Lee model and the Black-Karasinski model discussed in the next section, a binomial model of the one-period interest rate. The model assumes that

$$\Delta \log r(t_i) = \eta(t_i)\,\Delta t \pm \sigma(t_i)\sqrt{\Delta t},$$ (14.4)

where "+" and "−" have probability one-half each under the risk-neutral measure. The volatility $\sigma(t_i)$ has the interpretation of the *percentage* volatility of the one-period rate (rather than the absolute volatility, as in the Vasicek and Ho-Lee models). The model implies that the one-period rate will always be nonnegative.

A significant feature of the model is that the volatility is allowed to be time-varying. This would produce a non-recombining tree except that the drift $\eta(t_i)$ is allowed to vary across the date-t_{i-1} nodes, i.e., to be a random variable, depending on the level of the one-period rate at date t_{i-1}. To understand this, consider the following two-period tree. For convenience, we write a_i for $\eta(t_i)\,\Delta t$ and b_i for $\sigma(t_i)\sqrt{\Delta t}$.

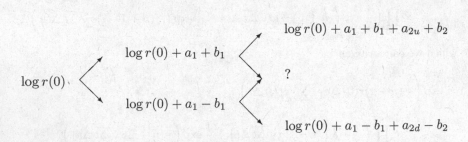

We have written a_{2u} and a_{2d} to demonstrate that the drift between date 1 and date 2 can vary, depending on whether we are at the top or bottom node at date 1. If we arrive at the node marked with a question mark via a down move from the top node at date 1, the value will be

$$\log r(0) + a_1 + b_1 + a_{2u} - b_2\,.$$

On the other hand, if we arrive at it via an up move from the bottom node at date 1, the value will be

$$\log r(0) + a_1 - b_1 + a_{2d} + b_2\,.$$

For the tree to be recombining, these values must be the same, which implies

$$a_{2d} = a_{2u} + 2b_1 - 2b_2\,.$$

In general, at each date there are two free parameters: the volatility and the drift at one of the nodes, the drifts at the other nodes then being determined by the requirement that the tree be recombining.

The dependence of the drift on the node can be expressed as a linear dependence on the logarithm of the one-period rate. In other words, it is possible (and convenient) to write the Black-Derman-Toy model (14.4) as

$$\Delta \log r(t_i) = \kappa(t_i)[\theta(t_i) - \log r(t_{i-1})] \Delta t \pm \sigma(t_i)\sqrt{\Delta t}, \qquad (14.5)$$

where now the functions $\kappa(t_i)$, $\theta(t_i)$, and $\sigma(t_i)$ are deterministic—i.e., depending on time but constant across nodes at each date.

In the two-period example above, we have

$$a_{2u} = \kappa(t_2)[\theta(t_2) - \{\log r(t_0) + a_1 + b_1\}]\Delta t,$$

and

$$a_{2d} = \kappa(t_2)[\theta(t_2) - \{\log r(t_0) + a_1 - b_1\}]\Delta t,$$

so the relation $a_{2d} = a_{2u} + 2b_1 - 2b_2$ is equivalent to

$$\kappa(t_2) = \frac{b_1 - b_2}{b_1 \Delta t} = -\frac{1}{\sigma(t_1)} \times \frac{\sigma(t_2) - \sigma(t_1)}{\Delta t} = -\frac{1}{\sigma(t_1)} \times \frac{\Delta\sigma(t_2)}{\Delta t}. \qquad (14.6)$$

This same relationship holds at each node at each date (just consider the two-period example as two periods extending from any node in the tree); thus, in general, we have

$$\kappa(t_i) = -\frac{1}{\sigma(t_{i-1})} \times \frac{\Delta\sigma(t_i)}{\Delta t}. \qquad (14.7)$$

Equations (14.5) and (14.7) define the Black-Derman-Toy model. The free parameters at each date are $\sigma(t_i)$ and $\theta(t_i)$, and the parameter $\kappa(t_i)$ is defined by (14.7). Alternatively, one can view $\kappa(t_i)$ and $\theta(t_i)$ as free parameters and define $\sigma(t_i)$ from (14.7). Note that $\kappa(t_i)$ can be interpreted as a mean-reversion parameter for $\log r(t_i)$, at least when it is positive (i.e., when $\Delta\sigma(t_i) < 0$). Because there are two free parameters at each date rather than one (as in the Ho-Lee model) it is possible to match both market bond prices and market yield volatilities or cap prices.

In continuous time, we would write (14.7) as

$$\kappa(t) = -\frac{\mathrm{d}\log\sigma(t)}{\mathrm{d}t}.$$

Therefore, the continuous-time version of the Black-Derman-Toy model is

$$\mathrm{d}\log r(t) = -\frac{\mathrm{d}\log\sigma(t)}{\mathrm{d}t}[\theta(t) - \log r(t)]\,\mathrm{d}t + \sigma(t)\,\mathrm{d}B(t),$$

with B being a Brownian motion under the risk-neutral measure.

14.3 Black-Karasinski

The Black-Karasinski [5] model is similar to the Black-Derman-Toy model—it assumes (14.5) for the changes in the logarithm of the one-period rate—but it removes the linkage (14.7) between the mean-reversion parameter and the volatility. It does this by allowing the lengths of the time steps to vary. Denote the length of the time-step $t_i - t_{i-1}$ by τ_i. Consider again the two-period example of the previous section. As a necessary condition for the tree to be recombining, we deduced (in (14.6)) that

$$\kappa(t_2) = \frac{b_1 - b_2}{b_1 \, \tau_2} \ .$$

In this model, we have $b_i = \sigma(t_i)\sqrt{\tau_i}$. Making this substitution, we obtain

$$\kappa(t_2) = \frac{\sigma(t_1)\sqrt{\tau_1} - \sigma(t_2)\sqrt{\tau_2}}{\sigma(t_1)\sqrt{\tau_1}\tau_2} \ ,$$

which we can rewrite as

$$\kappa(t_2)\sigma(t_1)\sqrt{\tau_1}\tau_2 + \sigma(t_2)\sqrt{\tau_2} - \sigma(t_1)\sqrt{\tau_1} = 0 \ .$$

This is a quadratic equation in the unknown $\sqrt{\tau_2}$ with the unique positive solution (assuming $\kappa(t_2) > 0$)

$$\sqrt{\tau_2} = \frac{\sqrt{\sigma(t_2)^2 + 4\kappa(t_2)\sigma(t_1)^2\sqrt{\tau_1}} - \sigma(t_2)}{2\kappa(t_2)\sigma(t_1)\sqrt{\tau_1}} \ .$$

This relation must hold at each date. Thus, squaring both sides, we obtain the general formula

$$\tau_i = \frac{\left[\sqrt{\sigma(t_i)^2 + 4\kappa(t_i)\sigma(t_{i-1})^2\sqrt{\tau_{i-1}}} - \sigma(t_i)\right]^2}{4\kappa(t_i)^2\sigma(t_{i-1})^2\tau_{i-1}}. \tag{14.8}$$

To summarize, the Black-Karasinski model is given by (14.5) with three free parameters—$\kappa(t_i)$, $\theta(t_i)$, and $\sigma(t_i)$—at each date. It is implemented in a recombining tree by defining the length of each time step $\tau_i = t_i - t_{i-1}$ for $i \geq 2$ according to the formula (14.8). The length of the first time step τ_1 can be chosen arbitrarily.

14.4 Cox-Ingersoll-Ross

Cox, Ingersoll, and Ross [19] introduced a continuous-time model[1] in which the short rate satisfies

---

[1] Cox, Ingersoll and Ross (hereafter CIR) also discuss a variety of other continuous-time models, but this particular model is so well known that it is often simply called *the* CIR model.

$$dr(t) = \kappa[\theta - r(t)]\,dt + \sigma\sqrt{r(t)}\,dB(t), \tag{14.9}$$

where κ, θ, and σ are positive constants and B is a Brownian motion under the risk-neutral measure. Like the Vasicek model, this short rate process has a long-run mean of θ. The difference between the CIR model and the Vasicek model is that the volatility in the CIR model is proportional to the square root of the short rate rather than being constant. Because of this fact, the short rate can never be negative. Intuitively, the reason is that the volatility $\sigma\sqrt{r(t)}$ is very small if $r(t)$ is near zero, so the drift will dominate the change in $r(t)$, pushing it upwards towards θ. This interest rate model was mimicked by Heston [34] in his stochastic volatility model discussed in Chap. 4. We will briefly discuss three topics in connection with this model: discount bond prices, calibrating the model to the current market, and pricing fixed-income derivatives.

Discount Bond Prices in the CIR Model

Discount bond prices can be most easily computed in the CIR model by solving the fundamental partial differential equation (pde) discussed in Chap. 10. Let $P(t, u)$ denote the price at date t of a discount bond maturing at date u, having remaining time to maturity of $\tau = u - t$. The discount bond price will depend on the remaining time to maturity and the short rate at date t, because, as in the Vasicek model, the short rate is the only random factor in this model. Thus, there must be some deterministic function f such that $P(t, u) = f(r(t), \tau)$. As in Chap. 10, the fundamental pde is obtained by applying Itô's formula to f to compute df in terms of the partial derivatives of f and then using the fact that the expected return of the discount bond (hence the drift of df/f) must equal the short rate under the risk-neutral measure.

From Itô's formula and the definition (14.9) of dr, we have

$$
\begin{aligned}
df &= \frac{\partial f}{\partial \tau}\,d\tau + \frac{\partial f}{\partial r}\,dr + \frac{1}{2}\frac{\partial^2 f}{\partial r^2}\,(dr)^2 \\
&= -\frac{\partial f}{\partial \tau}\,dt + \frac{\partial f}{\partial r}\left\{\kappa[\theta - r]\,dt + \sigma\sqrt{r}\,dB\right\} + \frac{1}{2}\frac{\partial^2 f}{\partial r^2}\sigma^2 r\,dt \\
&= \left(-\frac{\partial f}{\partial \tau} + \frac{\partial f}{\partial r}\kappa[\theta - r] + \frac{1}{2}\frac{\partial^2 f}{\partial r^2}\sigma^2 r\right)dt + \frac{\partial f}{\partial r}\sigma\sqrt{r}\,dB. \tag{14.10}
\end{aligned}
$$

Equating the drift to $rf\,dt$ gives us the fundamental pde:

$$-\frac{\partial f}{\partial \tau} + \frac{\partial f}{\partial r}\kappa[\theta - r] + \frac{1}{2}\frac{\partial^2 f}{\partial r^2}\sigma^2 r = rf. \tag{14.11}$$

This equation should be solved for the function f subject to the boundary condition that the value of the discount bond is one at maturity; i.e., $f(r, 0) = 1$ for all r.

The solution can be obtained by "guessing" a solution of the same form as the Vasicek bond pricing formula (13.6a), namely[2]

$$f(r, \tau) = \exp\left(-a(\tau) - b(\tau)r\right) \tag{14.12}$$

for deterministic functions a and b. The boundary condition

$$f(r, 0) = \exp\left(-a(0) - b(0)r\right) = 1$$

for all r implies $a(0) = b(0) = 0$, and it can easily be checked that the fundamental pde is equivalent to

$$b'(\tau) = 1 - \kappa b(\tau) - \frac{1}{2}\sigma^2 b^2(\tau) , \tag{14.13a}$$

and

$$a'(\tau) = \kappa\theta b(\tau) , \tag{14.13b}$$

where the "primes" denote derivatives. By differentiating, one can verify that the solution of (14.13a) (which is called a "Riccati equation") with the boundary condition $b(0) = 0$ is

$$b(\tau) = \frac{2\left(e^{\gamma\tau} - 1\right)}{c(\tau)} , \tag{14.14a}$$

where

$$\gamma = \sqrt{\kappa^2 + 2\sigma^2} \quad \text{and} \quad c(\tau) = (\kappa + \gamma)\left(e^{\gamma\tau} - 1\right) + 2\gamma . \tag{14.14b}$$

Integrating (14.13b) then gives

$$a(\tau) = -\frac{2\kappa\theta}{\sigma^2}\left[\frac{(\kappa + \gamma)\tau}{2} + \log\frac{2\gamma}{c(\tau)}\right] . \tag{14.14c}$$

To summarize,

The price at date t of a discount bond maturing at $u > t$ in the CIR model is

$$P(t, u) = \exp\left(-a(\tau) - b(\tau)r(t)\right) , \tag{14.15}$$

where $\tau = u - t$ and $a(\tau)$ and $b(\tau)$ are defined in (14.14).

Note that (14.12) implies $\partial f/\partial r = -b(\tau)f$. Substituting this into (14.10) gives us the discount bond return as

$$\frac{\mathrm{d}P(t, u)}{P(t, u)} = \frac{\mathrm{d}f}{f} = r(t)\,\mathrm{d}t - b(\tau)\sigma\sqrt{r(t)}\,\mathrm{d}B(t) . \tag{14.16}$$

---

[2] This guess works because the CIR model, like the Vasicek model, is an "affine model." See Sect. 14.5.

This is again similar to the Vasicek model except for the appearance of the $\sqrt{r}$ factor in the volatility. Because of this factor, the volatility is random. Thus, the option pricing formulas of previous chapters cannot be directly applied to price discount bond options (and hence caps, floors, coupon bond options, and swaptions). Nevertheless, the ideas underlying those formulas can be applied to obtain similar valuation formulas.

Hedge ratios depend on relative volatilities, so they are determined by the function b, just as discussed in Sect. 13.4 for the Vasicek model.

Calibrating the CIR Model to the Yield Curve

The CIR model can be calibrated to current market conditions by taking one or more of the parameters κ, θ and σ to be time-varying, as in the extended Vasicek model. This was suggested by Cox, Ingersoll and Ross. However, the simplest way to calibrate the model to discount bond prices, which was also suggested by Cox, Ingersoll and Ross, is to take the short rate to be the sum of a non-random function of time and a square-root process as defined in (14.9). Specifically, let

$$r(t) = \phi(t) + \hat{r}(t) ,$$

where ϕ is a non-random function and $\hat{r}$ satisfies

$$d\hat{r}(t) = \kappa[\theta - \hat{r}(t)] \, dt + \sigma \sqrt{\hat{r}(t)} \, dB(t) , \tag{14.17}$$

with $\hat{r}(0) = r(0)$. Then, as in the Hull-White model, discount bond prices are given by

$$P(t, u) = \exp\left(-\int_t^u \phi(s) \, ds\right) E_t^R\left[\exp\left(-\int_t^u \hat{r}(s) \, ds\right)\right] .$$

Moreover, the expectation in this equation is the discount bond pricing function calculated in the previous subsection, so we have

$$P(t, u) = \exp\left(-\int_t^u \phi(s) \, ds\right) \exp\left(-a(\tau) - b(\tau)\hat{r}(t)\right) . \tag{14.18}$$

In particular, discount bond prices at date 0 are

$$P(0, u) = \exp\left(-\int_0^u \phi(s) \, ds\right) \exp\left(-a(u) - b(u)r(0)\right) ,$$

so the model can be calibrated to market prices $P^{\text{mkt}}(0, u)$ by setting

$$\exp\left(-\int_0^u \phi(s) \, ds\right) = \exp\left(a(u) + b(u)r(0)\right) P^{\text{mkt}}(0, u) \tag{14.19}$$

for each u. Note that the calibration does not affect discount bond returns: the expected return under the risk-neutral measure must still be the short rate and the volatility is unaffected by a deterministic factor. Therefore, we have, as in (14.16),

$$\frac{\mathrm{d}P(t,u)}{P(t,u)} = r(t)\,\mathrm{d}t - b(\tau)\sigma\sqrt{\hat{r}(t)}\,\mathrm{d}B(t)\,. \tag{14.20}$$

This can be verified by applying Itô's formula to (14.18).

Pricing Fixed-Income Derivatives in the CIR Model

In the previous chapter, it was shown for the extended Vasicek model that pricing formulas for caps, floors, coupon bond options, and swaptions can be derived from a pricing formula for discount bond options. The same is true in the CIR model — caps and floors are of course portfolios of discount bond options and, in a single-factor model such as the Vasicek or CIR model, coupon bond options and swaptions can also be priced as portfolios of discount bond options. Here we will explain briefly how to price discount bond options in the CIR model.

Consider a call option maturing at date T with the underlying being a discount bond maturing at $u > T$. Let K denote the strike price. From our fundamental pricing formula (1.17), the value at date 0 of the option is

$$P(0,u) \times \mathrm{prob}^{u}\big(P(T,u) > K\big) - P(0,T) \times \mathrm{prob}^{T}\big(P(T,u) > K\big)\,, \tag{14.21}$$

where prob^{u} denotes the probability measure using the discount bond maturing at u as numeraire and prob^{T} denotes the probability measure using the discount bond maturing at T as the numeraire. Using the calibration of the previous subsection, the price of the underlying at date T will be, according to the model,

$$P(T,u) = \exp\left(-\int_{T}^{u} \phi(s)\,\mathrm{d}s - a(u-T) - b(u-T)\hat{r}(T)\right)\,.$$

Therefore, the option will finish in the money if and only if

$$\frac{-\int_{T}^{u} \phi(s)\,\mathrm{d}s - a(u-T) - \log K}{b(u-T)} > \hat{r}(T)\,. \tag{14.22}$$

Thus, to price discount bond options, we need to compute the probabilities that $\hat{r}(T)$ is less than a given number, using discount bonds as numeraires.

Consider the discount bond maturing at u. The calculation for the discount bond maturing at T can be done in the same way. We use the fact that

$$\frac{e^{\int_{0}^{t} r(s)\,\mathrm{d}s}}{P(t,u)}$$

is a martingale, when $P(t,u)$ is used as the numeraire, for $t \leq u$. Let $Z(t)$ denote this ratio and apply Itô's formula for ratios to derive

$$\frac{dZ(t)}{Z(t)} - r(t)\,dt - \frac{dP(t,u)}{P(t,u)} + \left(\frac{dP(t,u)}{P(t,u)}\right)^2 .$$

Substituting from (14.20) now gives us

$$\frac{dZ(t)}{Z(t)} = b(\tau)\sigma\sqrt{\hat{r}(t)}\,dB(t) + b(\tau)^2\sigma^2\hat{r}(t)\,dt$$

$$= b(\tau)\sigma\sqrt{\hat{r}(t)}\left[dB(t) + b(\tau)\sigma\sqrt{\hat{r}(t)}\,dt\right] .$$

Given that Z is a martingale, dZ/Z cannot have a drift, so it must be that B^* defined by $B^*(0) = 0$ and

$$dB^*(t) = dB(t) + b(\tau)\sigma\sqrt{\hat{r}(t)}\,dt$$

is a martingale, and hence a Brownian motion, when $P(t,u)$ is used as the numeraire. Substituting this into the definition (14.17) of $\hat{r}$, we have

$$d\hat{r}(t) = \kappa[\theta - \hat{r}(t)]\,dt + \sigma\sqrt{\hat{r}(t)}\left[dB^*(t) - b(\tau)\sigma\sqrt{\hat{r}(t)}\,dt\right]$$

$$= \kappa[\theta - \hat{r}(t)]\,dt + \sigma\sqrt{\hat{r}(t)}\,dB^*(t) - \sigma^2 b(\tau)\hat{r}(t)\,dt$$

$$= \kappa^*(t)[\theta^*(t) - \hat{r}(t)]\,dt + \sigma\sqrt{\hat{r}(t)}\,dB^*(t) , \tag{14.23}$$

where we define

$$\kappa^*(t) = \kappa + \sigma^2 b(u - t) \qquad \text{and} \qquad \theta^*(t) = \frac{\kappa\theta}{\kappa^*(t)} .$$

Thus, using a discount bond as numeraire, the process $\hat{r}$ is still a square root process, but now with a time-dependent long-run mean and mean-reversion rate.

The random variable $\hat{r}(T)$ defined by $\hat{r}(0) = 0$ and equation (14.23) for $t \le T$ is a transformation of a random variable having what is called a "non-central chi-square" distribution. See Appendix B.3 for further discussion and calculation of the probabilities $\text{prob}^u\big(P(T,u) > K\big)$ and $\text{prob}^T\big(P(T,u) > K\big)$.

14.5 Longstaff-Schwartz

Cox, Ingersoll and Ross suggest adding two independent square-root processes to obtain a two-factor model. This means that we would take $r(t) = x_1(t) + x_2(t)$, where

$$dx_i(t) = \kappa_i[\theta_i - x_i(t)]\,dt + \sigma_i\sqrt{x_i(t)}\,dB_i(t) , \tag{14.24}$$

where B_1 and B_2 are independent Brownian motions under the risk-neutral measure and κ_i, θ_i and σ_i are positive constants for $i = 1, 2$. Longstaff and

Schwartz [47] investigate this model further, including providing an "equilibrium" foundation, deriving discount bond option prices, and estimating the coefficients. The model is usually called the Longstaff-Schwartz model.

An important observation made by Longstaff and Schwartz is that the model can be rewritten so that the short rate and its volatility are the factors (rather than the unobservable x_1 and x_2).[3] Note that the instantaneous variance of $r = x_1 + x_2$ is

$$\left(\sigma_1\sqrt{x_1(t)}\,dB_1(t) + \sigma_2\sqrt{x_2(t)}\,dB_2(t)\right)^2 = \left(\sigma_1^2 x_1(t) + \sigma_2^2 x_2(t)\right)\,dt\;.$$

Define $V(t) = \sigma_1^2 x_1(t) + \sigma_2^2 x_2(t)$. We can solve for x_1 and x_2 in terms of r and V as

$$x_1 = \frac{\sigma_2^2 r - V}{\sigma_2^2 - \sigma_1^2} \tag{14.25a}$$

$$x_2 = \frac{V - \sigma_1^2 r}{\sigma_2^2 - \sigma_1^2}\;, \tag{14.25b}$$

provided $\sigma_1 \neq \sigma_2$. Making these substitutions for x_1 and x_2 on the right-hand side of (14.24) and noting that $dr = dx_1 + dx_2$ and $dV = \sigma_1^2\,dx_1 + \sigma_2^2\,dx_2$, we obtain the following equations presented by Longstaff and Schwartz: .

$$dr = \left(\alpha\gamma + \beta\eta - \frac{\beta\delta - \alpha\xi}{\beta - \alpha}r - \frac{\xi - \delta}{\beta - \alpha}V\right)dt$$
$$+ \alpha\sqrt{\frac{\beta r - V}{\alpha(\beta - \alpha)}}\,dB_1 + \beta\sqrt{\frac{V - \alpha r}{\beta(\beta - \alpha)}}\,dB_2\;, \quad (14.26a)$$

$$dV = \left(\alpha^2\gamma + \beta^2\eta - \frac{\alpha\beta(\delta - \xi)}{\beta - \alpha}r - \frac{\beta\xi - \alpha\delta}{\beta - \alpha}V\right)dt$$
$$+ \alpha^2\sqrt{\frac{\beta r - V}{\alpha(\beta - \alpha)}}\,dB_1 + \beta^2\sqrt{\frac{V - \alpha r}{\beta(\beta - \alpha)}}\,dB_2\;, \quad (14.26b)$$

where $\delta = \kappa_1$, $\xi = \kappa_2$, $\alpha = \sigma_1^2$, $\beta = \sigma_2^2$, $\gamma = \kappa_1\theta_1/\sigma_1^2$, and $\eta = \kappa_2\theta_2/\sigma_2^2$. Thus, this can be regarded as a two-factor model in which the factors are the short rate and its instantaneous variance, with the six parameters δ, ξ, α, β, γ, and η.

The simplest way to compute discount bond prices in this model is to return to the definition $r = x_1 + x_2$. Discount bond prices are

---

[3] Of course, the volatility is also not directly observable. Longstaff and Schwartz use a GARCH model to estimate it and then use the time series of estimated volatilities and the time series of short rates to estimate the parameters of the model.

$$P(t,u) = E_t \left[\exp \left(- \int_t^u r(s)\, ds \right) \right]$$

$$= E_t \left[\exp \left(- \int_t^u x_1(s)\, ds \right) \exp \left(- \int_t^u x_2(s)\, ds \right) \right]$$

$$= E_t \left[\exp \left(- \int_t^u x_1(s)\, ds \right) \right] E_t \left[\exp \left(- \int_t^u x_2(s)\, ds \right) \right],$$

due to the independence of x_1 and x_2. Moreover, these expectations have the same form as discount bond prices in the CIR model, namely

$$E_t \left[\exp \left(- \int_t^u x_i(s)\, ds \right) \right] = \exp \left(-a_i(\tau) - b_i(\tau) x_i(t) \right), \qquad (14.27)$$

where the functions a_i and b_i are defined in (14.14), using the parameters κ_i, θ_i and σ_i. The expectations (14.27) can be written in terms of $r(t)$ and $V(t)$ by substituting from (14.25).

The Vasicek, CIR and Longstaff-Schwartz models are examples of "affine models." An affine model is defined by a set of factors $x_1, \ldots, x_n$, where

- The short rate is an affine function of the factors;[4] i.e., $r(t) = \alpha_0 + \sum_{i=1}^n \alpha_i x_i(t)$ for constants α_i,
- The drift of each factor is an affine function of the factors.
- The instantaneous variance of each factor is an affine function of the factors.
- The instantaneous covariance of each pair of factors is an affine function of the factors.

In any affine model, discount bond prices are of the form

$$P(t,u) = \exp \left(-a(\tau) - \sum_{i=1}^n b_i(\tau) x_i(t) \right) \qquad (14.28)$$

for deterministic functions a and b_i for $i = 1, \ldots, n$, as we have seen is true for the Vasicek, CIR, and Longstaff-Schwartz models. Most, but certainly not all, of the continuous-time models studied in the finance literature are affine.

In any single-factor affine model, the short rate can be used as the factor. Thus, the general affine one-factor model is of the form

$$dr = \kappa(\theta - r)\, dt + \sqrt{\alpha + \beta r}\, dB,$$

for constants κ, θ, α and β, where B is a Brownian motion under the risk-neutral measure. The Vasicek model is the special case $\beta = 0$ (α being the same as the parameter σ^2). The CIR model is the special case $\alpha = 0$ (β being the same as the parameter σ^2).

---

[4] An affine function of a real variable x is a function $f(x) = a + bx$ for constants a and b. This is often called a linear function, but technically a linear function is of the form $f(x) = bx$. Thus, an affine function is a constant plus a linear function.

Because yields of discount bonds are affine functions of the factors in an affine model, as (14.28) shows, the short rate and yields at $n-1$ fixed times-to-maturity τ_i can be chosen to be the n factors (except in the rare case that the linear transformation from factors to the short rate and yields fails to be invertible). The transformation from factors to yields is analogous to the transformation from (x_1, x_2) to (r, V) in the Longstaff-Schwartz model. Important papers on affine models include Duffie and Kan [24] and Dai and Singleton [22].

14.6 Heath-Jarrow-Morton

Heath, Jarrow and Morton [33] propose an alternative framework for modelling. Rather than modelling the evolution of the short rate (and possibly other factors such as the volatility of the short rate or other yields), Heath, Jarrow and Morton (hereafter, HJM) propose modelling the evolution of instantaneous forward rates. They derive a formula for the drifts of instantaneous forward rates under the risk-neutral measure, in terms of the volatilities of the forward rates. A model is therefore completely defined by specifying the volatilities of forward rates. A model of this type is easily fit to market discount bond prices by simply using the initial term structure of forward rates as an input. By calibrating the volatility structure, the model can also be fit to other market prices. Any of the continuous-time models we have discussed can be written in the HJM form. The virtue of the HJM approach is that it facilitates the construction of new models. The disadvantage of the HJM approach is that models of this form will generally be path-dependent—bond prices and the prices of other fixed-income instruments at any point in time depend on the entire history of the forward rate processes, rather than depending only on the values of a small set of factors. This makes computation quite difficult, just as computation with non-recombining binomial trees is much more difficult than with recombining trees.

As discussed in Sect. 13.6, the forward rate at date t for an instantaneous loan at date $u \geq t$ is $f(t, u)$ defined by

$$f(t, u) = -\frac{\mathrm{d} \log P(t, u)}{\mathrm{d}u}. \tag{14.29}$$

The short rate at time t is the forward rate for maturity date t; i.e., $r(t) = f(t, t)$. By integrating (14.29), one can see that discount bond prices are written in terms of forward rates as:

$$P(t, u) = \exp\left(-\int_t^u f(t, s)\,\mathrm{d}s\right). \tag{14.29'}$$

Heath, Jarrow and Morton use the definition (14.29') and the fact that the expected return of a discount bond under the risk-neutral measure must be the short rate to derive a formula for the drifts of forward rates under the risk-neutral measure.

Assume, for the sake of simplicity, that there is only a single source of uncertainty (i.e., a single Brownian motion) driving the yield curve. Then, for each fixed u, the forward rate $f(t, u)$ at date $t < u$ will evolve as

$$\mathrm{d}f(t, u) = \mu(t, u)\,\mathrm{d}t + \sigma(t, u)\,\mathrm{d}B(t), \qquad (14.30)$$

for some μ and σ, where B is a Brownian motion under the risk-neutral measure. In general $\mu(t, u)$ and $\sigma(t, u)$ could depend on the entire history of the Brownian motion through date t. Heath, Jarrow and Morton show that

$$\mu(t, u) = \sigma(t, u) \int_t^u \sigma(t, s)\,\mathrm{d}s. \qquad (14.31)$$

One can show that $\int_t^u \sigma(t, s)\,\mathrm{d}s$ is the volatility of $P(t, u)$. Therefore, (14.31) states that the drift of the forward rate is the product of the volatilities of the forward rate and the discount bond return. A model is fully specified by specifying initial forward rates—i.e., $f(0, u)$ for all u—and the volatility processes $\sigma(t, u)$ for each u and $t \leq u$. The generalization to multiple Brownian motions is straightforward and allows for forward rates that are not instantaneously perfectly correlated. An important application of the HJM modelling framework is work by Brace, Gatarek and Musiela [9] and Miltersen, Sandmann, and Sondermann [52], who derive conditions on the volatility processes $\sigma(t, u)$ that guarantee forward LIBOR rates of a fixed maturity (e.g., quarterly or semi-annual rates) have deterministic volatilities, thus justifying the use of Black's formula in Sect. 12.4 for valuing caps and floors.

To see how some of the models we have discussed can be written in the HJM form, let us re-examine the Hull-White model. From the Hull-White bond price formula (13.14), the instantaneous forward rate in the Hull-White model is

$$f(t, u) = \phi(u) + \frac{\partial}{\partial u}a(u - t) + \hat{r}(t)\frac{\partial}{\partial u}b(u - t)$$

$$= \phi(u) - \frac{\sigma^2}{2\kappa^2}\left(1 + \mathrm{e}^{-2\kappa(u-t)} - 2\mathrm{e}^{-\kappa(u-t)}\right) + \mathrm{e}^{-\kappa(u-t)}\hat{r}(t).$$

Applying Itô's formula yields

$$\mathrm{d}f(t, u) = -\frac{\sigma^2}{\kappa}\left(\mathrm{e}^{-2\kappa(u-t)} - \mathrm{e}^{-\kappa(u-t)}\right)\mathrm{d}t + \sigma\mathrm{e}^{-\kappa(u-t)}\,\mathrm{d}B.$$

Therefore, in the HJM notation,

$$\mu(t, u) = -\frac{\sigma^2}{\kappa}\left(\mathrm{e}^{-2\kappa(u-t)} - \mathrm{e}^{-\kappa(u-t)}\right) \qquad \text{and} \qquad \sigma(t, u) = \sigma\mathrm{e}^{-\kappa(u-t)}.$$

A direct calculation shows that these functions μ and σ satisfy the HJM equation (14.31), as we knew they must, given that the HJM equation is based only on the assumption that expected returns of discount bonds equal the short rate under the risk-neutral measure. The initial forward rate curve in the Hull-White model is $f(0, u) = \phi(u)$ which is chosen to fit the market forward rate curve, as discussed in Sect. 13.6. Thus, rather than defining the Hull-White model as we did in Chap. 13, it could be defined alternatively as an HJM model in which the volatility process is the deterministic function $\sigma(t, u) = \sigma e^{-\kappa(u-t)}$ for positive constants σ and κ. The normal distribution of the short rate in the Hull-White model is a consequence of the volatility $\sigma(t, u)$ being non-random.

Likewise, the Cox-Ingersoll-Ross model fit to market bond prices as we discussed in Sect. 14.4 could be described as an HJM model. Calculations of the sort we have just done show that the volatility process in the CIR model is $\sigma(t, u) = \sigma b'(\tau)\sqrt{\hat{r}(t)}$, where $\tau = u - t$ and the function b is defined in (14.14). In this case, the volatility is random, but it depends only on the short rate at date t. Similarly, in any factor model, such as the Longstaff-Schwartz model, the volatilities $\sigma(t, u)$ of the forward rates will depend only on the factors at each date t. As mentioned at the beginning of this section, such a factor structure simplifies calculations considerably.

HJM models are sometimes written in a slightly different fashion than we have done here. If we define

$$\Sigma(t, u) = \int_t^u \sigma(t, s)\, \mathrm{d}s$$

then we have $\mathrm{d}\Sigma(t, u)/\mathrm{d}u = \sigma(t, u)$, and the HJM equation (14.31) can be written as

$$\mu(t, u) = \Sigma(t, u)\frac{\mathrm{d}\Sigma(t, u)}{\mathrm{d}u},$$

so the evolution of forward rates can be written as

$$\mathrm{d}f(t, u) = \Sigma(t, u)\frac{\mathrm{d}\Sigma(t, u)}{\mathrm{d}u}\, \mathrm{d}t + \frac{\mathrm{d}\Sigma(t, u)}{\mathrm{d}u}\, \mathrm{d}B(t).$$

For example, in the Hull-White model we have $\Sigma(t, u) = \sigma b(\tau)$ where b is defined in (13.14), and in the CIR model, we have $\Sigma(t, u) = \sigma b(\tau)\sqrt{\hat{r}(t)}$, where b is defined in (14.14).

14.7 Market Models Again

Many fixed-income derivatives (e.g., caps, floors, and swaps) have cash flows that depend on simple interest rates (e.g., LIBOR). In this chapter and the preceding chapter, we discussed valuation formulas for fixed-income derivatives based on (i) models of the short rate or one-period rate and possibly other factors, or (ii) models of instantaneous forward rates. However, it is possible, and simpler for many purposes, to model simple interest rates directly. For example, we observed in Chap. 12 that Black's formula can be applied to value caps and floors when the underlying simple interest rates have nonrandom volatilities. Models of this type are called "market models" or "LIBOR models" or, sometimes, "BGM models," the last name referring to the paper of Brace, Gatarek and Musiela [9].[5] This class of models has become quite popular in recent years. A thorough and very readable account is given by Rebonato [56].

As in Chap. 12, we will use "LIBOR" as a generic name for simple interest rates. One important fact about forward LIBOR rates that we have already essentially derived is that they are martingales under the corresponding forward measures. As was discussed in Sect. 7.7, a probability measure corresponding to a discount bond being the numeraire is called a "forward measure," because the forward price of any contract maturing at the same time as the discount bond is a martingale under that measure. We showed in Sect. 12.3 that forward LIBOR rates are forward prices of portfolios that pay spot rates. Specifically, considering a LIBOR rate of term (also called "tenor") Δt and the forward LIBOR rate corresponding to loans over a period u to $u + \Delta t$, the forward LIBOR rate is a martingale under the measure corresponding to the discount bond maturing at $u + \Delta t$.

To price derivatives other than caps and floors (e.g., swaptions), it is important to know the dynamics of forward LIBOR rates under other probability measures as well—for example, it is useful to know the dynamics under the forward measures corresponding to discount bonds maturing at dates $T \neq u + \Delta t$, or under the risk-neutral measure, or under the measure that uses as numeraire the portfolio consisting of rolling over an investment at spot LIBOR rates.[6] We will derive here the dynamics under different forward measures.

Consider dates $t_1 < t_2 < \cdots < t_N$ with $t_i - t_{i-1} = \Delta t$ for each i. At dates $t \leq t_i$, we denote the forward LIBOR rate for the time period (t_i, t_{i+1}) by $\mathcal{R}_i(t)$. The forward LIBOR rate satisfies equation (12.2), which we repeat here:

$$\frac{P(t, t_i)}{P(t, t_{i+1})} = 1 + \mathcal{R}_i(t)\,\Delta t \,. \tag{14.32}$$

Fix a date t_n. We will compute the drift of each rate $\mathcal{R}_i(t)$ when we use the discount bond maturing at t_n as the numeraire.

Let $\sigma_i(t)$ denote the volatility of rate $\mathcal{R}_i(t)$ at date t. This means that

$$\frac{\mathrm{d}\mathcal{R}_i(t)}{\mathcal{R}_i(t)} = \mu_i(t)\,\mathrm{d}t + \sigma_i(t)\,\mathrm{d}B_i(t) \,, \tag{14.33}$$

---

[5] Other important work on this topic includes Miltersen, Sandmann, and Sondermann [52] and Jamshidian [43].

[6] Actually, for this theory, it is not even necessary that the short rate exist, so the risk-neutral measure may not even be defined. The risk-neutral measure uses as numeraire the portfolio that consists of continuously rolling over an investment at the instantaneously risk-free rate, and the more natural object in a market model is the portfolio that consists of rolling over an investment at spot LIBOR rates.

for some μ_i, where the B_i are Brownian motions when $P(t, t_n)$ is used as the numeraire. The different rates should be correlated, so the Brownian motions will be correlated. Let ρ_{ij} denote the correlation of B_i and B_j. We will show that the drifts μ_i are determined by the volatilities and correlations, in analogy to the HJM result for instantaneous forward rates. Specifying the volatilities and correlations of forward rates, and inputting initial forward rates, must therefore determine the value of any security whose cash flows depend on the LIBOR rates of term Δt at the dates $t_1, \ldots, t_N$. Of course, this does not mean that there are simple formulas. Obviously, the easiest case is when the volatilities and correlations are nonrandom. In this case, we can use Black's formula to price caps and floors as in Chap. 12. However, even when the σ_i are nonrandom, forward swap rates will have random volatilities, as mentioned in Sect. 12.6.

If $i = n - 1$, then (14.32) implies

$$\mathcal{R}_i(t) = \frac{P(t, t_{n-1}) - P(t, t_n)}{P(t, t_n)\, \Delta t} \, .$$

Hence, it is the ratio of a non-dividend-paying asset (portfolio) price to the price of the numeraire asset. Consequently, it is a martingale, and we have $\mu_{n-1} = 0$. This is the case discussed in the second paragraph of this section. Consider now $i \neq n - 1$.

Define

$$Y(t) = \frac{P(t, t_{i+1})}{P(t, t_n)} \, , \tag{14.34}$$

and

$$Z(t) = \mathcal{R}_i(t) Y(t) \, . \tag{14.35}$$

Note that Y is the ratio of a non-dividend-paying asset price to the price of the numeraire asset and hence is a martingale. Furthermore, (14.32) gives us

$$Z(t) = \frac{P(t, t_i) - P(t, t_{i+1})}{P(t, t_n)\, \Delta t} \, ,$$

and hence Z is also the ratio of a non-dividend-paying asset (portfolio) price to the price of the numeraire asset and consequently a martingale. Itô's formula applied to (14.35) yields

$$\frac{\mathrm{d}Z}{Z} = \frac{\mathrm{d}\mathcal{R}_i}{\mathcal{R}_i} + \frac{\mathrm{d}Y}{Y} + \left(\frac{\mathrm{d}\mathcal{R}_i}{\mathcal{R}_i}\right)\left(\frac{\mathrm{d}Y}{Y}\right) \, .$$

Because both Z and Y are martingales, the drift of $\mathrm{d}\mathcal{R}_i / \mathcal{R}_i$ must cancel the product (covariance) term in this equation, implying

$$\mu_i \, \mathrm{d}t = -\left(\frac{\mathrm{d}\mathcal{R}_i}{\mathcal{R}_i}\right)\left(\frac{\mathrm{d}Y}{Y}\right) \, . \tag{14.36}$$

To compute the covariance, it is helpful to define

$$X_j(t) = 1 + \mathcal{R}_j(t)\,\Delta t \qquad\qquad (14.37)$$

for $j = 1, \ldots, N$. Then we have

$$\frac{\mathrm{d}X_j}{X_j} = \left(\frac{\mathcal{R}_j\,\Delta t}{1 + \mathcal{R}_j\,\Delta t} \right) \left(\mu_j\,\mathrm{d}t + \sigma_j\,\mathrm{d}B_j \right). \qquad\qquad (14.38)$$

We distinguish two cases. If $i < n - 1$, then the definitions (14.32), (14.34) and (14.37) imply

$$Y(t) = \frac{P(t, t_{i+1})}{P(t, t_{i+2})} \times \frac{P(t, t_{i+2})}{P(t, t_{i+3})} \cdots \times \frac{P(t, t_{n-1})}{P(t, t_n)}$$

$$= X_{i+1}(t) \times X_{i+2}(t) \times \cdots \times X_{n-1}(t).$$

In this case, (14.33) and (14.38) yield

$$\left(\frac{\mathrm{d}\mathcal{R}_i}{\mathcal{R}_i} \right) \left(\frac{\mathrm{d}Y}{Y} \right) = \sum_{j=i+1}^{n-1} \left(\frac{\mathrm{d}\mathcal{R}_i}{\mathcal{R}_i} \right) \left(\frac{\mathrm{d}X_j}{X_j} \right)$$

$$= \sum_{j=i+1}^{n-1} \left(\frac{\mathcal{R}_j\,\Delta t}{1 + \mathcal{R}_j\,\Delta t} \right) \sigma_i \sigma_j \rho_{ij}\,\mathrm{d}t.$$

On the other hand, if $i > n - 1$, then the definitions (14.32), (14.34) and (14.37) imply

$$\frac{1}{Y(t)} = \frac{P(t, t_n)}{P(t, t_{n+1})} \times \frac{P(t, t_{n+1})}{P(t, t_{n+2})} \times \cdots \times \frac{P(t, t_i)}{P(t, t_{i+1})}$$

$$= X_n(t) \times X_{n+1}(t) \times \cdots \times X_i(t).$$

In this case, (14.33) and (14.38) yield

$$\left(\frac{\mathrm{d}\mathcal{R}_i}{\mathcal{R}_i} \right) \left(\frac{\mathrm{d}Y}{Y} \right) = -\sum_{j=n}^{i} \left(\frac{\mathrm{d}\mathcal{R}_i}{\mathcal{R}_i} \right) \left(\frac{\mathrm{d}X_j}{X_j} \right)$$

$$= -\sum_{j=n}^{i} \left(\frac{\mathcal{R}_j\,\Delta t}{1 + \mathcal{R}_j\,\Delta t} \right) \sigma_i \sigma_j \rho_{ij}\,\mathrm{d}t.$$

We conclude:

When we use the discount bond maturing at t_n as the numeraire, the drift of (expected percentage change in) the forward rate $\mathcal{R}_i$ is

$$\mu_i(t) = \begin{cases} \sum_{j=i+1}^{n-1} \left(\frac{\mathcal{R}_j(t)\,\Delta t}{1 + \mathcal{R}_j(t)\,\Delta t} \right) \sigma_i \sigma_j \rho_{ij} & \text{if } i < n - 1\,, \\ 0 & \text{if } i = n - 1\,, \\ -\sum_{j=n}^{i} \left(\frac{\mathcal{R}_j(t)\,\Delta t}{1 + \mathcal{R}_j(t)\,\Delta t} \right) \sigma_i \sigma_j \rho_{ij} & \text{if } i > n - 1\,. \end{cases} \qquad (14.39)$$

Problems

14.1. Create an Excel worksheet demonstrating a four-period Ho-Lee model. Allow the user to input σ, Δt, and $P^{\mathrm{mkt}}(t_n)$ for $n = 1, \ldots, 5$. Compute $\phi(t_n)$ from (14.3) for $n = 1, \ldots, 4$. Create the one-period interest rate tree (starting from $r(0) = -\log P^{\mathrm{mkt}}(t_1)/\Delta t$) and the valuation tree for a discount bond maturing at t_5. Verify that the tree gives the price $P^{\mathrm{mkt}}(t_5)$. Note: to create a binomial tree in a spreadsheet, it is probably easiest to put the topmost (or bottommost) nodes along one row and the other nodes in a triangle below (or above).

14.2. Modify the preceding exercise to include the valuation tree for a caplet with t_4 as its reset date and t_5 as its payment date. Note that the payoff of the caplet at date t_5 is

$$\max\left(0, \mathcal{R}(t_4) - \bar{\mathcal{R}}\right)\Delta t,$$

where

$$\mathcal{R}(t_4) = \frac{1}{P(t_4, t_5)} - 1.$$

Allow the user to input $\bar{\mathcal{R}}$.

14.3. Create a function `Ho_Lee_Caplet` that values a caplet in the Ho-Lee model. Look up market discount bond prices from a function such as `DiscountBondPrice` in Prob. 12.1 and calibrate to the market from (14.3). The inputs to the function should be $\bar{\mathcal{R}}$, σ, T_1, T_2, N_1 and N_2, where T_1 is the reset date for the caplet, T_2 is the payment date for the caplet, N_1 is the number of periods between date 0 and T_1, and N_2 is the number of periods between T_1 and T_2. Note that the payoff of the caplet at date T_2 is

$$\max\left(0, \mathcal{R}(T_1) - \bar{\mathcal{R}}\right)' \times (T_2 - T_1),$$

where

$$\mathcal{R}(T_1) = \frac{1}{P(T_1, T_2)} - 1.$$

14.4. Create an Excel worksheet demonstrating a four-period Black-Derman-Toy model. Allow the user to input Δt, $r(0)$, and $\sigma(t_i)$ and $\theta(t_i)$ for $i = 1, \ldots, 4$. Create the one-period interest rate and the valuation tree for a discount bond maturing at t_5.

14.5. Modify the preceding exercise by including the valuation tree for a caplet with reset date t_4 and payment date t_5, as in Prob. 14.2.

14.6. Repeat Probs. 14.4 and 14.5 for the Black-Karasinski model, allowing the user to also input $\kappa(t_i)$ for $i = 1, \ldots, 4$.

14.7. Create a VBA function `CIR_Caplet_MC` that values a caplet in the CIR model using Monte Carlo, without calibrating the model to the current yield curve. Simulate the CIR process as described in Sect. 4.5 for the Heston model. The inputs should be $\bar{\mathcal{R}}$, $r(0)$, κ, θ, σ, T_1, T_2, N and M, where T_1 is the reset date for the caplet, T_2 is the payment date for the caplet, N is the number of periods between 0 and T_1, and M is the number of simulations. The payoff of the caplet is as in Prob. 14.3, where $P(T_1, T_2)$ is the function of $r(T_1)$ and $T_2 - T_1$ given in (14.18) with $\phi = 0$.

14.8. Modify the function in the preceding exercise to create a function `CIR_Calibrated_Caplet_MC` that values a caplet in the CIR model using Monte Carlo, with the model calibrated to the market. Look up market discount bond prices from a function such as `DiscountBondPrice` in Prob. 12.1. To compute $\phi(t_i)$ for $i = 1, \ldots, N$ from market bond prices, use (14.19) for dates t_i and t_{i+1} as in Sect. 13.8.

A

Programming in VBA

The purpose of this appendix is to provide an introduction to the features of Excel VBA that are used in the book. To learn more about VBA in a finance setting, Jackson and Staunton [40] is a good source.

A.1 VBA Editor and Modules

Subroutines and functions are created in the VBA editor, which is reached through Tools/ Macros/Visual Basic Editor. When the editor opens, click Insert/Module to open an editing screen in which subroutines and functions can be typed. If you have opened a new workbook and inserted a module, you should see on the left a small pane with the heading Project-VBA Project that lists the elements of the workbook, including Module 1, which is the default name for the collection of things you might type in the editing screen (if the pane is not present, click View/Project Window). You should also see on the left a pane with the heading Properties-Module 1 (if it is not there, click View/Properties). You can rename Module 1 to something more useful by highlighting Module 1 in the Properties Window and typing the new name. You can add another module by clicking Insert/Module again. If you save the Excel workbook, all of the modules (and hence all of the subroutines and functions created in them) are saved with the workbook.

If you open the workbook distributed with this book, you will see in the Project Window modules named Chapt2, Chapt3, ..., Chapter 13. Each of these modules contains the VBA code in the corresponding book chapter. To view the code in a particular module, right-click on its name in the Project Window and select View Code. You will see a collection of programs separated by gray lines (which are added by the VBA editor to make things more readable). Each subroutine starts with Sub and ends with End Sub and each function starts with Function and ends with End Function. You will also see "Option Explicit" at the top of each module—this will be discussed below.

The organization of subroutines and functions into modules is not important (except for globally defined variables, which are not used in this book). All of the subroutines and all of the functions in all of the modules in any open workbook are available to use in any open workbook. However, you may find it convenient to organize your work into separate modules, for example Homework1, Homework2, etc.

If you open multiple workbooks, and open the VBA editor with one of them, then the Project Window will list each workbook and the modules associated with each. When the workbooks are saved, each set of modules will be saved with the associated workbook.

If VBA catches an error (normally a syntax error or an undeclared variable if "Option Explicit" has been declared) when executing a subroutine or function, a message box will pop up to inform the user. If the "Debug" option is chosen in this box, the offending VBA code will be highlighted in the VBA editor. After correcting the error, you need to click Run/Reset in the editor (or the square button in the editor's toolbar).

VBA ignores everything written on a line following an apostrophe, so comments can be placed on any line by preceding them with an apostrophe. Including comments in your subroutines and functions is very important to make them understandable.

The underscore character indicates that a line is to be continued. For example,

```
y = x + 5
```

is the same as

```
y = x _
+ 5
```

This is useful for breaking long lines.

A.2 Subroutines and Functions

A subroutine is also called a macro. It is a way of automating tasks, including mathematical calculations, cell formatting, and outputting results to cells. The subroutines in this book simulate a random process and output the results to the active worksheet. The other programs in the book are user-defined functions.

To execute a macro, click Tools/Macros. Clicking the name of a macro and then clicking Run will execute it. A macro or function created in one workbook can be used in another. To execute a macro created in another workbook, simply open both workbooks at the same time, click Tools/Macros and choose the option "All Open Workbooks" for the macros to be displayed.[1]

---

[1] Since you will be running macros and using user-defined functions frequently, it is useful to add buttons to the toolbar to execute the keystrokes of clicking

To create a macro in the VBA editor, type

```
Sub WhateverNameYouWant()
...
list of commands
...
End Sub
```

You will notice that the editor automatically adds the parentheses () at the end of the subroutine name and adds the End Sub statement when you type Sub WhateverNameYouWant.

A user-defined function is executed just like any other Excel function—in a cell of the spreadsheet, type =FunctionName(arguments). The arguments supplied to the functions can be numbers or can be cell references, just as with any other Excel function. To see the user-defined functions that have been created, click Insert/Function and select the category User Defined. You may see a lot of functions created by Excel add-ins in addition to the functions that are in the modules. You can also execute a function by double-clicking on its name here.

To create a function in the VBA editor, type

```
Function AnotherName(argument1, argument2, ..., lastargument)
...
list of commands
...
AnotherName = WhateverTheAnswerMightBe
End Function
```

A.3 Message Box and Input Box

One way for a subroutine or function to deliver information is through the MsgBox function. In Module 1, type

```
Sub WhateverNameYouWant()
MsgBox("Whatever you want to type.")
End Sub
```

When you execute this macro, a message box will pop up, displaying the message. To close the message box, click OK. The message box function is useful primarily for displaying error messages. However, the message box can also return the results of mathematical operations, as the next example shows.

Tools/Macros and Insert/Function, if the buttons are not already there. To add the macro button, click Tools/Customize/Commands/Tools, scroll to Macros, and drag the "Macros ..." button to the toolbar. To add the function button, scroll to Insert and drag the "Insert Function" button to the toolbar.

One way for a subroutine or function to obtain information from the user is via the `InputBox` function. In Module 1, type

```
Sub AnotherSub()
x = InputBox("What is your favorite number?")
MsgBox("You said your favorite number was  " & x)
End Sub
```

When you execute this macro, a box will pop up displaying the text "What is your favorite number?" and providing a facility for inputting a number. When you hit Enter or click OK, the input box will disappear and the message box will appear, displaying the message and the number you chose.

A.4 Writing to and Reading from Cells

You can write a number, text, or formula to any cell in any worksheet of any open workbook. For example, executing the following macro

```
Sub WritingTest()
Workbooks("Book1.xls").Sheets("Sheet1").Range("B3").Value = 7
End Sub
```

will write the number 7 to cell B3 of Sheet1 of Book1.xls. The statement can be shortened to

```
Sheets("Sheet1").Range("B3").Value = 7
```

if you want to write to the active workbook, and it can be shortened to

```
Range("B3").Value = 7
```

if you want to write to the active sheet in the active workbook. To write text to the cell, enclose it in parentheses; for example, we could replace `Value = 7` with `Value = "some text"`. It is also possible to write a formula to a cell by replacing `Value = 7` with, for example, `Formula = "=A6"`. Running any macro of this sort will over-write anything that may already be in cell B3.

In the macros in this book, rather than writing to a particular cell, we write to the active cell of the active sheet of the active workbook (i.e., the cell in which the cursor is) and to cells surrounding the active cell. This is done as follows:

```
Sub WritingTest()
ActiveCell.Value = 7
ActiveCell.Offset(1,2) = 8
End Sub
```

This macro writes the number 7 to the active cell and the number 8 to the cell that is one row below and two columns to the right of the active cell.

A subroutine or function can also read directly from a cell in a workbook, though we do not use that feature in this book. The syntax is the same as for writing to a cell; for example, `x = ActiveCell.Value` assigns the value in the active cell to the variable x.

The formatting of cells (and ranges of cells) can also be changed in Excel macros. Moreover, the active cell/sheet/workbook can also be selected within a macro, and charts can be generated within macros, etc. We use VBA mainly as a computational engine in this book rather than as a means to create and modify worksheets, so we do not use many of the features of Excel VBA.

A.5 Variables and Assignments

Variable names must begin with a letter, be less than 256 characters long, and cannot include various special characters (in particular, they cannot contain blank spaces, hyphens or periods). Variable names are not case sensitive: a is the same variable as A (in fact, you may find the VBA editor changing the capitalization of names to maintain consistency across a project). It is of course a good idea to use names that mean something, so your programs are easier to read later. You cannot use any name already reserved by VBA or Excel; for example, attempting to create a variable with the name Sub will generate an error message.

An expression like y = x + 3 is an assignment statement (unless it is prefaced by an If, ElseIf or Do While—see below). The computer evaluates the right-hand side, by looking up the value already assigned to x, adding 3, and storing this value in the memory space reserved for y. A statement like x = x + 3 is perfectly acceptable. It simply adds 3 to the value of x. It doesn't matter whether you add spaces around the = and + signs; the VBA editor will automatically adjust the spacing.

It is optional whether you must specifically allocate memory space for a variable. If you type "Option Explicit" in a VBA module, then all variables must be declared. This is done with the keyword Dim at the beginning of the program (more on this below). If you do not type "Option Explicit," then you can create a new variable in the middle of a program simply by assigning it a value. For example, you can type y = x+3. If y has not been previously defined, then it will be created and assigned the value x+3. If x has not been defined, it will be created and given the value 0.

The main virtue of selecting "Option Explicit" is that it helps to avoid typographical errors. Suppose for example that you intend to assign a new value to a variable named HardToSpell. If you misspell the name in the assignment statement and have not declared "Option Explicit," then VBA will create a new variable with the misspelled name. The program will still execute, but it will not calculate what you intended it to calculate. Likewise, if you intend to perform some operation with HardToSpell and assign the result to another variable and you misspell HardToSpell, then a new variable will be created with the misspelled name, given a value of zero, and the operation will be performed with the value zero rather than with the value of HardToSpell. In both cases, with "Option Explicit" declared, VBA will generate an error message alerting you to the misspelling.

A.6 Mathematical Operations

The basic mathematical operations are performed in VBA in the same way as in Excel: addition, subtraction, multiplication (the asterisk symbol), division (/), and exponentiation (the caret symbol—3^2 is 3 squared). The natural exponential is also the same in VBA as in Excel: Exp(6) is e^6. The square root and natural logarithm functions are also available in VBA but with different names than in Excel. The name of the square root function is Sqr in VBA (rather than Sqrt as in Excel) and the name of the natural logarithm function is Log in VBA (rather than Ln as in Excel). It does not matter whether or not you capitalize the names; the VBA editor will automatically capitalize, converting for example exp to Exp.

Other mathematical functions are used in VBA by preceding their Excel names with Application. For example Application.Max(3,4) returns the larger of 3 and 4. Of course, VBA means "Visual Basic for Applications" and the application being used here is Excel, so the name Application.Max indicates that the Excel Max function is to be used. A function that we use frequently is Application.NormSDist(d), which returns the probability that a standard normal random variable is less than or equal to d.

A.7 Random Numbers

Computers do not behave in a random way (though of course it may seem like it when one crashes) but they can generate sequences of numbers that pass statistical tests for randomness. The basic construction is the generation of a random integer in some range $[0, N]$ with each integer in the range being "equally likely." Dividing by N gives a number between 0 and 1 that has the appearance of being uniformly distributed. This number can then be transformed to give the appearance of a normal distribution or other standard distributions. Random integers are generated sequentially by an algorithm of the type $I_j = aI_{j-1} + c \bmod N$, for constants a and c. "mod N" means the remainder after dividing by N (7 mod 5 is 2, 10 mod 3 is 1, etc.). In this construction, I_{j-1} is called the "seed," and each integer becomes the seed for the next. This is certainly not a random construction, but if the constants and N are suitable chosen (N must be very large) then the integers will have the appearance of unpredictability, both to the human observer and according to formal statistical tests.

VBA has a built-in function for generating random variables that are uniformly distributed between 0 and 1. This function is called Rnd(). The same function is in Excel but called Rand(). Applying the inverse of the standard normal cumulative distribution function to a random variable that is uniformly distributed between 0 and 1 will generate a random variable with the standard normal distribution (i.e, the normal distribution with mean 0 and variance 1). The inverse of the standard normal cumulative distribution function is provided in Excel as the `NormSInv` function, and hence it can be called in VBA as `Application.NormsInv`. Given the existence of the `NormSInv` function, this is the simplest, though not the fastest, way to transform a uniformly distributed random variable into a normally distributed one. To reduce typing, the following function is used throughout this book.

```
Function RandN()
    RandN = Application.NormSInv(Rnd())
End Function
```

A.8 For Loops

A loop is a command or set of commands that executes repeatedly either for a fixed number of times or until some condition is violated. To execute the commands for a fixed number of times, use a "for loop."

To add the first 10 integers together we can create the following macro:

```
Sub AddIntegers()
x = 1
For i = 2 To 10
    x = x + i
Next i
ActiveCell.Value = x
End Sub
```

In the above, we first initialized the value of x to be 1. The statement(s) between the `For` statement and the `Next` statement are executed repeatedly. In the first passage through the loop, the variable i has the value 2 and the statement x = x + i translates as x = 1 + 2, so x is given the value 3. In the next passage, i has the value 3 and x has the value 3, so the statement x = x + i translates as x = 3 + 3, and x is given the value 6, etc.

Any variable name (not just i) can be used as a counter. The indentation of the line x = x + i is optional and serves only to make the program easier to read.

The number of iterations need not be fixed when the program is written. We can use variables in the `For` statement like `For i = y To z`. The number of iterations will then be determined by the values of y and z when the for loop is encountered.

In the statement `For i = 2 To 10`, MATLAB increases i by one each time it reaches the statement `Next i`. This is the default, but it can be changed. If you want i to increase by two each time, you can write

```
For i = 2 To 10 Step 2.
```

Negative step sizes and non-integer step sizes are also acceptable. For example, the statement `For i = 10 To 1 Step -1` produces a loop that executes "backwards," starting from i = 10 and counting down until i = 2.

A.9 While Loops and Logical Expressions

A "while loop" executes a block of statements repeatedly until some condition is violated. For example a crude way to add the first 10 integers would be with the following macro:

```
Sub AddIntegers2()
x = 0
i = 1
Do While i <= 10
    x = x + i
    i = i + 1
Loop
ActiveCell.Value = x
End Sub
```

When the program first encounters the `Do While` statement, it checks whether the condition $i \leq 10$ is true. If it is, then the statements preceding the `Loop` statement are executed. The condition $i \leq 10$ is then checked again, and the statements are executed repeatedly in this way until the condition $i \leq 10$ is false. Be careful that the statements being executed will eventually cause the condition to be false.

The comparison operators that can be used in the `Do While` statement (and `If` and `ElseIf` described below) are less than (<), less than or equal to (<=), greater than (>), greater than or equal to (>=), and equal to (=).

The expression `Not(i > 10)` is equivalent to `(i <= 10)`. Multiple conditions can be combined: the expression `i <= 10 And y > 6` is true if (and only if) both $i \leq 10$ and $y > 6$ are true, and the expression `i <= 10 Or y > 6` is true if either or both of its component statements is true.

A.10 If, Else, and ElseIf Statements

You can cause a statement to execute only when a certain condition is satisfied by prefacing it with an `If` statement. The format is

```
If y <= 10 Then
    x = 2 * x
End If
```

which doubles x if $y \leq 10$ and does nothing otherwise. Rather than doing nothing otherwise, you can cause a different statement or block of statements to execute when the condition is violated by including an `Else`. For example,

```
If y <= 10 Then
    x = 2 * x
Else
    x = 3 * x
End If
```

In this case, if $y > 10$, the statements following the Else statement execute, tripling x. Finally, you can check multiple conditions sequentially with ElseIf. Consider the following:

```
If y <= 10 Then
    x = 2 * x
ElseIf y <= 20 Then
    x = 3 * x
ElseIf y <= 30 Then
    x = 4 * x
Else
    x = 5 * x
End If
```

The conditions are checked sequentially as follows. If $y \le 10$, then x is doubled and execution of of the If block ends. If $y > 10$, the condition $y \le 20$ is checked. If this is true, x is tripled. If it is not true, the next condition is checked, etc. The result is that x is doubled when $y \le 10$; it is tripled when $10 < y \le 20$; it is quadrupled when $20 < y \le 30$; and it is quintupled when $y > 30$.

A.11 Variable Declarations

As mentioned before, if "Option Explicit" is declared, each variable must be declared at the beginning of a subroutine or function. A variable can be declared to be of a specific type or the type can be left unspecified and VBA will choose what seems to be the appropriate type. For numerical calculations, the important types are Integer, Long, Double, and Variant. The Integer data type is for storage of integers between -32,768 and 32,769. The Long data type can store integers between plus or minus 2 billion (actually a bit more than 2 billion). The Double data type stores arbitrary (floating point) numbers, to sixteen digits of accuracy. The Variant data type is the default type for variables whose type is not specified, and it adjusts itself automatically to the data stored within it. To declare a variable to be of a particular type, there are two equally acceptable syntaxes. For example, the Double type can be declared either as Dim x As Double or Dim x#. The Integer type can be declared either as Dim x As Integer or Dim x%. Note that the syntax Dim i, j, k As Integer is acceptable but it declares only k as being of type Integer, with i and j still being of type Variant. On the other hand, Dim i%, j%, k% declares i, j and k as being of type Integer.

In this book, the data type is left unspecified (hence as Variant), with the exception that the type of large arrays is declared. The Variant data type requires more memory for storage, so this is a bit inefficient.

Variables declared within a function or subroutine are "local variables." They can only be accessed within the function or subroutine within which they are defined. To understand this, consider the following simple example of a function (TestFunction) calling another function (AddTwo).

```
Function TestFunction(x)
TestFunction = x * AddTwo(x)
End Function

Function AddTwo(x)
Dim y
y = x + 2
AddTwo = y
End Function
```

The result of TestFunction(3) is $3 \times 5 = 15$. Consider now the following (strange) change to TestFunction.

```
Function TestFunction(x)
Dim z
z = x * AddTwo(x)
TestFunction = y
End Function
```

The new feature is that TestFunction(x) attempts to return y, which is defined only in AddTwo. If TestFunction(3) is executed, then one of two things will happen: (i) if "Option Explicit" has been declared, an error message will appear with the information that the variable y has not been declared within TestFunction, or (ii) if "Option Explicit" has not been declared, the function will return a value of zero. The reason in both cases is that the variable y defined within AddTwo is not available to TestFunction—it is local to AddTwo. In case (ii), a new variable y is created within TestFunction and, like all new variables, is given a default value of zero. The error message is probably preferable in this circumstance, which points again to the value of the "Option Explicit" declaration.

It is possible to declare a variable so that it is available to (and can be modified by) all of the functions in a module, or all of the functions in a workbook, or even all of the functions in all open workbooks. Such variables are called "global variables." That facility is useful in some situations, but it is not used in this book.

A.12 Variable Passing

As we have seen, functions and macros can call other functions or macros to perform part of their work. For example, macros shown previously in this appendix call the MsgBox function. The default arrangement in VBA is that variables are passed to functions (or to macros—though variables are not passed to macros in this book) "by reference" rather than "by value." This means that the actual memory location of the variable is given to the function, and any changes made to a variable by a function will affect the use of the variable in a calling function. Consider, for example the following simple change to the function AddTwo:

```
Function AddTwo(x)
x = x + 2
AddTwo = x
End Function
```

This function still adds the number 2 to its input. Now when we execute

```
TestFunction(3)
```

and it reaches the line

```
TestFunction = x * AddTwo(x)
```

it will be multiplying x by 5 as before. However, now x has been changed in AddTwo from 3 to 5, so the result of TestFunction(3) is $5 \times 5 = 25$.

This may sometimes be what one wants, but it is more likely that it will produce mistakes. There are two possible solutions. One is to change the function AddTwo as follows:

```
Function AddTwo(ByVal x)
x = x + 2
AddTwo = x
End Function
```

This forces VBA to pass only the value of x and not the memory location. So when 2 is added to x and returned to TestFunction(3), the value of x in TestFunction is still 3.

The second solution is more straightforward: simply do not change input variables within a function. That is, we can use our first version of AddTwo, which created a new variable to store the sum of x and 2, rather than changing the value of x (or we could use the simpler one-line function AddTwo = x + 2). The functions in this book follow this second approach—**we avoid changing the values of input variables**.

A.13 Arrays

It is very useful to be able to use a single variable name to store multiple values. For example, we can write loops such as

```
For i = 1 To 10
    x(i) = whatever
Next i
```

An array variable must be declared, regardless of whether "Option Explicit" is declared. Normally, the declaration takes the form Dim x(10) if the largest index number of x is known (to equal 10) when the function or macro is written. The default in VBA is that the first element is indexed by 0.[2] Therefore,

---

[2] This can be changed so that the default is for the first element to be indexed by 1 with the statement "Option Base 1."

`Dim x(10)` creates a vector with 11 elements, which are accessed as `x(0)`, ..., `x(10)`. The type of each element is Variant, unless it is declared otherwise—for example, `Dim x(10) As Integer` reserves memory locations for 11 integers. Multidimensional arrays can also be used. For example, `x(10, 6, 12)` creates a 3-dimensional array, with $11 \times 7 \times 13$ elements. The first index does not have to be zero. The declaration `Dim x(1 To 10)` creates a vector with 10 elements, which are accessed as `x(1)`, ..., `x(10)`. Likewise, one can use, for example `Dim x(-6 To 3)` to start the indexing at -6 and end at 3.

If the dimension of the array is not fixed, which is often the case, then normally it must be declared with empty parentheses—for example, `Dim x()`. The dimension will depend on the input arguments, or on calculations based on the input arguments. Before the array is used, the program must include a statement specifying the dimension, of the form `ReDim x(N)`, or `ReDim x(1 To N)`, where the variable N is either an input argument to the function or has been calculated prior to the statement `ReDim x(N)`.

The exception to the above statements about declaring array variables, whether the number of elements is known in advance or not, is when an array is assigned to a variable by a call to a function. The `Array` function is one example of a function that creates an array. For example

```
Dim x
x = Array(3, 6, 7)
```

will create an array with elements `x(0)=3`, `x(1)=6`, and `x(2)=7`. Replacing `Dim x` with `Dim x(2)` in this context will not work.

Functions can take arrays as inputs and return arrays as outputs. Arrays can be input by (i) typing the array as an argument of the function, (ii) inputting the worksheet cells in which the array resides, or (iii) passing the array as an output from another function. An array created in one function is passed to another function in the same way that any other variable is passed. To type an array as an input, enclose it curly braces, separate items in each row with a comma, and separate rows with a semicolon—for example $\{3, 1, 2; 4, 6, 2\}$ is an array with two rows and three columns, the first row being $\{3, 1, 2\}$ and the second row being $\{4, 6, 2\}$. The same array might be input via cell references as `B3:C5`.

Arrays can also be output to Excel worksheets. Consider the following:

```
Function MyArray(x)
Dim y(3)
For i = 1 To 3
   y(i) = i * x
Next i
MyArray = y
End Function
```

Note that the array y has four elements. The program does not define element 0, so it is zero by default. If we execute MyArray(2), the other elements will be `y(1) = 2`, `y(2) = 4`, and `y(3) = 6`. If we execute the function by typing =MyArray(2) in a cell of a worksheet, the number 0 will appear. (To avoid this and have the output show up in three cells instead of four, we could have declared Dim y(1 To 3).) To see the rest of the output, highlight the active cell and the three cells immediately to the right on the same row. Click the function key F2 and then hold down the key combination CTRL-SHIFT-ENTER. This is the standard Excel procedure for displaying arrays returned by functions. For example, the output of Excel's matrix algebra functions, such as MMULT, is revealed in the same way.[3] Two-dimensional arrays can be output to worksheets in the same way.

A.14 Debugging

Errors (bugs) are inevitable. VBA will catch some types (for example, syntax errors) and inform you. The more troublesome errors are those that do not prevent the program from running but lead to incorrect results. It is essential therefore to debug each program carefully.

To debug a subroutine, put the cursor on the subroutine name in the Visual Basic editor. Click on Debug/Step Into (or the function key F8) to step through the subroutine one line at a time. Putting the cursor over any variable will show the value of the variable at that stage of the program. To observe the values of variables more systematically, you can include statements of the form Debug.Print x or Debug.Print "The value of x is " & x in the subroutine. The subroutine will then print to the Immediate Window. To view the Immediate Window, click View/Immediate Window. Click on Run/Reset (or the square button on the toolbar) to discontinue debugging.

To debug a function, one can rewrite it as a subroutine, defining values for the input variables in the beginning of the subroutine. The VBA debugger has many other features. Debug/Step Over is particularly useful for stepping over a line that does not need to be checked and will be time consuming to check, for example, a call to another function.

---

[3] Once this is done, the individual cells in which the array was output cannot be changed. Attempting to do so will generate an error message, and it may be necessary to hit the Escape key once or twice to allow any use of the worksheet after the error message appears.

B

Miscellaneous Facts
about Continuous-Time Models

B.1 Girsanov's Theorem

In Sect. 2.9, we were able to compute the expected return of an asset under different numeraires directly, by using Itô's formula and the fact that the ratio of a non-dividend-paying asset price to the numeraire asset price is a martingale under the measure associated with the numeraire. In other cases (e.g., Heston's stochastic volatility model and Vasicek's model) the drift of a process could not be computed directly when we changed numeraires, because the process (volatility in Heston's model and the short rate in Vasicek's model) was not an asset price. In general, the change in the drift of a process when we change numeraires (or, more generally, change probability measures) is given by Girsanov's theorem.

An heuristic explanation of Girsanov's theorem is as follows. Let λ be a constant, and let B be a Brownian motion under a probability measure that we will denote by prob. Let $B^*(t) = B(t) + \lambda t$; i.e., $\mathrm{d}B^* = \mathrm{d}B + \lambda\,\mathrm{d}t$. Girsanov's theorem shows how to change the probability measure so that the drift of B^* is zero, i.e., how to change the probability measure to make B^* a martingale and hence (by Levy's theorem) a Brownian motion.

Consider discrete time periods of length Δt and approximate B by a binomial process that steps up or down by $\sqrt{\Delta t}$ in each time period, with up and down being equally likely. This approximation implies that the changes ΔB of the binomial process have mean equal to zero and variance equal to Δt, just as for a true Brownian motion. We have $\Delta B^* = \lambda\,\Delta t \pm \sqrt{\Delta t}$. If we change the probability of the up move to $(1 - \lambda\sqrt{\Delta t})/2$ and the probability of the down move to $(1 + \lambda\sqrt{\Delta t})/2$, then the expected change in B^* will be

$$\left(\frac{1 - \lambda\sqrt{\Delta t}}{2}\right)\left(\lambda\,\Delta t + \sqrt{\Delta t}\right) + \left(\frac{1 + \lambda\sqrt{\Delta t}}{2}\right)\left(\lambda\,\Delta t - \sqrt{\Delta t}\right) = 0\,.$$

Therefore, B^* is a martingale under these revised probabilities.

Changing the probabilities of each "branch" of the binomial tree in this way implies that the probability of a path through the tree is changed as follows. The probability of a path is the product of the probabilities of the branches, so, letting prob* denote the revised probabilities, we have

$$\frac{\text{prob}^*(\text{path through time } t)}{\text{prob}(\text{path through time } t)}$$
$$= \frac{\text{prob}^*(\text{path through time } t-\Delta t)}{\text{prob}(\text{path through time } t-\Delta t)} \times \frac{\text{prob}^*(\text{branch at } t)}{\text{prob}(\text{branch at } t)} .$$

Note that our definitions imply

$$\frac{\text{prob}^*(\text{up branch at } t)}{\text{prob}(\text{up branch at } t)} = \frac{\frac{1}{2}\left(1 - \lambda\sqrt{\Delta t}\right)}{1/2} = 1 - \lambda\,\Delta B(t) ,$$

and

$$\frac{\text{prob}^*(\text{down branch at } t)}{\text{prob}(\text{down branch at } t)} = \frac{\frac{1}{2}\left(1 + \lambda\sqrt{\Delta t}\right)}{1/2} = 1 - \lambda\,\Delta B(t) .$$

Therefore,

$$\frac{\text{prob}^*(\text{path through time } t)}{\text{prob}(\text{path through time } t)}$$
$$= \frac{\text{prob}^*(\text{path through time } t-\Delta t)}{\text{prob}(\text{path through time } t-\Delta t)} \times \left(1 - \lambda\,\Delta B(t)\right) .$$

If we let $Y(t)$ denote the ratio of path probabilities through time t, this shows that the percent change in Y at time t is $-\lambda\,\Delta B(t)$, i.e.,

$$Y(t) = Y(t - \Delta t) \times \left(1 - \lambda\,\Delta B(t)\right) \implies \frac{Y(t) - Y(t - \Delta t)}{Y(t - \Delta t)} = -\lambda\,\Delta B(t) .$$

A continuous-time formulation of this equation is

$$\frac{\mathrm{d}Y(t)}{Y(t)} = -\lambda\,\mathrm{d}B(t) .$$

This equation implies that Y is a geometric Brownian motion with explicit solution (given that the ratio of path probabilities at date 0 is $Y(0) = 1$)

$$Y(t) = \exp\left(-\lambda^2 t/2 - \lambda B(t)\right) . \tag{B.1}$$

The above heuristic argument suggests that the process (B.1) defines a ratio of path probabilities, prob* to prob, such that B^* is a martingale under prob*. Because B^* is continuous and its quadratic variation through each date t is equal to t (because the addition of λt to B does not alter the quadratic variation of B), Levy's theorem implies that B^* must in fact be a Brownian motion relative to the measure prob*. This is the content of Girsanov's theorem. In the formal statement, there is no reference to ratios of path probabilities, because individual paths actually have zero probability under either prob or prob*. Instead, the theorem states that B^* is converted to a Brownian motion by multiplying the probability of any event (set of paths) by the conditional expectation of Y, given the event.

There is no need to assume λ is a constant, provided the random process λ is sufficiently regular that the general form of (B.1), i.e.,

$$Y(t) \equiv \exp\left\{ -\frac{1}{2} \int_0^t \lambda^2(u)\, du - \int_0^t \lambda(u)\, dB(u) \right\}, \tag{B.2}$$

is a martingale.[1]

Girsanov's Theorem: Let B be a Brownian motion on a time horizon $[0, T]$ and let λ be a stochastic process such that Y defined by (B.2) is a martingale. Define

$$B^*(t) = B(t) + \int_0^t \lambda(u)\, du, \tag{B.3}$$

and define a new probability measure prob* by setting prob$^*(A) = 0$ for each event A such that prob$(A) = 0$, and by defining

$$\text{prob}^*(A) = E\big[Y(T)|A\big] \times \text{prob}(A) \tag{B.4}$$

for each event A such that prob$(A) > 0$. Then B^* is a Brownian motion on the time horizon $[0, T]$ relative to prob*.

The definition of prob* in the boxed statement emphasizes the ratio of probabilities aspect. It is equivalent to the definition

$$\text{prob}^*(A) = E\left[1_A Y(T) \right] \tag{B.5}$$

for each event A. Thus, it is consistent with the definition (1.11) of the probability of an event A when we use a non-dividend-paying asset price S as the numeraire. The relation between the two is that the "ratio of path probabilities" $Y(T)$ equals $\phi(T)S(T)/S(0)$, where $\phi(T)$ denotes the random state price at date T.

---

[1] The process (B.2) is an Itô process with zero drift. A sufficient condition for it to be a martingale is that

$$E\left[\exp\left\{ \frac{1}{2} \int_0^T \lambda^2(u)\, du \right\} \right] < \infty \,.$$

This is called "Novikov's condition." See, e.g., Karatzas and Shreve [45].

Note also that for any random variable X (for which the mean exists) the mean of X under prob*, which we denote by $E^*[X]$, is given by

$$E^*[X] = E[Y(T)X] . \tag{B.6}$$

In some cases we may be given (perhaps by equilibrium arguments) the random variable Y defining the change of measure, and we wish to compute the change in the drift of a Brownian motion (in order to compute, for example, the drift of a volatility or an interest rate). Thus, we need to reverse the above process, in which we started with the change of drift λ and computed Y. This is straightforward. Given $Y(T)$, define $Y(t) = E_t[Y(T)]$, i.e., the expectation of $Y(T)$ under the original measure, given information at date t. Equation (B.2) shows that

$$\frac{dY}{Y} = -\lambda \, dB .$$

Therefore,

$$-(dB)\left(\frac{dY}{Y}\right) = \lambda \, dt .$$

It follows that the definition

$$dB^* = dB - (dB)\left(\frac{dY}{Y}\right)$$

gives us a Brownian motion B^* relative to the measure prob*. In other words, the drift of B under the measure prob* is $(dB)(dY)/Y$.

B.2 Distribution of the Minimum of a Geometric Brownian Motion

Here we will give an explanation of formulas used in Chap. 8 for valuing barrier and lookback options. From a mathematical point of view, our discussion will be decidedly informal.

Consider an asset price S satisfying

$$d \log S = \mu \, dt + \sigma \, dB ,$$

for constants μ and σ, where B is a Brownian motion. Consider constants $K \geq L$ with $L < \log S(0)$. Define $z = \min_{0 \leq t \leq T} S(t)$. Define

$$x = \begin{cases} 1 & \text{if } S(T) > K \text{ and } z > L , \\ 0 & \text{otherwise} . \end{cases}$$

To price a down-and-out call, we need to compute prob$(x = 1)$. As in Sect. 8.6, define

$$y = \begin{cases} 1 & \text{if } S(T) > K \text{ and } z \le L \,, \\ 0 & \text{otherwise} \,. \end{cases}$$

The event $S(T) > K$ is the union of the disjoint events $x = 1$ and $y = 1$, so we have

$$\text{prob}(x = 1) = \text{prob}(S(T) > K) - \text{prob}(y = 1)$$
$$= N(d) - \text{prob}(y = 1) \,,$$

where

$$d = \frac{\log\left(\frac{S(0)}{K}\right) + \mu T}{\sigma\sqrt{T}} \,. \tag{B.7}$$

Thus, the remaining task is to compute $\text{prob}(y = 1)$.

To price lookback options, it is necessary to know the cumulative distribution function of z, i.e., we need to know $\text{prob}(z \le L)$ for arbitrary L. The event $z \le L$ is the union of the disjoint events $S(T) \le L$ and $y = 1$, where we specialize to the case $K = L$ in the definition of y. Thus,

$$\text{prob}(z \le L) = \text{prob}(S(T) \le L) + \text{prob}(y = 1)$$
$$= N(-d) + \text{prob}(y = 1) \,,$$

where again we take $K = L$ in the definition of d. Thus, for pricing lookbacks also, the key task is to compute $\text{prob}(y = 1)$.

Assume first that $\mu = 0$, so $\log S$ is a Brownian motion with zero drift. We want to compute the probability of the paths of $\log S$ that dip below $\log L$ and end above $\log K$. Each such path has a "twin" defined by reflecting the path (as in a mirror image) through the horizontal line $x(t) = \log L$ after the first time $\log S$ hits $\log L$. The original path increases by at least $\log K - \log L$ after hitting $\log L$ (otherwise, it could not end above $\log K$). So, the twin decreases by at least $\log K - \log L$ after hitting $\log L$. This means that it ends below $2\log L - \log K$. Moreover, each path ending below $2\log L - \log K$ is the twin in this sense of a path hitting $\log L$ and then ending above $\log K$. Because $\log S$ has no drift, the "twins" are equally likely. Therefore, when $\mu = 0$,

$$\text{prob}(y = 1) = \text{prob}\left(\log S(T) \le 2\log L - \log K\right)$$
$$= \text{prob}\left(\frac{B(T)}{\sqrt{T}} \le \frac{2\log L - \log K - \log S(0) - \mu T}{\sigma\sqrt{T}}\right)$$
$$= N(d^*) \,,$$

where

$$d^* = \frac{\log\left(\frac{L^2}{KS(0)}\right)}{\sigma\sqrt{T}} \,. \tag{B.8}$$

Now consider the case $\mu \ne 0$, the case in which we are really interested. By Girsanov's theorem, the process B^* defined by $B^*(0) = 0$ and

$$dB^* = dB + \frac{\mu}{\sigma}\,dt$$

is a Brownian motion under the measure prob* defined by (B.1) and (B.4), where we take $\lambda = \mu/\sigma$ in the definition of $Y(T)$. The purpose of this definition is that we have

$$d\log S = \mu\,dt + \sigma\left(dB^* - \frac{\mu}{\sigma}\,dt\right) = \sigma\,dB^*\,.$$

Letting E denote expectation relative to the measure under which B is a Brownian motion and E^* denote expectation relative to prob*, we have from (B.6) that

$$\text{prob}(y=1) = E[y] = E\left[Y(T)\frac{y}{Y(T)}\right]$$

$$= E^*\left[\frac{y}{Y(T)}\right]$$

$$= E^*\left[\exp\left(\frac{1}{2}\lambda^2 T + \lambda B(T)\right)y\right]$$

$$= E^*\left[\exp\left(\frac{1}{2}\lambda^2 T + \lambda[B^*(T) - \lambda T]\right)y\right]$$

$$= E^*\left[\exp\left(-\frac{1}{2}\lambda^2 T + \lambda B^*(T)\right)y\right]\,. \qquad (B.9)$$

Because $\log S$ has no drift under prob*, the twin paths described before are equally likely under prob*. However, the reflection leads to low values of $\log S(T)$ and hence to low values of $B^*(T)$ rather than high values, and we must compensate for this in (B.9). Specifically, for a path of $\log S$ that ends above $\log K$, we have

$$B^*(T) = \frac{\log K - \log S(0) + \varepsilon}{\sigma} \qquad (B.10)$$

for some $\varepsilon > 0$ and the reflection of this path has

$$B^*(T) = \frac{2\log L - \log K - \log S(0) - \varepsilon}{\sigma} \qquad (B.10')$$

for the same ε. Therefore, to use the reflected path, we compute

$$\varepsilon = 2\log L - \log K - \log S(0) - \sigma B^*(T)$$

from (B.10$'$) and substitute this into the right-hand side of (B.10) to obtain

$$\frac{\log K - \log S(0) + 2\log L - \log K - \log S(0) - \sigma B^*(T)}{\sigma}$$

$$= \frac{2\log L - 2\log S(0)}{\sigma} - B^*(T)$$

as the value that should replace $B^*(T)$ in (B.9) when we use the reflected paths. As in the case $\mu = 0$, using the reflected paths means replacing the random variable y with y' defined as

$$y' = \begin{cases} 1 & \text{if } \log S(T) \leq 2\log L - \log K , \\ 0 & \text{otherwise .} \end{cases}$$

Substituting into (B.9) and employing some algebra gives us

$$\text{prob}(y = 1) = E^* \left[\exp\left(-\frac{1}{2}\lambda^2 T + \lambda \left[\frac{2\log L - 2\log S(0)}{\sigma} - B^*(T) \right] \right) y' \right]$$

$$= \left(\frac{L}{S(0)} \right)^{2\mu/\sigma^2} E^* \left[\exp\left(-\frac{1}{2}\lambda^2 T - \lambda[B(T) + \lambda T] \right) y' \right]$$

$$= \left(\frac{L}{S(0)} \right)^{2\mu/\sigma^2} E^* \left[\exp\left(-\frac{3}{2}\lambda^2 T - \lambda B(T) \right) y' \right]$$

$$= \left(\frac{L}{S(0)} \right)^{2\mu/\sigma^2} E \left[\exp\left(-2\lambda^2 T - 2\lambda B(T) \right) y' \right] , \tag{B.11}$$

where for the last equality we used (B.6) again.

Now we will define another change of measure. Set $\delta = 2\lambda$,

$$Z(T) = \exp\left(-\delta^2 T/2 - \delta B(T) \right)$$

and $\text{prob}^{**}(A) = E[1_A Z(T)]$ for each event A. From the definition of δ and (B.6) we have

$$E \left[\exp\left(-2\lambda^2 T - 2\lambda B(T) \right) y' \right] = E \left[\exp\left(-\frac{1}{2}\delta^2 T - \delta B(T) \right) y' \right]$$

$$= E^{**}[y']$$

$$= \text{prob}^{**}(y' = 1) . \tag{B.12}$$

Moreover, Girsanov's theorem states that $dB^{**} = dB + \delta\, dt$ defines a Brownian motion B^{**} under the measure prob^{**}. The event $y' = 1$ is equivalent to

$$\log S(0) + \mu T + \sigma B(T) \leq \log\left(\frac{L^2}{K} \right)$$

$$\iff \log S(0) + \mu T + \sigma[B^{**}(T) - \delta T] \leq \log\left(\frac{L^2}{K} \right)$$

$$\iff \log S(0) - \mu T + \sigma B^{**}(T) \leq \log\left(\frac{L^2}{K} \right)$$

$$\iff \frac{B^{**}(T)}{\sqrt{T}} \leq d' , \tag{B.13}$$

where we define

$$d' = \frac{\log\left(\frac{L^2}{KS(0)}\right) + \mu T}{\sigma\sqrt{T}} \, . \tag{B.14}$$

Combining (B.11), (B.12) and (B.13) yields

$$\mathrm{prob}(y = 1) = \left(\frac{L}{S(0)}\right)^{2\mu/\sigma^2} \mathrm{N}(d') \, .$$

Summarizing, we have

Assume $d\log S = \mu\,dt + \sigma\,dB$ where B is a Brownian motion. Define $z = \min_{0 \le t \le T} S(t)$. For $K \ge L$ and $L \le \log S(0)$,

1. The probability that $S(T) > K$ and $z > L$ is

$$\mathrm{N}(d) - \left(\frac{L}{S(0)}\right)^{2\mu/\sigma^2} \mathrm{N}(d') \, ,$$

   where d is defined in (B.7) and d' is defined in (B.14).

2. The probability that $z \le L$ is

$$\mathrm{N}(-d) + \left(\frac{L}{S(0)}\right)^{2\mu/\sigma^2} \mathrm{N}(d') \, ,$$

   where d is defined in (B.7) and d' is defined in (B.14), substituting $K = L$ in both.

B.3 Bessel Squared Processes and the CIR Model

This section will present additional results regarding the CIR square-root short rate process discussed in Sect. 14.4. The ideas described here are one way (though not the only way) to derive the CIR discount bond option pricing formula. We begin with the following simpler process

$$dx(t) = \delta\,dt + 2\sqrt{x(t)}\,dZ \tag{B.15}$$

for a Brownian motion Z and constant $\delta > 0$. This is called a Bessel-squared process with parameter δ. The parameter δ determines whether x can ever reach zero. If $\delta \ge 2$, then with probability one, $x(t)$ is strictly positive for all t; whereas, if $\delta < 2$, then with positive probability, x will sometimes hit zero (but will never go negative).

In the particular (rare) case that δ is an integer, the squared length of a δ-dimensional vector of independent Brownian motions is a process x satisfying (B.15). To see this, let $B_1, \ldots, B_\delta$ be independent Brownian motions starting at given values b_i; i.e., $B_i(0) = b_i$. Define $x(t) = \sum_{i=1}^{\delta} B_i(t)^2$. Then Itô's formula gives us

$$dx(t) = \sum_{i=1}^{\delta} 2B_i(t)\, dB_i(t) + \sum_{i=1}^{\delta} dt$$

$$= \delta\, dt + 2\sqrt{x(t)} \sum_{i=1}^{\delta} \frac{B_i(t)}{\sqrt{x(t)}}\, dB_i(t).$$

The process Z defined by $Z(0) = 0$ and

$$dZ = \sum_{i=1}^{\delta} \frac{B_i(t)}{\sqrt{x(t)}}\, dB_i(t)$$

is a Brownian motion (because it is a continuous martingale with $(dZ)^2 = dt$); thus, we obtain (B.15).

Continuing to assume that δ is an integer and that $x(t) = \sum_{i=1}^{\delta} B_i(t)^2$, note that, for any t, the random variables ξ_i defined as $\xi_i = [B_i(t) - B_i(0)]/\sqrt{t}$ are independent standard normals, and we have

$$x(t) = \sum_{i=1}^{\delta} \left[b_i + B_i(t) - B_i(0)\right]^2$$

$$= t \times \sum_{i=1}^{\delta} \left(\frac{b_i}{\sqrt{t}} + \xi_i\right)^2.$$

A random variable of the form $\sum_{i=1}^{\delta} (\gamma_i + \xi_i)^2$, where the γ_i are constants and the ξ_i are independent standard normals, is said to have a non-central chi-square distribution with δ degrees of freedom and noncentrality parameter $\sum_{i=1}^{\delta} \gamma_i^2$. Thus, $x(t)$ is equal to t times a non-central chi-square random variable with δ degrees of freedom and noncentrality parameter

$$\sum_{i=1}^{\delta} \frac{b_i^2}{t} = \frac{x(0)}{t}.$$

The noncentral chi-square distribution can be defined for a non-integer degrees of freedom also, and a process x satisfying (B.15) for a non-integer δ has the same relation to it, namely,

---

If x satisfies (B.15), then for any t and $\alpha > 0$, the probability that $x(t) \leq \alpha$ is equal to the probability that $z \leq \alpha/t$, where z is a random variable with a non-central chi-square distribution with δ degrees of freedom and noncentrality parameter $x(0)/t$.

Now consider the CIR process (14.9). Define $\delta = 4\kappa\theta/\sigma^2$ and define x by (B.15), with $x(0) = r(0)$. Set[2]

$$h(t) = \frac{\sigma^2}{4\kappa}\left(e^{\kappa t} - 1\right) ,$$

and

$$r(t) = e^{-\kappa t}x(h(t)) .$$

Then it can be shown[3] that r satisfies the CIR equation (14.9), namely

$$dr = \kappa(\theta - r)\,dt + \sigma\sqrt{r}\,dB \tag{B.16}$$

for a Brownian motion B. For any t and $\alpha > 0$, the probability that $r(t) \leq \alpha$ is equal to the probability that $x(h(t)) \leq e^{\kappa t}\alpha$. In view of the previous boxed statement, this implies:

If r satisfies the CIR equation (B.16) where κ, θ and σ are positive constants, then, for any $t > 0$ and any α, the probability that $r(t) \leq \alpha$ is the probability that $z \leq e^{\kappa t}\alpha/h(t)$, where z is a random variable with a non-central chi-square distribution with $\delta = 4\kappa\theta/\sigma^2$ degrees of freedom and noncentrality parameter $r(0)/h(t)$.

To derive the discount bond option pricing formula for the CIR model, we need to know the distribution of $r(T)$ when the parameters κ and θ are time-dependent. Let w denote either u (the maturity of the underlying) or T (the maturity of the option). Using the discount bond maturing at w as the numeraire, we repeat here (14.23), dropping now the "hat" on $\hat{r}$:

$$dr(t) = \kappa^*(t)[\theta^*(t) - r(t)]\,dt + \sigma\sqrt{r(t)}\,dB^*(t) , \tag{B.17}$$

where

$$\kappa^*(t) = \kappa + \sigma^2 b(w - t) \qquad \text{and} \qquad \theta^*(t) = \frac{\kappa\theta}{\kappa^*(t)} .$$

Because $\kappa^*(t)\theta^*(t) = \kappa\theta$, we again define $\delta = 4\kappa\theta/\sigma^2$ but now set

$$h^*(t) = \frac{\sigma^2}{4}\int_0^t \exp\left(\int_0^s \kappa^*(y)\,dy\right) ds$$

---

[2] I learned this transformation from unpublished lecture notes of Hans Buehlmann.

[3] The key to this calculation is the fact that if Z is a Brownian motion and h is a continuously differentiable function with $h'(s) > 0$ for all $s > 0$ then B defined by

$$B(t) = \int_0^t \frac{1}{\sqrt{h'(s)}}\,dZ_{h(s)}$$

is a Brownian motion also.

and

$$r(t) = \exp\left(-\int_0^t \kappa^*(s)\,\mathrm{d}s\right) x(h^*(t))\,.$$

Then it can be shown that r satisfies (B.17) for a Brownian motion B^*. Thus, as in the previous paragraphs, the probability that $r(T) \le \alpha$, where r satisfies (B.17), is the probability that

$$z \le \frac{\exp\left(\int_0^T \kappa^*(s)\,\mathrm{d}s\right)\alpha}{h^*(T)}\,,$$

where z has a non-central chi-square distribution with δ degrees of freedom and noncentrality parameter $r(0)/h^*(T)$.

Straightforward calculations, using in particular the fact that $b(\tau) = a'(\tau)/(\kappa\theta)$ and

$$\int \frac{\mathrm{e}^{\gamma t}}{c(t)^2}\,\mathrm{d}t = -\frac{1}{(\kappa+\gamma)\gamma}\int \frac{\mathrm{d}}{\mathrm{d}t}\left(\frac{1}{c(t)}\right)\,\mathrm{d}t = -\frac{1}{(\kappa+\gamma)\gamma c(t)}$$

give us:

$$\exp\left(\int_0^T \kappa^*(s)\,\mathrm{d}s\right) = \frac{\mathrm{e}^{-\gamma T}c(w)^2}{c(w-T)^2}$$

and

$$h^*(T) = \frac{\sigma^2 \mathrm{e}^{-\gamma w}c(w)}{4(\kappa+\gamma)\gamma}\left[\frac{c(w)}{c(w-T)} - 1\right]\,,$$

where γ and c are defined in (14.14). This simplifies somewhat in the case $w = T$ because $c(0) = 2\gamma$. Thus, the probabilities in the CIR option pricing formula (14.21), which are the probabilities of the event shown in (14.22), are as follows:

- $\mathrm{prob}^u\big(P(T,u) > K\big)$ is the probability that

$$z \le -\frac{\mu_u}{\lambda_u}\left(\frac{\int_T^u \phi(s)\,\mathrm{d}s + a(u-T) + \log K}{b(u-T)}\right)\,,$$

where z has a non-central chi-square distribution with $4\kappa\theta/\sigma^2$ degrees of freedom and noncentrality parameter $r(0)/\lambda_u$, and

$$\mu_u = \frac{\mathrm{e}^{-\gamma T}c(u)^2}{c(u-T)^2}\,,$$

$$\lambda_u = \frac{\sigma^2 \mathrm{e}^{-\gamma u}c(u)}{4(\kappa+\gamma)\gamma}\left[\frac{c(u)}{c(u-T)} - 1\right]\,.$$

- $\text{prob}^T \left(P(T, u) > K \right)$ is the probability that

$$z \leq -\frac{\mu_T}{\lambda_T} \left(\frac{\int_T^u \phi(s)\, \mathrm{d}s + a(u - T) + \log K}{b(u - T)} \right),$$

where z has a non-central chi-square distribution with $4\kappa\theta/\sigma^2$ degrees of freedom and noncentrality parameter $r(0)/\lambda_T$, and

$$\mu_T = \frac{\mathrm{e}^{-\gamma T} c(T)^2}{4\gamma^2},$$

$$\lambda_T = \frac{\sigma^2 \mathrm{e}^{-\gamma T} c(T)}{4(\kappa + \gamma)\gamma} \left[\frac{c(T)}{2\gamma} - 1 \right].$$

List of Programs

List of Symbols

e, exp	natural exponential
cov	covariance
log	natural logarithm function
max	maximum function
min	minimum function
n	normal density function
prob, prob*	probability
prob^R, prob^S, ...	probability using an asset as numeraire
var	variance
N	cumulative normal distribution function
M	bivariate normal distribution function
a, b, c	constants or functions of time
d	real number, or, as subscript, indicator of down state
d_1, d_2	real numbers
f, g, h, Σ	functions
i, j, k, ℓ, n	integers
p	probability
q	dividend yield (probability in Chap 1)
r	risk-free rate or continuously-compounded return
r_f	foreign risk-free rate
s, t, w, T	real numbers representing time
u	real number, or, as subscript, indicator of up state
v	squared volatility
x	real number, function of time, or random variable
y, z, ξ, ε	real numbers or random variables
A	set of states of the world, or a constant
1_A	random variable that equals 1 when the event A occurs and zero otherwise
B, B^*	Brownian motion
C	call option price

E, E^*	expectation
E_t, E_t^*	conditional expectation at date t
$E^R, E^S, \ldots$	expectation using an asset as numeraire
$E_t^R, E_t^S, \ldots$	conditional expectation at date t using an asset as numeraire
F	forward price
F^*	futures price
K	exercise price
L	boundary for a barrier option
M, N	integers
P	discount bond or put option price
P^{mkt}	market discount bond price
R	risk-free accumulation factor
$\mathcal{R}$	simple interest rate (LIBOR)
S	asset price
V, W	portfolio values
X, Y, Z	random processes
$\alpha, \beta, \kappa, \mu, \nu, \theta$	constants or functions of time
$\delta, \Gamma, \Theta, \mathcal{V}$	option Greeks (delta, gamma, theta, vega)
λ	constant, function of time, or random process
ϕ	constant or function of time (or, in Chap 1, a state price density)
π	irrational number pi (or, in Chap 1, a state price)
ρ	correlation (or option derivative with respect to r)
σ	volatility
τ	time to maturity
Δt	length of time period
ΔB	change in B over a discrete time period
$\sqrt{}$	square root
$\times$	multiplication
$!$	factorial
d	differential
∂	partial differential
$\int$	integral
$\sum$	sum
$\prod$	product
$\Longrightarrow$	implies
$\Longleftrightarrow$	is equivalent to

References

1. Arrow, K.J.: The role of securities in the optimal allocation of risk bearing. Review of Economic Studies, **31**, 91–96 (1964). Translation of Le rôle des valeurs boursières pour la repartition la meillure des risques. Econometrie (1952)
2. Bielecki, T. , Rutkowski, M.: Credit Risk: Modeling, Valuation and Hedging. Springer, Berlin Heidelberg New York (2002)
3. Black, F.: The pricing of commodity contracts. Journal of Financial Economics, **3**, 167–179 (1976)
4. Black, F., Derman, E., Toy, W.: A one-factor model of interest rates and its application to Treasury bond options. Financial Analysts Journal, January/February, 33–39 (1990)
5. Black, F., Karasinski, P.: Bond and option pricing when short rates are lognormal. Financial Analysts Journal, July-August, 52–59 (1991)
6. Black, F., Scholes, M.: The pricing of options and corporate liabilities. Journal of Political Economy, **81**, 637–654 (1973)
7. Bollerslev, T.: Generalized autoregressive conditional heteroskedasticity. Journal of Econometrics, **31**, 307–327 (1986)
8. Boyle, P.: Options: a Monte Carlo approach. Journal of Financial Economics, **4**, 323–338 (1977)
9. Brace, A., Gatarek, D. , Musiela, M.: The market model of interest rate dynamics. Mathematical Finance, **7**, 127–154 (1996)
10. Brandimarte, P.: Numerical Methods in Finance: A MATLAB-Based Introduction, Wiley, New York (2002)
11. Brennan, M., Schwartz, E.: Finite difference methods and jump processes arising in the pricing of contingent claims: a synthesis. Journal of Financial and Quantitative Analysis 13, 461–474 (1978)
12. Brigo, D., Mercurio, F.: Interest Rate Models, Theory and Practice, Springer, Berlin Heidelberg New York (2001)
13. Broadie, M., Detemple, J.: American option valuation: new bounds, approximations, and a comparison of existing methods. Review of Financial Studies, **9**, 1211–1250 (1997)
14. Broadie, M., Glasserman, P.: Estimating security price derivatives using simulation. Management Science, **42**, 269–285 (1996)

15. Broadie, M., Glasserman, P.: Pricing American-style securities using simulation. Journal of Economic Dynamics and Control, **21**, 1323–1352 (1997)

16. Broadie, M., Kaya, O.: Exact simulation of stochastic volatility and other affine jump diffusion processes. Operations Research (forthcoming)

17. Clewlow, L., Strickland, C.: Implementing Derivatives Models, Wiley, New York (1998)

18. Conze, A., Viswanathan.: Path dependent options: the case of lookback options. Journal of Finance, **46**, 1893–1907 (1991)

19. Cox, J., Ingersoll, J., Ross, S.: A theory of the term structure of interest rates. Econometrica, **53**, 385–408 (1985)

20. Cox, J., Ross, S.: The valuation of options for alternative stochastic processes. Journal of Financial Economics, **3**, 145–166 (1976)

21. Cox, J., Ross, S., Rubinstein, M.: Option pricing: a simplified approach. Journal of Financial Economics, **7**, 229–263 (1979)

22. Dai, Q., Singleton, K.: Specification analysis of affine term structure models. Journal of Finance, **55**, 1943–1978 (2000)

23. Drezner, Z.: Computation of the bivariate normal integral. Mathematics of Computation, **32**, 277-279 (1978)

24. Duffie, D., Kan, R.: A yield-factor model of interest rates., Mathematical Finance, **6**, 379–406 (1996)

25. Duffie, D., Singleton, K.: Credit Risk: Pricing, Measurement and Management, Princeton University Press, Princeton, New Jersey (2003)

26. Epps, T. W.: Pricing Derivative Securities, World Scientific Publishing, Singapore (2000)

27. Geman, H., El Karoui, N., Rochet, J.-C.: Changes of numeraire, changes of probability measure and option pricing. Journal of Applied Probability, **32**, 443–458 (1995)

28. Geske, R.: The valuation of compound options. Journal of Financial Economics, **7**, 63–81 (1979)

29. Glasserman, P.: Monte Carlo Methods in Financial Engineering, Springer, New York Berlin Heidelberg (2004)

30. Goldman, M., Sosin, H., Gatto, M.: Path dependent options: 'buy at the low, sell at the high.' Journal of Finance, **34**, 1111–1127 (1979)

31. Harrison, J.M., Kreps, D.: Martingales and arbitrage in multiperiod securities markets. Journal of Economic Theory, **20**, 381–408 (1979)

32. Haug, E.G.: The Complete Guide to Option Pricing Formulas, McGraw-Hill, New York (1998)

33. Heath, D., Jarrow, R., Morton, A.: Bond pricing and the term structure of interest rates: a new methodology for contingent claims valuation. Econometrica, **60**, 77–105 (1992)

34. Heston, S.: A closed-form solution for options with stochastic volatility with applications to bond and currency options. Review of Financial Studies, **6**, 327–344 (1993)

35. Heston, S., Nandi, S.: A closed-form GARCH option valuation model. Review of Financial Studies, **13**, 585–625 (2000)

36. Ho, T., Lee, S.: Term structure movements and pricing interest rate contingent claims. Journal of Finance, **41**, 1011–1029 (1986)

37. Hull, J.: Options, Futures, and Other Derivatives, Prentice-Hall, 5th ed, Upper Saddle River, New Jersey (2002)

38. Hull, J., White, A.: Pricing interest-rate-derivative securities. Review of Financial Studies, **3**, 573–592 (1990)

39. Jäckel, P.: Monte Carlo Methods in Finance, Wiley, New York (2002)

40. Jackson, M., Staunton, M.: Advanced Modelling in Finance Using Excel and VBA, Wiley, New York (2001)

41. James J., Webber, N.: Interest Rate Modelling, Wiley, New York (2000)

42. Jamshidian, F.: An exact bond option formula., Journal of Finance, **44**, 205–209 (1989)

43. Jamshidian, F.: LIBOR and swap market models and measures. Finance and Stochastics, **1**, 293–330 (1997)

44. Jarrow, R., Rudd, A.: Option Pricing, Dow Jones-Irwin, Homewood, Illinois (1983)

45. Karatzas, I., Shreve, S.: Brownian Motion and Stochastic Calculus, Springer, New York Berlin Heidelberg (1988)

46. Leisen, D.P.J., Reimer, M.: Binomial models for option valuation—examining and improving convergence., Applied Mathematical Finance, **3**, 319–346 (1996)

47. Longstaff, F., Schwartz, E.: Interest rate volatility and the term structure: a two-factor general equilibrium model., Journal of Finance, **47**, 1259–1282 (1992)

48. Longstaff, F., Schwartz, E.: Valuing American options by simulation: a simple least-squares approach. Review of Financial Studies, **14**, 113–147 (2001)

49. McDonald, R.: Derivatives Markets, Addison Wesley, Boston (2003)

50. Margrabe, W.: The value of an option to exchange one asset for another. Journal of Finance, **33**, 177–186 (1978)

51. Merton, R.: Theory of rational option pricing. Bell Journal of Economics and Management Science, **4**, 141–183 (1973)

52. Miltersen, K.R., Sandermann, K., Sondermann, D.: Closed form solutions for term structure derivatives with log-normal interest rates., Journal of Finance, **52**, 409–430 (1997)

53. Mina, J., Xiao, J.: Return to RiskMetrics, The Evolution of a Standard (2001)

54. Musiela, M., Rutkowski, M.: Martingale Methods in Financial Modeling, Springer, Berlin Heidelberg New York (1997)

55. Rebonato, R.: Interest-Rate Option Models, 2nd ed., Wiley, New York, (1998)

56. Rebonato, R.: Modern Pricing of Interest-Rate Derivatives, The LIBOR Market Model and Beyond, Princeton University Press, Princeton, New Jersey (2002)

57. Schönbucher, P.: Credit Derivatives Pricing Models: Models, Pricing and Implementation, Wiley, New York (2003)

58. Stulz, R.: Options on the minimum or maximum of two risky assets: analysis and applications, Journal of Financial Economics, **10**, 161–185 (1982)

59. Tavakoli, J.: Credit Derivatives and Synthetic Structures, Wiley, New York (2001)

60. Tavella, D.A.: Quantitative Methods in Derivatives Pricing, Wiley, New York (2002)

61. Trigeorgis, A.: A log-transformed binomial analysis method for valuing complex multi-option investments. Journal of Financial and Quantitative Analysis, **26**, 309–326 (1991)

62. Vasicek, O.: An equilibrium characterization of the term structure. Journal of Financial Economics, **5**, 177–188 (1977)

63. Wilmott, P.: Paul Wilmott on Quantitative Finance, Vol. 2, Wiley, New York (2000)

64. Wilmott, P., Dewynne, J., Howison, S.: Option Pricing: Mathematical Models and Computation, Oxford Financial Press, Oxford (2000)

65. Zhang, P.G.: Exotic Options, A Guide to Second Generation Options, 2nd ed., World Scientific Publishing, Singapore (1998)

Index